WIND

❄ **POWER**

FOR HOME & BUSINESS

The Real Goods Solar Living Books

Paul Gipe, *Wind Power for Home & Business: Renewable Energy for the 1990s and Beyond*

Michael Potts, *The Independent Home: Living Well with Power from the Sun, Wind, and Water*

Edward Harland, *Eco-Renovation: The Ecological Home Improvement Guide*

Real Goods Solar Living Sourcebook: The Complete Guide to Renewable Energy Technologies, Ninth Edition, edited by John Schaeffer

Athena Swentzell Steen, Bill Steen, and David Bainbridge, with David Eisenberg, *The Straw Bale House*

Nancy Cole and P.J. Skerrett, Union of Concerned Scientists, *Renewables Are Ready: People Creating Renewable Energy Solutions*

David Easton, *The Rammed Earth House*

Sam Clark, *The Independent Builder: Designing & Building a House Your Own Way*

James Kachadorian, *The Passive Solar House: Using Solar Design to Heat & Cool Your Home*

Real Goods Trading Company in Ukiah, California, was founded in 1978 to make available new tools to help people live self-sufficiently and sustainably. Through seasonal catalogs, a newspaper (*The Real Goods News*), and the *Solar Living Sourcebook,* as well as retail outlets, Real Goods provides a broad range of renewable-energy and resource-efficient products for independent living.

"Knowledge is our most important product" is the Real Goods motto. To further its mission, Real Goods has joined with Chelsea Green Publishing Company to co-create and co-publish the Real Goods Solar Living Book series. The titles in this series are written by pioneering individuals who have firsthand experience in using innovative technology to live lightly on the planet. Chelsea Green books are both practical and inspirational, and they enlarge our view of what is possible as we enter the next millennium.

Ian Baldwin, Jr.
President, Chelsea Green

John Schaeffer
President, Real Goods

WIND POWER

FOR HOME & BUSINESS

Renewable Energy
for the 1990s and Beyond

Paul Gipe

A Real Goods Solar Living Book

Chelsea Green Publishing Company
White River Junction, Vermont

To my family and friends, whose encouragement has given me the freedom to choose my own path.

The installation and operation of small wind turbines entails a degree of risk. Always consult the manufacturer, applicable building codes, and the National Electrical Code™ before installing or operating your wind power system. For wind turbines that will be interconnected with the electric utility, always check with your local utility first. When in doubt, ask for advice. Recommendations in this book are not a substitute for the directives of wind turbine manufacturers, or federal, state, and local regulatory agencies.

The author assumes no liability for personal injury, property damage, or loss from using information contained in this book.

The views expressed are those of the author, and do not necessarily reflect those of any organization with which he may be associated.

Printed in the United States of America.
5 6 7 8 9 10   97 98 99

First printing 1993.

Chelsea Green Publishing Company
P.O. Box 428
White River Junction, Vermont
05001

Gipe, Paul.
    Wind power for home and business : renewable energy for the 1990s
    and beyond / Paul Gipe.
        p. cm.
    Includes bibliographical references and index.
    ISBN 0-930031-64-4 (pbk.)
    1. Wind power   I. Title
TJ820.G57 1993
621.4'5—dc20   93-9437

Contents

Tables

Preface

I find myself in the unique position of being at the same time an observer of the wind industry and a participant. As an observer I've traveled throughout the United States and Europe reporting on the technology and trends in the industry. As a participant I've installed anemometers in Pennsylvania and hunted windchargers in Montana. I know firsthand the danger and excitement of prowling the range for a 600-pound beast perched precariously atop 60 feet of rusted steel. And as an erstwhile dealer of small wind turbines I've felt the cold and exhaustion that come after working dawn to dusk in subfreezing weather installing a wind turbine.

From those experiences I prepared a day-long seminar on the promise and pitfalls of wind energy that was held at colleges and universities across the country. An earlier version of this book, which appeared in the early 1980s under the title *Wind Energy: How to Use It*, grew out of those seminars. At that time there was a chasm between the books written for the backyard tinkerers who wanted to build their own wind turbines, and those books surveying the entire field of wind energy. There was no book that answered the questions people raised in my seminars about how they could obtain a working wind system and not an experimenter's toy. *Wind Energy* was written to meet that need.

The book was unique because it didn't simply look at the technology. It gathered tips and advice from leaders in the field and offered practical guidance on how to select, buy, and install small wind machines, and how to do so safely. Since its publication in 1983, the demand for small wind turbines has declined dramatically following the collapse in the price of oil and the decrease in national interest in renewable energy. The technology, fortunately, continued to advance during the decade.

Times have again changed. With the heightened concern about our environment, there's a resurgence of interest in small wind turbines. From the growing number of inquiries that I and my colleagues in the wind industry have fielded during the past year, I've realized that there is again a need for frank information on how to use small wind machines.

This version of my original book has been extensively rewritten around the modern, integrated wind turbines that have come to dominate small wind machine technology. Further, *Wind Power for Home and Business* includes two new chapters not found in the original: one on remote power systems and another on using the wind to pump water. Both subjects represent expanding markets for small wind turbines.

Wind Energy sought to help newcomers to wind energy avoid the mistakes I and others had made and to spur greater use of wind energy. *Wind Power for Home and Business* seeks the same ends.

Paul Gipe
Tehachapi, California
December 1992

Acknowledgments

No one can write a book on a subject such as wind energy that crosses so many disciplines without the help of numerous contributors. I am greatly indebted to the folks at Bergey Windpower Company and Northern Power Systems. I'd like to particularly thank Mike and Karl Bergey. I've borrowed liberally from Mike's excellent technical publications as well as his thorough installation manual. Karl's perspective on wind technology from his years of work with private aviation has been invaluable.

I appreciate the quick responses by Woody Stevens at Northern Power Systems and Mick Sagrillo at Lake Michigan Wind & Sun to my frequent questions on stand-alone power systems. Mick is a fount of practical, hands-on knowledge of small wind turbines. His observations are peppered throughout the text.

The help provided by David Calley of Wind Baron, David MacKay of SOMA Ltd., Janet Tarling of L.V.M. Ltd., and Marlec Engineering added greatly to the discussion of micro wind turbines, a growing segment of the small machine market.

I am thankful for the assistance Dave Blittersdorf of NRG Systems, Kevin Cousineau of Zond Systems, and Susan Giordano of Second Wind provided on modern wind measuring instruments.

Randy Swisher, Tom Gray, and the staff of the American Wind Energy Association deserve a special note of thanks for their patience as the months dragged on and my contributions to AWEA's newsletter dropped to a trickle.

My thanks also to Mark Haller at SeaWest's Altamont Operations for his commentary; Bill Hopwood at Springhouse Energy Systems for his contributions on siting; Dean Kilgore at Dempster Industries; the Aermotor Windmill Company and Ken O'Brock of O'Brock Windmills for their help with mechanical windmills; Alan Wyatt at the Center for International Development for his analysis of wind-driven water pumping; Richard Perez at *Home Power* magazine, Jim Davis at Energy Transfer Systems, and Wes Edwards for their insights into remote power systems; Neil Kelley at the National Renewable Energy Laboratory, James Dubois at Southern California Edison Company, and Jos Buerskens at ECN for their help interpreting noise criteria; Vaughn Nelson and Ken Starcher at West Texas State University's Alternative Energy Institute for their review and suggestions; Dennis Elliott at Battelle Pacific Northwest Laboratory for up-dated maps of the U.S. and world wind resource; Phil Metcalf at Unarco Rohn for information on guyed lattice towers; Henry Dodd at Sandia National Laboratory, Oscar Holst Jensen at Vestas DWT, and Rolf Heier at Roheico, SINCO Products, Klein Tools, and Solarex for use of their illustrations; Photron for use of their DC Source Center trademark; Carol White for her production of the line drawings; Capitola Reece, Gene Heisey, and Art and Maxine Cook for sharing their experiences; Gil Morrissey for his tutelage to a sometimes dim-witted apprentice; Paul White for ably managing my office while I worked on the manuscript; Susan Nelson for her faith in the future; and Nancy Nies for her encouragement.

 1

Introduction

Wind works. It's reliable. It's economical. It makes environmental sense. And it's here now. Wind machines are not tomorrow's technology. Whether it's on a giant wind farm in California, in a small village in Morocco, or in the backyard of a Kansas wheat farmer, wind energy works today in a variety of applications around the world. You too can put this renewable resource to work. The following chapters explain how to go about doing just that: how to select and install the small wind power systems on the market today.

Wind technology has come a long way since the mid-1970s when the only wind turbines available were 1930s-era machines salvaged from ranchers on the Great Plains. During the past decade wind technology has come of age with the development of advanced small wind turbines. These rugged yet extremely simple designs have greatly improved the reliability and performance of small wind machines. But as you'll see in the chapters ahead, wind machines are not for everyone.

To use the wind successfully you must have a good site, have enough wind, and select the right machine. You also need something else. Using wind energy takes courage. Wind machines are not cheap, and whether you install it yourself or contract a dealer to do it, the installation of a wind machine is an undertaking fraught with risk and uncertainty. At some point, after considering all the pros and cons, a decision must be made that only you can make. You must weigh the options, then act. The people who use wind energy are prudent, but they're doers.

People use wind machines for many reasons: economic, environmental, and philosophical. The knowledge that you're saving money—in some cases earning it—is often sufficient reward for plunging into wind energy. Yet for many there's more to it than that. Windmills have fascinated us for centuries and will continue to do so. Like campfires or falling water, they're mesmerizing, indeed, entrancing. People respond almost instinctively. Few escape the excitement created by a sleek turbine whirring in the wind.

Working with the wind is more than just a means to cheap electricity. It becomes a way of life, a way of living in closer harmony with the world around us. Harnessing the wind for energy enables us to regain some sense of responsibility for meeting our own needs, and for reducing our impact on the environment. By generating our own electricity cleanly and with a renewable resource we can reduce the need for distant power plants and their attendant ills.

If you're fortunate enough to install a wind machine with your own hands, you'll experience sensations few others can share. You'll know what it's like to hang from the top of an 80-foot tower on a warm sunny day and gaze at the countryside spread out below you. You'll know the feeling of seeing your wind machine spinning overhead for the first time. You'll rediscover the sense of accomplishment one gets from a job well done. You and your friends may also find the hearty, backslapping camaraderie that grows among people after several arduous days working together. There's nothing quite like it.

However, installing a wind machine will never be risk free. Generating electricity isn't easy; it never has been. Your utility may make it look simple, but it's a tough and sometimes dangerous job. This book is designed to help you minimize the risk, to ensure—as much as possible—that you'll succeed in erecting a wind turbine that works reliably and safely.

Wind Power for Home and Business was written for those who ask questions, who want to know what's going on around them, who want to do what they can themselves. It's for those who want to make a difference. Yet this isn't another how-to-build-your-own windmill book, though much of what it contains is essential for building and installing one safely. Instead, *Wind Power* gives you what you need to make intelligent choices (whether or not to build one yourself, for example). It's about how to proceed in a logical and methodical manner to determine if you can use a wind machine and, if you can, the kind that's right for you.

This book differs from others on the subject in both describing the technology and explaining how to evaluate what's important and what's not. It also differs from others in primarily describing small wind turbines. These are machines that generate from a few watts to tens of kilowatts and use rotors spanning 1-7 meters (3-21 feet) in diameter. Larger wind turbines are included where necessary to illustrate development of today's technology.

How This Book Is Organized

Before we plunge into the text, there are a few preliminaries we need to deal with: how to navigate the chapters ahead, and the nomenclature that will be used.

Chapter 2 is an introduction to the varied applications of wind energy. It briefly discusses how wind energy is used today.

The wind itself is the subject of Chapter 3. You'll learn the importance of wind speed and how to find out what you have, as well as how to use the most up-to-date maps on wind resources available in the United States. You'll also learn when it's necessary to measure wind speed at your site and when it's not.

Chapter 4 describes two techniques for estimating the amount of energy that a wind turbine may capture, as well as how to interpret the information published by manufacturers. With estimates of annual production in hand, you're then ready for evaluating whether your wind investment makes economic sense in Chapter 5.

Chapter 6 explores wind technology: where we've been, where we are today, and where we're headed. The advantages and disadvantages of the various technologies in use today are explained. You'll also get a feel for what's on the market and how to evaluate it. Towers, necessary companions to wind machines, are the subject of Chapter 7.

How to cut costs is the subject of Chapter 8. If you think that building the wind machine yourself is the answer, this chapter asks you some hard questions. If you're a tinkerer, it suggests finding a used wind turbine. If you're looking for a trouble-free wind machine, this chapter narrows your options to installing a small commercial wind machine yourself from a kit.

Once you've decided what you need, Chapter 9 explains how to put it all together, how to buy a wind machine by evaluating the manufacturer, what to include in a contract, and what to expect from the dealer.

You'll learn how you can sell power back to the utility company in Chapter 10. And you'll find out why the utility may have some legitimate concerns about your doing so.

If instead of generating electricity in parallel with the utility you prefer to declare your energy independence, Chapter 11 examines the components you'll need. This chapter explains why hybrid power systems using wind, solar, and a backup generator make more sense for remote sites than either technology alone. It's followed by a chapter on another nonutility role for wind energy: pumping water. Chapter 12 puts a modern twist on one of the oldest applications of wind energy.

Chapter 13 examines where you can and can't install a wind turbine, and why. You'll also learn about potential zoning conflicts, how to avoid them, and—where you can't—how to deal with them.

Chapter 14 was designed for those who want to install the wind turbine themselves. But it's not only for them; it's useful to anyone planning to erect a wind system. This extensive chapter offers tips on installation, anchoring, and wiring.

Once your wind machine is installed you'll need to operate it properly to ensure that it serves you well for many years to come. Chapter 15 reviews some simple startup procedures and suggests how to operate, maintain, and monitor the performance of your wind system.

Unlike many other books on solar and wind technology, Chapter 16 takes a close look at a taboo subject: safety. Chapter 17 completes *Wind Power for Home and Business* by looking at the future and where the technology may be headed.

Nomenclature—What Are We Talking About?

They've been called many things. Some unprintable. Most know them as windmills. Whether we call them wind machines, wind turbines, or just windmills, the subjects of this book are kinematic devices intended to capture the wind and put it to work.

Wind Machines

The terms *wind machine* and *wind turbine* are simple and unpretentious. They do the job with the least fuss and are used interchangeably in the following chapters. You'll find no reference to fancier terms like *SWECS* (for small wind energy conversion systems) or *WTG* (wind turbine generators) here. Few people who work with wind energy for a living use such bureaucratese.

The term *wind machine*, as used here, shouldn't be confused with the huge electric fans found in some orchards. These fans go by the same name but are used to stir still air during cold winter nights to protect valuable fruit. They move air, they're not moved by the air as in a wind turbine.

For references to the wind machine, tower, and ancillary equipment as a whole, *wind system* works well. *Windmill* is reserved for the multiblade, water-pumping wind machine (American farm windmill) and for the European wind machine (Dutch windmill), because that's what nearly everyone calls them.

Conventional wind turbines are comprised of three essential components: rotor, nacelle, and tower. The rotor, the spinning part of a wind turbine, is the most important. Because the rotor determines how much work a wind turbine can perform, the size of a wind turbine will often be described by the diameter of its rotor. The rotor attaches to the wind turbine's drive train, which typically sits atop the tower housed inside the nacelle. The tower simply raises the whole machine above the ground where it can better catch the wind.

Power and Energy—There Is a Difference

In casual conversation, we use the terms interchangeably. But there is a difference. When we describe what a wind machine can do for us—what it can produce—we must become more specific. Power and energy, though closely related, have separate and distinct meanings. They are both related to work.

In the technical sense, work is performed when a force acts through some distance. When you push a stalled car, for example, you are applying a force. If the car moves 10 feet, we say the force you applied acted through a distance of 10 feet. For work to be done, something must be accomplished: an object moved, lifted, or turned. If the object does not move, no work is accomplished. If the car did not move, no matter how hard you grunted and groaned to move it, no work was performed.

Energy is defined as the ability to do work or the amount of work actually performed. Both use the same units and are given in the same terms. When the wind strikes the blades of a wind turbine, it imparts a thrust or force that turns the rotor. A finite amount of energy in the flowing wind has been converted to rotational energy in the spinning rotor. When a force does work on an object, energy is transferred from one form to another.

Now couple that spinning rotor to a generator. Work is accomplished when electrons flow from the generator to a load, such as heating a wire in a toaster, or turning a motor in a fan. Now take a closer look at the electric motor. The flow of electrons, what we call electricity, transfers the rotational energy imparted by the wind to the wind turbine rotor through wires or power lines to spin the shaft of the motor, giving us, once again, rotational energy. Because the motor spins a fan, work is accomplished. We've gotten something out of it.

The conversion from one form of energy to another is never 100 percent efficient; that is, you can't get out as much as you put in. There's always some energy lost in the process. Friction's the chief culprit. Though no one has ever built a machine that can convert 100 percent of energy in one form to another, people keep trying. Their perpetual motion machines appear every so often as a "startling discovery" in the popular press. Wind machines are a favorite target of this breed. Such machines never operate as claimed, nor do they usually produce useful work.

As someone who uses energy, you're concerned about the actual work accomplished and the amount of energy transferred. It's the bottom line of the technical balance sheet. You measure the performance of a wind machine by what it does for you, the work it performs, the energy it transfers and puts to use.

Where does power come in? Power is the rate at which work's performed, the rate at which energy is transferred (changed or released), the rate at which the energy in the wind passes through a unit of area. Power is given in watts *W*, or in units of 1000 watts, 1 kilowatt *kW*.

Homes demand power, and wind turbines can provide it. But what's most important isn't power, it's energy. Homeowners seldom pay for power. They pay for energy, given in units of kilowatt-hours *kWh*. When you buy a kilowatt-hour of electricity, you're paying for the energy delivered, not the power.

Watt's More

In an electrical circuit, *current* is the flow of electrons. It's given in units of amperes or amps. We perceive voltage as the pressure trying to force the electrons to flow through a wire. But no flow takes place and no work is done unless a load like a toaster is attached to the circuit. (No electricity flows out of a receptacle until an appliance is plugged into it.) Power in watts is the product of voltage times current.

This represents the instantaneous rate of work being done when a force (voltage) moves electrons some distance through a wire (current). A toaster, for example, will operate at 110 volts *V* and will draw 10 amps *A*, or

$$\text{Power} = 110 \text{ V} \times 10 \text{ A} = 1100 \text{ W} = 1.1 \text{ kW.}$$

The toaster's demand for power is equivalent to eleven 100-W light bulbs. Yet in the average household, a toaster uses much less energy (power x time) than lighting because it's used so infrequently (only a few minutes every morning). Lights, on the other hand, are used for hours on end. One 100-W light bulb operating 11 hours will burn as much electricity—use as much energy—as a toaster operated for 1 hour.

100-W Light bulb: 100 W x 11 hours = 1100 Wh = 1.1 kWh
Toaster: 1100 W x 1 hour = 1100 Wh = 1.1 kWh

The distinction between kilowatts and kilowatt-hours is critically important. Knowing the difference can keep you from being confused by a wind turbine's size, in kilowatts, and how much energy, in kilowatt-hours, it will actually produce.

Why Equations?—They're Informative

There are numerous tables and more than a few equations used in the following chapters. They're there for a reason. They express the relationship between quantities. If we're calculating the power in the wind, the equation for power explains why wind speed plays such a critical role. But don't fret. Wherever an equation appears in the text, there'll be an accompanying table to summarize the results. You needn't use the equations if you don't want to; just use the tables. Tables are helpful because they are easier to read than equations. But trends are hard to spot. Where necessary, graphs are used to illustrate trends and general relationships, such as in the distribution of wind speeds over time.

False Precision—How to Avoid It

A long string of digits in a calculation gives a sense of precision that may or may not exist. Don't be fooled. Estimating the annual energy output of a wind turbine is an inexact process. The art of number crunching for technologies dependent on a natural resource, whether it's farming, mining, or drilling for oil, is finding the best approximation.

In the real world, natural phenomena seldom follow the orderly relationships shown in textbooks. If you plotted a graph of the relationship between wind speed and the instantaneous power generated by a wind turbine, you'd find that the points representing your measurements didn't fall neatly in line. Rather, they would be scattered. Some points would lie above and below where you'd expect the line should be. When your job is to interpret this information, you draw a line that best fits the data. The resulting line is an approximation of what happened; it doesn't say exactly what happened.

Knowing this, you should be on guard for indicators of false precision. Take the potential output from a wind turbine as an example. It's absurd to say that a wind machine will generate 495 kilowatt-hours per month when the data used in the estimate suggests the wind turbine could deliver from 450 to 550 kilowatt-hours per month. Why not simply say 500 kilowatt-hours per month and be done with it? By rounding off the calculation you indicate uncertainty. Better yet, present the estimate as a range of values from 450 to 550 kilowatt-hours. Then you know that the estimates are only a gross approximation of what could occur.

Since the advent of pocket calculators the results of calculations are often presented in meaningless detail. In the days of the slide rule, every student learned that a calculation was only as accurate as the divisions on the rule— as accurate as the least accurate number in the calculation. You couldn't carry a number out to 10 decimal places if you wanted to. Though slide rules have gone the way of the dodo, the concept remains valid: The results of a calculation are only as accurate as the least accurate number used.

Consider average wind speed as an example. It's normally presented to one decimal place, for example, 10.5 mph. It has three significant figures: two to the left, and one to the right of the decimal. If we want to use this average speed in a calculation to estimate the energy production from a wind machine, the results should likewise be presented to no more than three significant figures. Say the calculation resulted in 22,525.49 kilowatt-hours on your calculator. You'd need to round off the number to three figures instead of the seven indicated. The result (22,500 kilowatt-hours) is more realistic considering the accuracy of the numbers used to derive it. Ignoring the concept of significant figures leads to false precision.

Most scientists and engineers are accustomed to approximate arithmetic. They round off numbers every chance they get. It makes their work easier. More importantly, it allows them to get quickly to the heart of a calculation without wasting time on needless detail. The use of approximate arithmetic is how an engineer can take a seemingly complex problem, such as estimating the potential energy production from a wind turbine, and solve it mentally or scratch it out on the back of an envelope. Where appropriate, the calculations presented in the following chapters have been liberally rounded off.

Units—Yes, Metrics Too

Purists of either the metric or English system of measurements should be forewarned. The wind industry in the United States uses an unholy mix of both. Towers installed in the United States are built here so they're sold in feet. For this reason tower height will generally be given in feet. Wind turbines, on the other hand, are produced around the world. (Nearly 50 percent of the turbines installed in California were built outside the United States.) The convention among wind turbine manufacturers abroad, and now those in the United States as well, is to describe the diameter of their turbines in the metric system. Consequently, the size of a wind turbine will be given in meters, not feet.

The description of wind speeds isn't as clear-cut. Both metric, in meters per second (m/s), and English, in miles per hour (mph), are used. In general metric is preferred everywhere but the United States. Because the wind industry in the United States is gradually adopting the metric system for wind speed, metric units frequently appear in the text.

If you have an aversion for either system, don't panic. Often the English or metric equivalent will be included within parentheses. Conversion tables for common metric and English units can also be found in Appendix A.

In the next chapter we'll look at how wind energy is used today.

 **2**

Applications—

How to Use the Wind

Look around you. How do you currently use energy? To run your appliances, to heat and light your home? In almost every case, a wind machine can be used to supplement or replace your consumption of conventional energy. In this chapter we'll glance at some of the ways wind energy can be used today. More details are provided in subsequent chapters.

Supplying Stand-Alone Power at Remote Sites

Outside of mechanically pumping water, wind turbines are best known for their ability to generate power at remote sites. They've distinguished themselves in this role for decades. During the 1930s, when only 10 percent of the nation's farms were served by electricity, literally thousands of small wind turbines were in use, primarily on the Great Plains. These home light plants provided the only source of electricity to homesteaders in the days before the Rural Electrification Administration (REA) brought electricity to all. Electric utilities now serve 98 percent of rural families in the United States.

That's not true elsewhere. There may be as many as 110,000 small wind turbines in use by nomadic herdsmen in northwestern China. These small turbines (so small they can be carted on horseback from one encampment to another) are the sole source of power available on the Asian Great Plains that stretch from China to the Soviet Union. To meet the demand for these portable windchargers, China alone builds 10,000 per year.

What's a remote site? There's no hard and fast rule. Utilities will build a power line almost anywhere if someone will pay for it. As a rule, anyone more than a half mile (or 1 kilometer) from existing utility service will find it cheaper to install an independent power system than to bring in utility service.

Even today three-fourths of all small wind turbines built are destined for stand-alone power systems at remote sites. Some find their way to remote homesteads in Canada and Alaska far from the nearest village. Others serve mountaintop telecommunications sites where utility power could seldom be justified. Surprisingly, an increasing number are being put to use in the lower 48 states by homeowners determined to produce their own power even though they could just as easily buy their electricity from the local utility.

Because the wind is an intermittent resource, remote applications generally require some form of storage. For remote homes, where a fairly steady supply of electricity is expected, battery storage becomes a necessity. Batteries store surplus energy during windy days for later use during extended calms. In battery-charging wind systems direct current (DC) from the batteries can be used, as is, in DC appliances or inverted to alternating current (AC) similar to that from the utility.

At any remote site with an average wind speed greater than 9 miles per hour (mph), or about 4 meters per second (m/s), small wind turbines produce power at less cost than gasoline or diesel generators. At such sites, says Mike Bergey of Bergey Windpower, wind systems are also more cost-effective than solar photovoltaics (PVs) alone. If these are your only other sources of power, and if you have the wind, a battery-charging wind system may be practical for you (see Figure 2-1).

Hybrids

A decade ago solar and wind proponents wore blinders. If you wanted a wind system for a remote site the dealer would happily oblige, sizing the wind turbine and the batteries to carry the entire load. If you'd asked a solar dealer to do the same, they would have covered your roof with solar cells (PV modules). Never would the twain have met. Fortunately, that's not so today. Both Bergey Windpower and Northern Power Systems, two small wind turbine pioneers, have successfully demonstrated wind and solar hybrids that capitalize on each technology's assets. In many locales wind and solar resources complement each other with strong winds in winter balanced by strong sun in summer, thus enabling designers to reduce the size

Figure 2-1. *Small wind turbines, such as these in operation at the U.S. Department of Agriculture's experiment station near Bushland, Texas, can be used to charge batteries, pump water, or generate electricity in parallel with that from an electric utility.*

Figure 2-2. Extremely small wind machines capable of producing only a few watts have been developed for charging batteries on sailboats and recreational vehicles. They are so small that they can be erected and taken down by one person. (L.V.M. Ltd.)

of each component. They've found that these hybrids perform even better when coupled with small backup generators to reduce the amount of battery storage needed.

Low-Power Applications

There are numerous applications for low-power systems where storage isn't required. One is cathodic protection on pipelines. A small wind turbine, often a micro turbine (the smallest on the market), provides an electric charge to the surface of the metal pipe to counteract galvanic corrosion in highly reactive soils. Storage isn't needed during calm winds because corrosion is a slow process that occurs over long periods. Eventually the wind returns and again protects the exposed metal. At one time all cathodic protection in rural areas was provided by wind turbines. Today pipelines primarily use small PV modules for cathodic protection, but wind machines are making a comeback with the development of rugged micro turbines that produce only a few watts (see Figure 2-2).

Another option is to simply use a micro turbine to charge your auto battery or to power security lighting. Photovoltaics are being deployed extensively for such uses. There are few technical reasons why micro turbines, which can be installed for less than $1500, can't be similarly used.

Village Electrification

One-third of the world's people live without electricity. In China alone half the population lives without access to utility power. Many Third World nations are scrambling to expand their power systems to meet the demand for rural electrification. Most are following the pattern set by the United States during the 1930s: build new power plants and extend power lines from the cities to rural areas. However, this approach to rural electrification doesn't make sense today with the advent of reliable hybrid power systems using wind and solar energy.

Developing nations will find it more cost-effective, says Mike Bergey, to install hybrid power systems than to stretch heavily loaded, and often unreliable, central-station power from the large cities. Though these hybrid systems generate little power in comparison to central power plants, Third World villages use little power. One kilowatt-hour of electricity provides 10 times more services in India than it does in Indiana.

Hybrid power systems that feature small wind turbines, because of their relative low cost, enable strapped governments to get power into villages quickly. As the central power system expands to these villages, the hybrid systems can be removed and sent on to even more remote villages.

Interconnecting with the Utility

Some small wind turbines generate electricity identical to that supplied by the electric utility: constant-voltage, constant-frequency alternating current. With passage of the Public Utility Regulatory Policies Act (PURPA) in 1978, homeowners and businesses in the United States have been permitted to connect these wind turbines with the utility network. Since then some 4500 small wind turbines have been interconnected with local electric lines across the country (see Figure 2-3).

There are disadvantages to this application. First, you must deal with the utility company, a task that may try your patience. Second, if there's a power outage your wind machine will be idled as well, though there are ways around this (see Chapter 11). And because your wind turbine will be competing with electricity supplied by the utility, the economic requirements are more stringent than in a stand-alone system.

Electric utilities have had more than 100 years of practice producing electricity. They've had ample time to learn how to do it cheaply. In general you should only consider an interconnected wind system if you have an

Figure 2-3. With modern electronics, small wind turbines can perform several tasks simultaneously. This wind turbine at West Texas State University's Wind Test Center charges electric-vehicle batteries and produces utility compatible power for use at the test center or for sale back to the utility. The wind turbine, a Bergey Excel, is powerful enough to supply the needs of an all-electric home.

average wind speed greater than 10 mph (about 5 m/s) and electricity costs approaching 10 cents per kilowatt-hour. To compete with the utility you'll also need a bigger wind turbine than those typically used for stand-alone systems. Most interconnected small wind turbines in the United States are found in the range from 1 kilowatt (about 3 meters in diameter) to 25 kilowatts (about 10 meters in diameter).

How It Works

There are two types of wind turbine suitable for producing utility-compatible power: those that use induction generators, and those that use synchronous inverters. (These will be discussed more fully in Chapter 6.) Regardless of which is used, their interconnection with the utility is the same. Cable from the wind turbine's control box connects it to terminals within your utility service panel. In effect, the wind turbine becomes a part of your home's electrical circuit not unlike an electric stove does.

The wind machine reduces your consumption of utility-supplied electricity whether it's for lighting, appliances, or electric heat. The wind turbine will power your clock, your stereo, your refrigerator, or your lights. If you're a farmer, it will run your milkers or your feeders. In short, you'll use wind-generated electricity wherever you presently use utility power.

When the wind is blowing, the wind turbine produces electricity that flows to your service panel. From there it seeks out those circuits where electricity is being consumed. With electricity it's first come, first served. Electricity will flow to the first circuit where it's needed. If more energy is being generated than can be used by the first circuit, it will flow to the next, and so on. When the wind machine can't deliver as much energy as needed, the utility makes up the difference. There are no fancy electronics controlling which circuit gets what. It's all accomplished silently and effortlessly. That's the beauty of electricity.

If you're not using electricity when the wind machine is operating, or if you are not using as much as it's generating, the excess flows from your service panel through your electric meter out to the utility's lines and on down the road. In some cases the utility will permit you to run your kilowatt-hour meter backward. It seems mysterious, but it works. And it does so neatly, cleanly, and without fuss.

Farming the Wind

Under the right conditions the utility will pay a fair price for any surplus electricity you sell back to them. If so, you might find it profitable to install several wind turbines and sell as much power to the utility as you can. When you do, you're farming the wind for profit. More than 16,000 turbines have been installed in California just for this purpose (see Figure 2-4). These commercial wind farms, or wind power plants, are nothing more than a large-scale version of a small wind turbine interconnected with a residential customer's electric utility. But rather than meeting the domestic demand of a home or business, all the electricity generated by these wind power plants is delivered for sale to the utility.

In the early 1980s, some small wind turbines originally designed for homes or small businesses found their way to California's wind farms. Literally thousands of such wind machines were installed. These were 10-, 25-, and 40-kilowatt wind turbines that just a few years prior were being installed in backyards across the United States. Today the average size of wind turbines installed in California wind plants exceeds 250 kilowatts with rotors spanning 25 meters (80 feet) or more in diameter.

While the bulk of wind power generation in the United States is found in California's wind plants, Denmark has a far different story. Most of the 3500 wind turbines in Denmark are used by homeowners, farmers, and small businesses. While Americans were erecting 10-kilowatt wind turbines in their backyards during the early 1980s, the Danes were installing 55-kilowatt machines in theirs. Today the size of the average wind turbine installed for nonutility applications in Denmark is more than 150 kilowatts. As explained in Chapter 8 on cutting costs, Danes often join cooperatives and buy what many Americans would consider a medium-sized wind turbine and install it in the backyard of one of the co-op members. They then share in the revenues from the sale of electricity to the utility. Others install similarly sized machines for their own domestic use and sell the excess to the utility. As Denmark has decisively shown, medium-sized wind turbines need not be limited solely to commercial wind power plants; they may be successfully used to power homes, farms, and businesses.

Heating

For much of the country, heating comprises most of a home's energy demand. In many areas there is a good correlation between the availability of wind energy and the demand for heat. It's not surprising then that there have been numerous attempts to use wind machines strictly for heating. Advocates of the wind furnace concept believe that the same winter winds that rob a house of its warmth could be used for heating. Backers of the wind furnace

Figure 2-4. *Arrays of multiple wind turbines, such as these near Tehachapi, California, form wind power plants. One of the wind machines shown here produces enough electricity to supply the residential needs of 30 North American homes. The same turbine on the west coast of Denmark could supply the needs of 60 homes.*

approach argue that producing low-grade energy with either a generator or a mechanical churn is less costly than producing the same amount of utility-compatible electricity. It's simpler, proponents say, than trying to generate the high-grade electricity demanded by the utility. It hasn't worked out that way. Both Danish and American companies tried to commercialize the concept during the 1970s and early 1980s. All failed.

Experience has shown that in most cases it's cheaper and easier to interconnect the wind system directly with the utility than to generate heat and try to store it. It's more cost-effective to produce a high-grade form of electricity that can be used for all purposes, including home heating if desired, than to build a wind turbine that can be used only for one function. The sole wind turbine that has been used successfully for home heating is Bergey Windpower's Excel. This 10-kilowatt turbine was designed to meet the needs of *all-electric* homes on the Great Plains and to generate utility-compatible electricity (see Figure 2-3).

Water Pumping

Wind machines have been used to pump water throughout history, and pumping water remains an important application today both in the United States and in the developing world. The American farm windmill, which dependably pumps low volumes of water from shallow wells, is still extensively used for watering remote stock tanks on the Great Plains and Argentinean Pampas (see Figure 2-5). There are probably more than a million of these wind pumps still in use around the world today.

Researchers at West Texas State University and the U.S. Department of Agriculture have made major advances in water-pumping technology, first with wind-assisted irrigation and more recently with wind-electric pumping systems. In cooperation with Bergey Windpower they have developed pumping systems that use modern, electricity-producing wind machines. In almost all cases today these machines will pump more water at lower cost than the familiar American farm windmill.

Whether for pumping water, heating your house, or selling power to the utility, successfully harnessing the wind is largely determined by the wind speeds at your site. The speed of the wind and its relationship to the energy that a wind turbine can capture are the subjects of the next two chapters.

Figure 2-5. Wind machines have been traditionally used to pump water. Hundreds of thousands of mechanical wind pumps are still used throughout the world. Note the tank used to store domestic water.

3

Measuring the Wind

"Hi, I want a windmill."

"You sure?"

"Yeah, I can't wait to tell the utility to go you know where."

"Hmm. I'll bet they'll be glad to hear that. How much wind do you have?"

"Wind? Oh, We've got lots of wind. It's always windy here."

More than one new owner of a wind system has learned an expensive lesson—one that may seem patently obvious. Wind generators with little wind are like dams with little water. They don't work. There's wind everywhere, but not everywhere has enough. Ample wind is a prerequisite for successfully siting a wind machine, and the more the better. But just how much is enough? How do you know whether the wind over your site is sufficient? If you're living on the coast of Alaska or on the Texas Panhandle, you probably have enough wind. Unfortunately, few people live where the wind has such a well-deserved reputation for being so fierce. Most of us need a better description of the wind than "It's always windy here."

In this chapter we'll discuss the wind, what it is, how local climate and terrain affect it, and how it changes over time. We'll explore the meaning of wind power and how wind speed and power increase with height. You will learn where to find wind information for your area and how to determine the winds at your site. For those with an aversion to math, easy-to-use tables are presented wherever a formula appears. As mentioned earlier, our objective is to determine if we have enough wind to put wind energy to work. For stand-alone or remote applications we're seeking sites with average wind speeds above 9 mph (4 m/s); for offsetting utility generation we'll need average wind speeds above 10 mph, preferably above 12 mph (5 m/s) at the height of the wind turbine.[1]

Wind—What Is It?

The atmosphere is a huge, solar-fired engine that transfers heat from one part of the globe to another. Large-scale convective currents set in motion by the sun's rays carry heat from lower latitudes to northern climes. The rivers of air that pour across the surface of the earth in response to this global circulation are what we call wind, the working fluid in the atmospheric heat engine.

When the sun strikes the earth, it heats the soil near the surface. In turn, the soil warms the air lying above it. Warm air is less dense than cool air and, like a helium-filled balloon, rises. Cool air flows in to take its place and is itself heated. The rising warm air eventually cools and falls back to earth completing the convection cell. This cycle is repeated over and over again, rotating like the crankshaft in a car, as long as the solar engine driving it is in the sky. Thermals, rising currents of warm air that boil up over land during bright daylight hours, are as much sought after for soaring by humans as by hawks. The cumulus clouds of summer are a sign of the convective circulation that causes winds to strengthen in late afternoon. If you're a pilot, you probably prefer to fly in the early morning hours when winds are light. On the other hand, if you are making a trip to inspect a wind machine, the mid-afternoon when you're more likely to find it running, is better.

Winds are also stronger and more frequent along the shores of large lakes and along the coasts because of differential heating between the land and the water. During the day, the sun warms the land much quicker than it does the surface of the water. (Water has a higher specific heat and can store more energy than soil without a change in temperature.) The air above the land is once again warmed and rises. Cool air flows landward, replacing the warm air, creating a large convection cell. At night the flow reverses as the land cools more quickly than the water. In late afternoon, when the sea breezes are strongest, winds can reach 10-15 mph (5-7 m/s) on an otherwise

---

1. The cost of competing sources determines the actual wind speed at which wind energy becomes economically attractive. Higher wind speeds are needed for wind energy to compete with cheap utility power than to compete with the more costly sources of energy used in remote applications.

calm day. Land-sea breezes are most pronounced when winds due to large-scale weather systems are light. The influence of land-sea breezes diminishes rapidly inland and is insignificant more than 2 miles (3 km) from the beach.

The winds along the shore are also higher because of the long unobstructed path (fetch) that the wind travels over the water. Hills, trees, and buildings block the wind on land. The shores of the Great Lakes and the seacoasts of many nations have average wind speeds approaching 12-15 mph (6-7 m/s), partly due to these effects.

The mountain-valley breeze is another example of local winds caused by differential heating (see Figure 3-1). Mountain-valley breezes occur when the prevailing wind over a mountainous region is weak and there's marked heating and cooling. These breezes are found principally in the summer months when solar radiation is the strongest. During the day the sun heats the floor and sides of the valley. The warm air rises up the slopes and moves upstream. Cooler air is drawn up from the plains below, causing a valley breeze. At night the situation reverses and the mountains cool more quickly than the lowlands. Cool air cascades down the slopes and is channeled through the valley to the plains. Nighttime mountain breezes are generally stronger than valley breezes, with winds reaching speeds of 25 mph (11 m/s) in valleys with steeply sloping floors located between high ridges or mountain passes.

VALLEY WIND   **MOUNTAIN WIND**

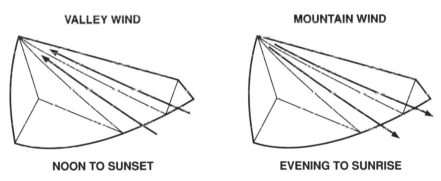

NOON TO SUNSET   **EVENING TO SUNRISE**

Figure 3-1. Mountain-valley winds. (Battelle PNL)

Mountain-valley breezes can be reinforced by the prevailing winds when the two flow in the same direction. The effects of channeling are most pronounced on North America's Pacific Coast where onshore winds are funneled into narrow gorges through the mountains. When heating of interior deserts causes large differences between temperatures along the coast and temperatures inland, convective flow reinforces the funneling effect. Average wind speeds above 20 mph (9 m/s) are typical, with winds in some seasons averaging well above that.

Many of the windy passes through mountain chains on the West Coast of the United States lie east to west. The tremendous wind potential in the Columbia River Gorge east of Portland, Oregon, is one such example. Another is the Altamont Pass east of San Francisco where cool air off the Pacific Ocean rushes over the low pass into the hot interior of the San Joaquin Valley. The San Gorgonio Pass near Palm Springs, California, is similar. This sea-level pass through the San Bernadino Mountains channels cool coastal air onto the blazing Sonoran Desert. Similar flows occur from the San Joaquin Valley across the Tehachapi Mountains onto the Mojave Desert. Winds through the Tehachapi Pass are more subject to the passage of cold fronts than in California's other passes. The storms associated with the cold fronts push masses of dense air through the Tehachapi Pass sometimes causing winds up to 100 mph (50 m/s).

Long ridges across the path of the wind, like Cameron Ridge in the Tehachapi Mountains, also enhance the flow over the summit (see Figure 3-2). Wind speeds may double as the flow accelerates up the gradual slope of a long ridge. As many California wind farmers have learned the hard way, this enhancement occurs only in the last third of the slope near the crest. Turbines lower down the slope perform dismally, as do those on the leeward side.

A similar terrain enhancement of the wind has been found on the island of Oahu in Hawaii. There the northeasterly winds sweep around the end of a long ridge that runs the length of the island from north to south. The winds accelerate as they pass over the ridge, but more so when they pass

INCREASE IN WIND SPEED OVER A RIDGE

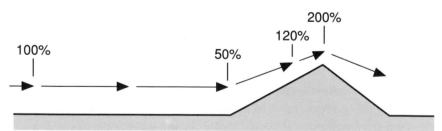

Figure 3-2. Variation in wind speed over a ridge. Wind speed increases near the summit of a long ridge lying across the wind's path. (Battelle PNL)

around the ends. Kahuku Point, at the northernmost end of the ridge, has long been a prime site for wind development because of this local effect. One of the world's largest wind turbines, the 3.2 MW Boeing Mod-5B, is in operation there.

Mount Washington, in the White Mountains of New Hampshire, is one of the few areas in the eastern United States where terrain enhancement is well known. In fact, this phenomenon was first observed on Mount Washington and in the Green Mountains of Vermont. Mount Washington has an average speed of approximately 38 mph (17 m/s), the highest average wind speed on the East Coast. On April 12, 1934, an observatory on the summit recorded the highest wind speed ever measured, 231 mph (103 m/s).

The effect of terrain on wind speeds within the region has been known for some time and has been used to advantage in the past. The Green Mountains of Vermont form a long northeasterly ridge that sits astride the prevailing winds just east of the long unobstructed fetch across Lake Champlain. Because of the potential speed enhancement, Palmer Putnam chose the Green Mountains as the site for the experimental Smith-Putnam wind turbine during the 1940s. Unfortunately, the terrain features that enhance wind speeds also create turbulence that can wreak havoc on wind machines. The 1500-kilowatt Smith-Putnam turbine eventually succumbed to just such turbulence atop Grandpa's Knob after only a few years of operation.

Mountains and ridges offer higher winds for reasons other than channeling. Prominent peaks often pierce temperature inversions that can blanket valleys and low-lying plains. Temperature inversions cause a stratification of the atmosphere near the surface. Above the inversion layer normal air flow prevails, but below the winds are stagnant. The air beneath an inversion layer may be completely cut off from the air circulation of the weather system moving through the region. Temperature inversions are common in hilly or mountainous terrain such as southern California and western Pennsylvania, both areas notorious for their air pollution episodes. In northern latitudes inversions are common during the fall and winter.

The inversion layer itself may accelerate the wind. The wind above a temperature inversion, essentially a giant lake of stagnant air, blows across the surface of the inversion unimpeded by the hills, trees, buildings, and other features that often hinder the wind. Ridge tops may not only possess more frequent winds, they may have stronger winds as well. The wind can skip across the inversion layer as though across the surface of a lake.

There are numerous regional winds around the globe that can also have a powerful influence on successfully siting a wind turbine. These winds, like the powerful chinook that roars down the east side of the Rocky Mountains, are due to infrequent local meteorological and geographic anomalies. The Santa Ana in southern California, for example, results from high pressure systems that occasionally move over the Basin and Range provinces of the American Southwest. But whether it's the foehn in the Alps, the scirocco sweeping across the Mediterranean from North Africa, the legendary mistral of southern France, or the tramontana howling out of the eastern Pyrenees, these winds are a force to be reckoned with. They can power wind turbines or destroy them. In the early days of the U.S. Department of Energy's wind program, chinooks with winds sometimes above 100 mph (50 m/s), wreaked havoc on the flimsy experimental turbines at the Rocky Flats test center near Denver, Colorado. Today, wind farmers in Tehachapi harness the once-feared Santa Ana winds. And in southern France operators of two small wind plants eagerly look forward to the tramontana that sends tourists scurrying for cover at nearby resorts.

Wind Speed and Time

The wind is an intermittent resource: calm one day, howling the next. Wind speed and direction vary widely over almost all measuring periods. Because wind speed fluctuates, it becomes necessary to average wind speed over a period of time, usually over an entire year.

The average annual wind speed itself is not constant. It varies from year to year. The average speed can change as much as 25 percent from one year to the next. This can amount to more than 2 mph (1 m/s) in a moderate wind regime where an average of 10 mph (5 m/s) is the norm. Meteorologists say that they gather 10 years of data or more before they feel comfortable that all the yearly cycles have been recorded.

Average wind speeds vary by season and by month. "March roars in like a lion and goes out like a lamb" is a popular adage signifying that early spring is windy, while summer is not. For much of North America's interior, winds are light during summer and fall and increase during the winter, reaching their maximum in the spring (see Figure 3-3).

When we looked at the differential heating of the earth's surface and its effects on local winds, we saw that wind speeds often increased during late afternoon after convective circulation had been set in motion. This tells us that wind speeds vary by time of day, not only because of changing weather

MONTHLY AVERAGE WIND SPEED

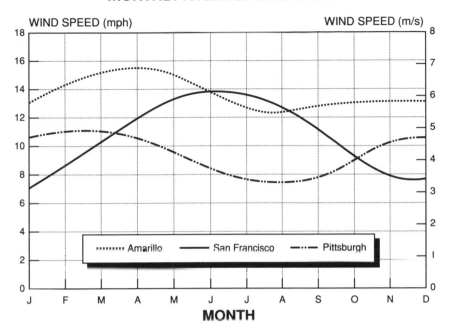

Figure 3-3. Historical monthly wind speed. For the interior sites of Pittsburgh, Pennsylvania, and Amarillo, Texas, wind speeds increase during the winter and spring. Wind speeds are strongest near San Francisco during the summer when seasonal flows of cool marine air rush toward the blistering San Joaquin Valley. It's this seasonal flow that drives the wind power plants in the Altamont Pass east of San Francisco. The annual average wind speed at Amarillo is 13.7 mph (6.1 m/s); Pittsburgh, 9.3 mph (4.2 m/s); and San Francisco, 10.5 mph (4.7 m/s).

but also because of convective heating. The effects of local convective winds are greater during the summer when winds are light and solar heating is strong. Convective circulation leads to a dramatic difference between wind speeds during daylight hours and those at night.

The diurnal difference in wind speeds is less marked during winter because there's less convective circulation. During winter and spring, winds are dominated by storm systems. It's the recurring storms of winter that push up the average wind speeds across the Midwest and northeastern United States. Storms formed in the Gulf of Alaska are the source for the winds that funnel through the Tehachapi Pass.

Power in the Wind

One of the most important tools in working with the wind, whether designing a wind turbine or using one, is a firm understanding of what factors determine the power in the wind. For the sake of thoroughness, we'll start right at the beginning. Wind is air in motion. (Good so far.) The air about us has mass (think of it as weight if you're unfamiliar with the term). Though extremely light, it has substance. A gallon of air is similar to a gallon of water, but the gallon of air is lighter. It has less mass than that of water because air is less dense. It's more diffuse. Like any other moving substance, whether it's a gallon of water plummeting over Niagara Falls or a car speeding down the highway, this moving air contains kinetic energy. This energy of motion gives the wind its ability to perform work.

When the wind strikes an object, it exerts a force in an attempt to move it out of the way. Some of the wind's kinetic energy is given up or transferred, causing the object to move. When it does, we say the wind has performed work. We can see this when leaves skitter across the ground, trees sway, or the blades of a wind turbine move through the air.

The amount of energy in the wind is a function of its speed and mass. At higher speeds more energy is available, in much the same way a car on the highway contains more energy than a car of equal size it passes. It takes more effort—energy—to stop a car driven at 70 mph than it does one at 50 mph. Likewise, heavy cars contain more energy than light cars traveling at the same speed. This relationship between mass, speed, and energy is given by the equation for kinetic energy where m represents the air's mass and S its speed.[2]

$$\text{Kinetic Energy} = \tfrac{1}{2} \, mS^2$$

The air's mass can be derived from the product of its density d and its volume. Because the air is constantly in motion, the volume must be found by multiplying the wind's speed S by the area A through which it passes during a given period of time t.

$$m = dASt$$

When we substitute this value for mass into the earlier equation, we can find the kinetic energy in the wind.

---

2. Technically the term is V for velocity. Even though speed and velocity have separate and distinct meanings, they'll be used synonymously here.

$$\text{Wind Energy} = \tfrac{1}{2}\,(dASt)\,S^2$$
$$= \tfrac{1}{2}\,dAtS^3$$

We've gone through this derivation for a reason. Equations are the language of science, and in that terse, compact script the fundamentals of wind energy are precisely stated. But before we go over each of them, let's complete one more step.

Power, as you may remember, is the rate at which energy is available, or the rate at which energy passes through an area per unit of time.

$$P = \tfrac{1}{2}\,dAS^3$$

Power *P*, we've now learned, is dependent upon air density, the area intercepting the wind, and wind speed. Increase any one of these factors and you increase the power available from the wind.

Air Density

The wind is a diffuse source of power because air is less dense than most common substances. (Water, for example, is 800 times denser than air. This is why so much more effort, comparatively, has been expended to harness hydropower than wind power.)

Air density *d* decreases with increasing temperature and increasing altitude. Air is less dense in summer than in winter, varying 10-15 percent from one season to the next. On a yearly average, though, seasonal changes in temperature have such a slight influence on power that we can safely disregard them. We'll assume for all calculations that we're in an area where conditions approximate those at a standard temperature for our values of air density.[3] Of course, if you plan to install a wind machine on the North Pole or on the sands of the Sahara, average temperature can make such a difference in air density that it should be taken into account. If this is the case, see Appendix B for temperature corrections to air density.

Increasing altitude has a more substantial effect on air density than average temperature. For example, at the same wind speed there is more wind power at a coastal site than at Denver, Colorado (elevation 5000 feet above sea level). There's 15 percent less wind power at Denver (see Table 3-1). For our calculations, we're going to assume conditions at sea level.

---

3. Standard temperature is 59°F or 15°C.

Table 3-1

Change in Air Density with Elevation

Elevation (ft)	(m)	Relative Change	Elevation (ft)	(m)	Relative Change
0	0	100%	5,000	1,524	85%
500	152	99%	6,000	1,829	82%
1,000	305	97%	7,000	2,134	79%
2,000	610	94%	8,000	2,439	76%
3,000	915	91%	9,000	2,744	73%
4,000	1,220	88%	10,000	3,049	70%

Swept Area

Unlike changes in air density, changes in the area swept by a wind turbine rotor influence the power available significantly. Wind turbines with large rotors intercept more wind than those with smaller rotors and, consequently, capture more power. Doubling the area swept by a wind turbine rotor, for example, will double the power available to it. This principle is fundamental to understanding wind turbine design. Knowing this, you can quickly size up any wind machine by noting the dimensions of its rotor (see Figure 3-4).

Consider a conventional wind turbine whose blades spin about a horizontal axis. The rotor sweeps a disc the area of which can be found from the equation for the area of a circle, where A is the area and R is the radius of the rotor (approximately the length of one blade).

$$A = \pi R^2$$

This tells us that the area swept by the rotor of a conventional wind turbine is proportional to the square of the rotor's radius. Relatively small increases in blade length produce a substantial increase in swept area. Doubling the length of each blade—doubling the rotor diameter—increases the wind turbine's swept area four times.

$$A/\pi = R^2 = (2/1)^2 = \frac{2 \times 2}{1 \times 1} = 4$$

Figure 3-4 illustrates this concept graphically.

Wind Speed and Speed Distributions

No other factor is more important to the amount of wind energy available to a wind turbine than the speed of the wind. Because the power in the wind is a cubic function of wind speed, changes in speed produce a profound effect on power, even more so than does the swept area.

Consider an example where the wind speed at two sites differs by only 20 percent. At one site the wind speed is 10 mph, at the other, 12 mph. How much difference does this cause in the power available?

From an earlier equation we learned that power is proportional to speed cubed. Therefore the ratio of powers is proportional to the ratio of speeds cubed.

$$P_2/P_1 = (S_2/S_1)^3 = (12/10)^3 = 1.73. \text{ Therefore } P_2 = 1.73P_1.$$

Although there's only a 20 percent difference between the wind speeds at the two sites, there's 73 percent more power available at the windier

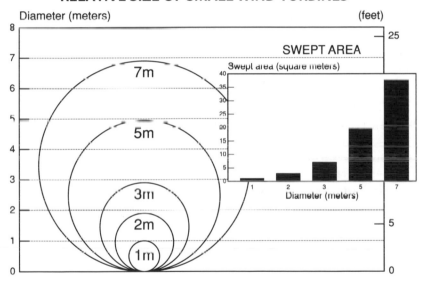

Figure 3-4. The relative size of commercially available small wind turbines. Marlec Engineering's Rutland wind turbine uses a molded plastic rotor about 1 meter in diameter. New Zealand manufacturer SOMA Power's 400-watt turbine uses a rotor spanning 2 meters in diameter. The rotor on the SOMA turbine sweeps four times more area than that of the Marlec, one of the smallest wind turbines on the world market. With a 7-meter rotor, the Bergey Excel intercepts more than five times the wind energy than that of the Northern Power Systems' HR1 with a rotor 3 meters in diameter. The Northern Power Systems' HR3, at 5 meters in diameter, can capture 25 times more wind energy than the 1-meter diameter Marlec.

location. This is why there's such a fuss concerning the proper siting of a wind machine: small differences in wind speed caused by bordering trees or buildings can drastically reduce the power a wind turbine can potentially capture. To grasp the full effect, consider what happens when the wind speed doubles from one site to the next. Doubling wind speed does not simply double the power available. Instead, power increases a whopping eight times.

$$P_2/P_1 = (S_2/S_1)^3 = (20/10)^3 = 2^3 = 8. \text{ Therefore } P_2 = 8P_1.$$

We can summarize the power equation with these general rules:
• Power is not significantly affected by changes in air density except for sites in arctic and desert environments, or sites above 3000 feet in elevation.
• Power is proportional to the area intercepted by the wind turbine. Double the area intercepting the wind and you double the power available.
• Power is a cubic function of wind speed. Double the speed, and power increases eight times.

At this point an important question arises. What wind speed are we talking about? The average wind speed? If so, what average wind speed? The annual average? Whichever we use determines the results we get.

Using the average annual wind speed alone in the power equation will not give us the right results; our calculation could differ from the actual power in the wind by a factor of two or more. To understand why, remember that wind speeds vary over time. The average speed is composed of winds above and below the average.

To illustrate, let's calculate the power density *P/A*, the rate at which energy passes through a unit of area, for an annual average wind speed of 15 mph.

$$P/A = \frac{1}{2} dS^3$$

Power density is a term frequently used by wind energy experts because it's a convenient shorthand for how energetic the winds are during a period of time, typically a year. Power density is normally given in units of watts per square meter (W/m^2). So let's take the value for air density at sea level and a temperature of 60 degrees F and plug it into the equation and see what we have.

$$P/A = 0.05472 \ S^3, \text{ where } S \text{ is in mph}^4$$
$$P/A = 0.6125 \ S^3, \text{ where } S \text{ is in meters per second}$$

---

4. *P/A* = 0.08355 *S*^{3}, where *S* is in knots; 1 knot = 1.15 mph.

At 15 mph,

$$P/A = 0.05472 \ (15)^3 = 0.05472 \ (3375) = 184 \ W/m^2.$$

Now, what happens if the wind blows half the time at 10 mph and half the time at 20 mph? The average speed remains 15 mph.

$$\frac{10 + 20}{2} = 15$$

Right? Yes, but watch what happens to the average power density using these two wind speeds.

At 10 mph, $P/A = 0.05472 \ (10)^3 = 55 \ W/m^2$
At 20 mph, $P/A = 0.05472 \ (20)^3 = 438 \ W/m^2$
Average $P/A = \dfrac{55 \ W/m^2 + 438 \ W/m^2}{2} = 247 \ W/m^2$

The average power density P/A equals 247 W/m²! How can this be? Both have the same average speed. The answer rests with the cubic relationship between power and speed.

Grab a cup of coffee, sit back, and ponder this statement. *The average of the cube of many different wind speeds will always be greater than the cube of the average speed.* Or stated another way, *the average of the cubes is greater than the cube of the average.* In this case, the average of the cube for two classes of wind speed (10 and 20 mph) is 1½ times the cube of the average.

The reason for this paradox is that the single number representing the average speed ignores the amount of wind above as well as below the average. It's the wind speeds above the average that contribute most of the power.

If we plotted a graph of the number of times, or frequency, with which winds occur at various speeds throughout the year, we'd find that there are few occurrences of no wind, and few occurrences of winds above hurricane force. Most of the time wind speeds fall somewhere between extremely low and extremely high.

The distribution of winds at various speeds differs from one site to the next but in general follows a bell-shaped curve for a family of Rayleigh distributions at sites with different average speeds (see Figure 3-5). The Rayleigh distribution is a hypothetical distribution of wind speeds based on a mathematical formula that attempts to approximate the real world. (A table of Rayleigh distributions at different average speeds can be found in Appendix C.)

The power density calculated from the Rayleigh distribution for a given

RAYLEIGH AVERAGE ANNUAL WIND SPEED DISTRIBUTION

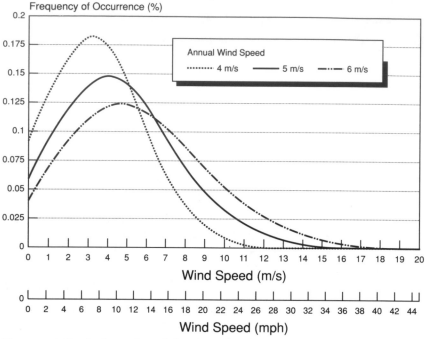

Figure 3-5. Rayleigh wind speed distribution. The average annual wind speed defines the shape of the Rayleigh distribution. As the average annual wind speed increases, the curve shifts toward the higher wind speeds on the right of the chart.

average wind speed is almost twice that derived from the average wind speed alone. This relationship holds for many sites with moderate to strong average annual wind speeds, but it will underestimate the potential at some and overestimate the potential at others. The real world is never as tidy as the mathematical models portray it. In a study of small wind turbine performance, Wisconsin Power & Light found that the Rayleigh distribution produced reasonably good estimates when used to project annual generation. Their estimates were only about 5 percent less than the power actually produced. Monthly estimates were less reliable.

The relationship between the power density derived from the average speed alone, and that from a speed distribution, whether an actual distribution measured at a site or the distribution created by a formula, is commonly called the *cube factor* or the *energy pattern factor*. For example, the cube factor for the Raleigh distribution is 1.9.

Table 3-2

*Effect of Speed Distribution on Wind Power Density
for Sites with Same Average Speed*

Site	Annual Average Wind Speed (m/s)	(mph)	Wind Power Density (W/m²)	EPF or Cube Factor	Wind Power Class (at 10 m)
Culebra, Puerto Rico	6.3	14	220	1.4	4
Tiana Beach, New York	6.3	14	285	1.9	5
San Gorgonio, California	6.3	14	365	2.4	6

Source: Battelle, PNL

Sites in the trade winds of the Caribbean often have high average wind speeds. But the winds are steady. They have few occurrences of extremely high winds (see Table 3-2). At trade wind sites, the Rayleigh distribution tends to overestimate potential generation by 10-20 percent.

On the other hand, sites in mountain passes, such as the San Gorgonio Pass in southern California, are subjected to funneling effects where there are more occurrences of winds at high speed than would be predicted by the Rayleigh distribution. Despite this, FloWind Corporation, a developer of wind power plants, found that wind speeds on top of Cameron Ridge near Tehachapi, California, closely approximate those of the Rayleigh distribution.

Because the speed distribution plays such an important role in determining power, it's always preferable to use an actual measured distribution whenever possible. Battelle Pacific Northwest Laboratory notes that the measured distribution for a site near Ellensburg, Washington, produces a power density of 320 W/m², twice that from a Rayleigh distribution (160 W/m²) for the same average speed of 5.3 m/s (12 mph).

Despite its limitations, the Rayleigh distribution remains a useful tool for most sites. Meteorologists often use a more flexible mathematical formula, the Weibull distribution, which can more closely model the wind speeds at a wide range of sites than the Rayleigh distribution. The Rayleigh distribution is a member of the Weibull family of speed distributions.

Where do we stand now? We can calculate power density in two ways. We can sum a series of power density calculations for each wind speed and its frequency of occurrence (the number of hours per year the wind blows at that speed) for the site's distribution of wind speeds. Or we can use the average wind speed and the cube factor for the Rayleigh distribution of 1.9.

Average Annual P/A = 0.05472 S^3 x 1.9, where S is in mph
Average Annual P/A = 0.6125 S^3 x 1.9, where S is in m/s

Once we know the annual power density at a site, we can quickly estimate how much power a typical wind turbine will intercept.

Estimating the power available to the wind turbine becomes simply the product of power density *P/A*, and the turbine's swept area *A* in square meters.

$$P = P/A \times A$$

We'll discuss this equation more thoroughly in the next chapter. We're not quite ready to estimate how much of this power a wind machine is capable of capturing. There's one step remaining, because the wind turbine will be mounted atop a tower that's typically two to three times taller than the tower used to measure wind speed. How wind speed and power change with height is the subject of the next section.

INCREASE IN POWER WITH HEIGHT
ABOVE 30 FT (10 M)

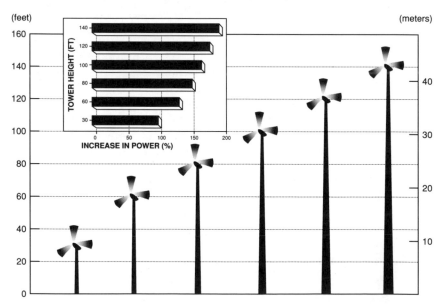

Figure 3-6. *The power available to a wind machine increases with height. The power available at 80 feet (24 meters) above the ground is 150% of that at 30 feet.*

Wind Speed, Power, and Height

Wind speed, and hence power, varies directly with height above the ground (see Figure 3-6). Wind moving across the earth's surface encounters friction caused by the turbulent flow over and around mountains, hills, trees, buildings, and other obstructions in its path. These effects decrease with increasing height above the surface until unhindered air flow is restored. Consequently, as friction and turbulence decrease, wind speed increases.

As you can imagine, frictional effects differ from one surface to another, depending on surface roughness. Friction is higher around trees and buildings than it is over the smooth surface of a lake. In the same manner, the rate at which wind speed increases with height varies with the degree of surface roughness. Wind speeds increase with height at the greatest rate over hilly or mountainous terrain, and at the least rate over smooth terrain like that of the Great Plains. Because of this, it's more important to use a tall tower when siting in hilly terrain than it is on the Texas Panhandle.

At low wind speeds, the change in speed with height or *wind shear* is less pronounced and more erratic. In light or calm winds, as may be encountered during a temperature inversion, wind speeds may increase slightly between the ground and a certain height, and then begin to decrease. Real-world experience has shown that changes in wind speed with height are not constant. In the Altamont Pass east of San Francisco, Pacific Gas & Electric found that above 200 feet (60 meters) wind speeds decreased with increasing height. The utility found this effect after they installed a Boeing Mod-2, the rotor of which was 300 feet (91 meters) in diameter. At times the uppermost part of the rotor would extend above the layer of fast-moving air. On average, however, wind shear is positive, and wind speed increases with height.

Hills, mountains, and numerous buildings often influence wind speed in unpredictable ways. With this caveat in mind, we can use the following formula to estimate the increase in wind speed with height, where S_o is the wind speed at the original height (the height of the anemometer, for example), S is the wind speed at the new height, H_o is the original height, and H is the new height. The exponent α varies with surface roughness and is a measure of the friction encountered by the wind as it moves across the ground.

$$S/S_o = (H/H_o)^\alpha$$

Table 3-3

Changes in Wind Speed with Height

Site	Speed at Height (m/s)		Speed Increase	Approx. α
	9.1 m (30 ft)	45.7 m (150 ft)		
San Gorgonio Pass	6.2	7.7	1.24	0.13
Livingston, Montana	6.8	8.4	1.24	0.13
Clayton, New Mexico	5.4	7.3	1.35	0.18
Minot, North Dakota	6.5	8.4	1.29	0.16
Amarillo, Texas	6.3	8.1	1.29	0.16

Source: Battelle, PNL

Over smooth, level, grass-covered terrain the *one-seventh power law* applies; that is, the exponent for wind shear, α, equals 0.14 or 1/7. This is a widely used rule of thumb for the increase in wind speed with height. It's valid for much of the Great Plains as seen in Table 3-3. For the sites in Table 3-3, which were candidates in the U.S. Department of Energy's large wind turbine program, the increase in speed after a five-fold increase in height was 1.24–1.35 times the original speed.

Once again, the one-seventh power law is just a guide. The actual increase in wind speed with height in Table 3-3, relative to the wind speed at 30 feet, doesn't exactly correspond to the one-seventh power law. But it's close. FloWind also found a wind shear exponent approximating the one-seventh power law (0.15-0.20) at its wind plant in the Tehachapi Pass.

Although wind shear often follows the one-seventh law, it doesn't always. Obstructions significantly reduce wind speeds near the ground. Over row crops such as corn, or over hedges and a few scattered trees, wind speed increases more dramatically with height than the one-seventh law predicts. Wind shear, α, rises to 1/5 (0.20). When the surface is rougher still, say with more trees and a few buildings, α increases further to ¼ (0.25). The speed profile becomes even steeper over woods and clusters of buildings.

The increase in wind speed with height holds true only for the height above the *effective* ground level. The wind rushing over a field of corn sees the top of the corn, not the soil on which it grows, as the *effective* ground level. For a woodlot this is the uppermost point where the branches of the trees are touching, not necessarily the tops of the trees.

Let's see what all this means by examining the effect a tall tower will have on the average wind speed at a site in Kansas where the one-seventh

Table 3-4

Increase in Wind Speed with Height

Above 30 ft (10 m)*

Height		Surface Roughness Exponent (α)			
		1/10	1/7	1/5	1/4
(ft)	(m)	(0.10)	(0.14)	(0.20)	(0.25)
30	9	1.00	1.00	1.00	1.00
60	18	1.07	1.10	1.15	1.19
80	24	1.10	1.15	1.22	1.28
100	30	1.13	1.19	1.27	1.35
120	37	1.15	1.22	1.32	1.41
140	43	1.17	1.25	1.36	1.47
150	46	1.17	**1.26**	1.38	1.50
160	49	1.18	1.27	1.40	1.52

*30 feet is approximately equivalent to a 10-meter anemometer mast.

power law applies. In our example the wind speed was measured at 30 feet (10 meters), H_o and we want to install our wind turbine on a 150-foot (45-meter) tower H. If you don't like working with formulas, all is not lost. This calculation and others have already been made in Table 3-4 for a range of tower heights and wind shear.

$$S/S_o = (H/H_o)^{1/7} = (150/30)^{1/7} = 1.26$$

The wind speed at the height of the wind turbine increased 26 percent by installing it on a 150-foot tower.

Here's how to use Table 3-4 in the previous example. Find the 150-foot row under "Height" and follow along the row until it intersects with the column under the roughness exponent for 1/7. The result, 1.26, indicates that the wind speed at the new height will be 1.26 times that at 30 feet, or 26 percent greater.

But we can't stop here. Power is a cubic function of speed. Where P_o is the power at the original height of 30 feet and P is the power at the new height, the increase in power on the 150-foot tower in our example is given in the following formula. As before, there's no need to calculate the change in power density with height if you're averse to formulas. Just use Table 3-5 with the wind shear exponent appropriate for your site.

$$P/P_o = (H/H_o)^{3\alpha} = (150/30)^{0.43} = 1.99$$

Table 3-5

Increase in Wind Power Density with Height

Above 30 ft (10 m)*

Height		Surface Roughness Exponent (α)			
		1/10	1/7	1/5	1/4
(ft)	(m)	(0.10)	(0.14)	(0.20)	(0.25)
30	9	1.00	1.00	1.00	1.00
60	18	1.23	1.35	1.52	1.68
80	24	1.34	1.52	1.80	2.09
100	30	1.44	1.68	2.06	2.47
120	37	1.52	1.81	2.30	2.83
140	43	1.59	1.94	2.52	3.18
150	46	1.62	1.99	2.63	3.34
160	49	1.65	2.05	2.73	3.51

*30 feet is approximately equivalent to a 10-meter anemometer mast.

This brings us to another rule of thumb: Increasing tower height five-fold increases wind power density nearly two times at sites where the one-seventh power law applies.

Like all rules of thumb, this is only an approximation of the real world. In Table 3-6 the increase in power with height doesn't exactly follow this rule. The reason? The distribution of wind speeds changes slightly at the new heights.

Clearly you can't talk about wind speed without also referring to the height at which it's measured. The two always go together, though sometimes

Table 3-6

Changes in Power Density with Height

Site	Power Density at Height (W/m²)		Power Increase
	9.1 m (30 ft)	45.7 m (150 ft)	
San Gorgonio Pass	351	712	2.03
Livingston, Montana	457	794	1.74
Clayton, New Mexico	162	334	2.06
Minot, North Dakota	271	533	1.97
Amarillo, Texas	228	464	2.04

Source: Battelle, PNL

the height is assumed. In general the average wind speed at a specific site refers to the speed at the height of the anemometer (from 20 to 40 feet in the United States or 10 meters elsewhere). Confusion arises when you start talking to wind turbine manufacturers. The wind speeds they use may be either at the hub height of the wind machine or at some other height. It makes a big difference. Most rate their products at hub height; some don't. Reputable manufacturers and their dealers will clearly state which method they use.

After two decades of working with the wind, we've learned an important lesson—the hard way. No amount of historical wind data can substitute for measuring the wind at the specific site where you want to put your wind turbine. This includes measuring the wind at the proposed height of the turbine to avoid extrapolations that may or may not reflect actual conditions. More on this later. Next we will look at what historical wind data is available.

Sources of Wind Data

This section covers who has wind data, the kind of data they have, and its format. We will also discuss how this information has been updated and presented in easy-to-use maps of wind power density.

In the United States the National Weather Service (NWS) and the Federal Aviation Administration have collected the most extensive records. This data is stored at the National Climatic Center (NCC) in Asheville, North Carolina. The NCC also stores wind data from other federal agencies, including the Civil Aeronautics Administration, the Forest Service, and the Department of Defense.

At some of the stations, wind speed and direction have been recorded for over 30 years. Periodically, this information has been tabulated into a "Summary of Hourly Observations" that provides the long-term average wind speed, as well as the speed distribution. These summaries can be helpful when evaluating the winds at your site.

The NCC data, however, isn't always accurate. Much of the data has been collected at major airports, which are often sheltered from the wind. In general, wind data has been gathered near centers of population. People congregate in areas sheltered from storms and severe weather. (Cities are built more often in valleys than on windswept mountaintops.) Consequently, the data from the airports, military bases, and weather stations may not reflect the winds that exist at more exposed sites. Using data from these stations alone may lead to underestimating the potential wind power in an area. On the other hand, airports offer a vast clearing with minimal trees and buildings. In heavily wooded or developed areas, the open expanse at an airport may offset any sheltering effects of its location.

The data may be unreliable for other reasons as well: The wind-measuring instruments may have been inaccurate or poorly placed. The instruments at many airports were not properly maintained and frequently were located on or adjacent to the terminal building. At these stations, the data reflects the turbulence around the building better than anything else.

Data from remote sites is even more problematic. In some cases wind data was collected only during daylight hours or for a few hours during the summer months. In either case, the data doesn't represent what could be expected throughout the day or throughout the year. At some stations, so few observations of wind speed were recorded that the observer was literally noting whether it was *windy* or not.

In Pennsylvania, for example, data from several military bases was collected during World War II. From their summaries, the areas around these bases look attractive for wind development. Power densities over 200 W/m^2 are common, with some as high as 300 to 400 W/m^2. The truth, unfortunately, is less encouraging. Most of the original data was collected only during the day. Modern measurements reveal that average power densities of 100 to 150 W/m^2 are more realistic.

Historical average speeds are also available from air quality monitoring stations at both conventional and nuclear plants and from some industries. The limits on data from these sources are the same as those on data from airports. We measure air quality where wind speeds are low and where pollutants concentrate, such as in urban canyons and narrow mountain valleys. These are less than ideal locations for a wind machine, and the wind speeds are unrepresentative of better sites.

There are also numerous sources of short-term wind data: state energy offices, universities, nonprofit organizations, and wind system dealers. One or more of them might have collected wind data in your area. Many states (Pennsylvania, Texas, Oregon, Minnesota, California, Wisconsin, and New York are a few) have collected data from a network of instruments. Check around and find out what work has been done and who has the results.

While looking about for local sources of wind data, don't be smitten by the discovery of an abandoned airfield outside town and dash off to North Carolina for a search through NCC's archives. Much of the data from your find will not be of any value. Secondly, in the United States the work has already been done for all federal sources of data, and for many private ones too.

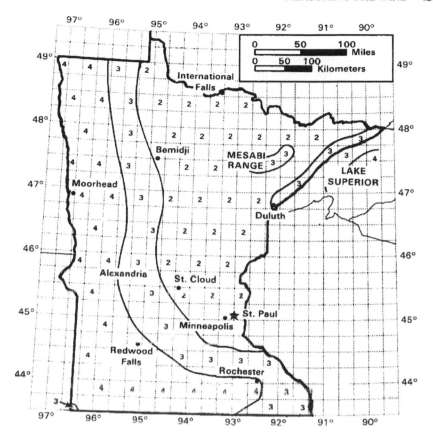

Figure 3-7. Battelle map of annual average wind power for Minnesota. Battelle Pacific Northwest Laboratory's survey of the wind resource in the United States is presented as maps of wind power density in W/m² at 10 meters (33 feet) above the ground. Since these maps were made in 1987, researchers have found that southwestern Minnesota has significantly better winds than those shown. The same may be true elsewhere in the United States.

Battelle Pacific Northwest Laboratory, under contract to the U.S. Department of Energy, has studied NCC's summarized data. They also have gone further and analyzed data from numerous new sources, as well as data from the many short-term recording stations installed specifically to measure wind energy. They have carefully reviewed the records for accuracy and presented their results as maps of average annual wind power density. (See Figure 3-7. Battelle's maps for the United States and the rest of the world are reprinted in Appendix D.)

Battelle's analysis also included adjusting the wind speed data for an-emometer height. Prior to the 1960s, the anemometer could have been mounted anywhere. Since then the height has been standardized at roughly 20 feet (6 meters) above ground in the United States. They present all the collected data at an international standard height of 10 meters.

The Battelle maps of average annual wind power do not give wind power density directly. Instead, they identify portions of the state or region with a numerical rating that corresponds to one of seven *wind power classes*. Each class represents a range of power density. For example, Class 4 repre-sents wind power density from 200 to 250 W/m², or wind speeds from 5.5 to 6 m/s (12.3-13.4 mph). The classes and their corresponding wind speeds and power density are shown in Appendix D.

These maps are derived from computer modeling of the historical wind data, the terrain, and regional weather patterns. The values shown represent only those sites such as hilltops, ridge crests, and mountain summits that are free of obstructions and are well exposed to strong prevailing winds. Esti-mates for ridge crests or mountain summits are shaded to indicate that the actual wind power there could be significantly greater if the terrain accen-tuates local wind speeds, or less if there's terrain interference. By giving a range of possible values rather than a single number, Battelle doesn't lure users into the mistaken notion that they can estimate wind speed with pre-cision.

Denmark's Risoe National Laboratory has done similar work in Eu-rope. Their *European Wind Atlas* (See Appendix I) provides a detailed look at the wind resources of the 12 countries in the European Community: Ireland, Great Britain, Denmark, Germany, France, Belgium, Luxem-bourg, Spain, Italy, Portugal, Greece, and the Netherlands. Like Battelle, Risoe presents the data as a range of values. But it goes further and suggests the wind speeds likely at sites with differing roughness, such as along coastlines. The European data is also available as a computer program for professional meteorologists.

Since its publication in the late 1980s, the European atlas has been used as a model by Finland, Sweden, Algeria, and Jordan to produce similar re-ports. Atlases for Poland, Canada, Egypt, Syria, and Turkey are also being prepared.

Surveying Your Site

To evaluate the potential at your site, begin by asking yourself, what is it that you want? Is it the instantaneous wind speed, the average annual speed, or the distribution of wind speeds? If you want to know when it's too windy to go sailing, instantaneous wind speed will suffice. If you want to estimate the annual energy output from a wind machine, then at least the average wind speed is necessary. Preferably you'll want the wind speed distribution as well. If you plan to use the wind turbine as your sole source of power at a mountaintop retreat, for example, then more detailed information may be required, such as the number of calm days and the time between them.[5]

In most cases you will be interested in how much energy a wind machine can produce in your area: more specifically, at your site. To estimate annual energy output, the speed distribution is preferred. The speed distribution gives you the most accurate results, but it isn't always necessary. Average speed will often suffice.

Now that you know what's needed, find out if someone has already done the work for you. Start by locating anyone nearby who has installed a wind turbine. "Talk to oldtimers," says Mick Sagrillo. They may remember someone who once used a windcharger. Pay them a visit.

Or find someone with an anemometer. What's his estimate of the average speed or power? Before you take any of his data to heart, determine if your site and his are comparable. Is the data typical of what you could expect? If so, you've saved a lot of time and effort. Normally you won't be so lucky and you'll have to make your own survey.

Start by looking at the vegetation. Trees and shrubs are frequently touted as a good qualitative indicator of wind speed. High winds and a harsh environment of ice and snow will deform them. The severity of the deformation, whether the tree is slightly flagged or completely bent to the ground, can be used as a rough gauge of wind speed. The types of deformity are,

Brushing: branches and twigs sweep downwind. This can be observed on both conifers and deciduous trees.

Flagging: branches sweep downwind, upwind branches are cropped short.

Throwing: trunk sweeps downwind.

Carpeting: trunk bends to the ground. Found frequently in Alpine or severe environments where trees grow only a few feet above the ground.

Use the scale of deformity in Figure 3-8 to find the range of wind speeds represented.

There are limitations to this technique. First, don't get excited by one or two odd-shaped trees. One flagged tree is insufficient to make a

---

5. An excellent guide through this process is Battelle's *A Siting Handbook for Small Wind Energy Conversion Systems*. See Appendix I for further information.

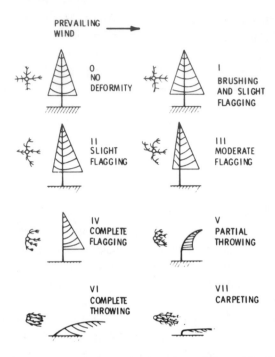

Figure 3-8. The degree to which conifers have been deformed by the wind can be used as a rough gauge of average annual wind speed. (Battelle PNL)

Griggs-Putnam Index of Deformity

Wind Speed	Index I	II	III	IV	V	VI	VII
mph	7-9	9-11	11-13	13-16	15-18	16-21	22+
m/s	3-4	4-5	5-6	6-7	7-8	8-9	10

judgment. There are many other causes for tree deformity besides the wind. You'll need to note several of the same species with an equal amount of deformation to determine if the wind is the cause. (Each species varies in its susceptibility to flagging.) Conifers, especially pine and fir trees are more reliable indicators of wind strength than are deciduous trees. Moreover, deformation is more obvious where freezing salt spray or ice frequently accompanies high winds. Such conditions are often found along coastlines and on mountaintops.

If you can't find any deformation, don't despair. The absence of flagging doesn't necessarily indicate a low speed. Too often the value of trees as a wind speed indicator is overplayed. At best, it gives only a crude range of possible wind speeds, and even then, it's most useful only where conifers dominate.

Next, find the nearest airport or other station where long-term records have been kept. Convert the wind data to power density and compare that with wind atlases for your area. Are they similar? Is the airport representative of your site, or is it better exposed or more sheltered? If the airport lies down on the valley floor and your site is on a plateau overlooking the valley, your site may experience much higher winds.

Now, put the pieces together. Estimate what you think is your average wind speed and power. Give yourself room for error. Avoid pinning your hopes on one number alone. Instead of saying 10 mph, give a range of 9 to 10 mph (or 10 to 11 mph, if you think the site might be a little windier). In other words, present your estimate as a wind power class. The 9-10 mph range falls within Battelle Class 2, for example. Be conservative. Most people overestimate the amount of wind available.

With these numbers in hand, use the techniques in the next section to estimate the output from several different wind machines. Choose the one that delivers the amount of energy you need and determine its cost. Call the manufacturer or wind system dealer for realistic estimates. Look at the economics.

Next, ask yourself how much risk you're willing to assume. Even though you may have done an admirable job of estimating the wind regime in your area from published sources, obstructions and terrain features can greatly reduce the actual wind energy available. If you don't mind this uncertainty, or if you're on the Great Plains and there isn't a tree for miles, then there's little need to go to the trouble of conducting a full-fledged wind speed survey. Measuring the wind at your site only becomes necessary when you're unwilling to take the risk that there's sufficient wind to produce what you expect.

For commercial wind power plants the economic risk is too great to proceed without on-site measurements. Professional wind developers always measure the wind first. But for residential and small commercial users this isn't always necessary. In Denmark and the Netherlands the wind resource is well known, and seldom do those installing individual machines measure the wind first. However, the United States, Canada, and Mexico span a continent with vast differences in terrain from one locale to the other. The Danish system doesn't work well on such a grand scale.

Assuming that wind machines generally make sense where average wind speeds are 12 mph (6 m/s) or greater and are uneconomic where wind speeds are less than 9 mph (4 m/s), an extended site survey is not needed in the following cases:

1. When a nearby airport has an average wind speed of 12 mph (6 m/s) or greater, or a wind power density of 200 W/m^2, and your site is within 10 miles over smooth terrain.

2. When a nearby airport has an average wind speed of 12 mph (6 m/s) or greater, or a wind power density of 200 W/m^2, and your site is more exposed, e.g. on a mountaintop.

3. When the average wind speed is less than 9 mph (4 m/s), or the power density is less than 75 W/m^2.

4. When there are numerous tall trees and buildings surrounding your site.

In general, a site survey is useful for small wind turbines when the estimated average wind speed is 10 mph (5 m/s) or less.

In the United States, site surveys may also be necessary when the estimated average wind speed is less than 12 mph (6 m/s) and the application requires a wind turbine 10 meters in diameter or larger. This is necessary in some areas because the wind turbine will compete with electricity that declines in price with increased consumption. Small wind turbines, in contrast, will offset the most expensive electricity.

Measurements, if they are to be made at all, should be taken at the intended location of the wind machine and at its proposed height. This is particularly important in rough terrain or where there are obstructions. Reliable measurements can only be made when the anemometer extends well above nearby trees and buildings.

Many past site surveys in the United States were often performed with 30-foot masts. The 30-foot mast is comparable to the 20-foot anemometer height at most U.S. airports and is close to the World Meteorological Organization's standard anemometer height of 10 meters. Data from a 30-foot mast, for example, could be compared directly to airport data without adjusting for height.

Wind machines shouldn't be installed on towers less than 80 feet tall. Consequently, it doesn't make any sense to install an anemometer on a 30-foot (10-meter) mast if the site is sheltered by trees, buildings, or the terrain. Even tall corn and low-lying shrubs raise the effective ground level, severely reducing the wind speed measured by the anemometer. If you need to measure the winds at your site, take a survey of the trees and buildings nearby and estimate their heights. You may find that the anemometer, and eventually the wind turbine, should be erected elsewhere.

Estimating the Height of Obstructions

Remember those comic pictures of a ragged artist thrusting her thumb at the world. There's a good reason for it. Artists use the technique to gauge proportions. You can use it to estimate the height of nearby trees and buildings. A pencil works better though. Here's how to use it.

Identify an object of known height at about the same distance from you as the tree or building you wish to measure. TV antennas, telephone poles, and houses work well. Hold the pencil at arm's length and sight along it. Line up the top of the pencil with, for example, the top of a tree. Slide your thumb down the pencil until it lines up with the bottom of the tree. Now turn to the object of known height and again sight along the pencil. While keeping the pencil at arm's length, move your arm up or down until your thumb lines up with the bottom of the object. Judge the proportions by noting how much of the pencil extends above the object. Is it twice the height, one-third greater, or the same?

Say you're using a TV antenna for comparison. Antennas in the United States are often assembled from two 10-foot sections (20 feet, or about 7 meters). The tree appears to be twice the height of the antenna, or 40 feet (12 meters) tall. Reverse the procedure and see if you get the same result. This method is quick, easy, and will give you a good idea how tall a tower you'll need.

Assume that the wind data you've examined is unconvincing. You want to measure the wind at your site to get a better picture of what's there. What next? Anemometers measure wind speed. Wind vanes indicate direction. That's simple enough. More complex is the kind of recording equipment you'll need. In the next section we'll go over the equipment that's on the market and discuss what probably meets your needs best.

Measuring Instruments

To perform a wind resource assessment you'll need an anemometer and recorder (see Figure 3-9). Instruments designed for the wind turbine market are generally less expensive, easier to use, and give you more of the information you seek than do those designed for meteorological use.

A wind-measuring instrument is composed of two parts: the sensor (the anemometer head), and a means for displaying the data it measures. The sensor generates an electrical signal that's proportional to wind speed.

Figure 3-9. Anemometer, mast, and electronic recorder. Measuring the wind can be as simple as pushing a button and writing down the data.

The NWS's sensor, for example, uses a three-bladed propeller to drive a DC generator (a miniature wind machine). Changes in wind speed cause a change in DC voltage that is then read by a calibrated voltmeter. The needle swings to and fro like the speedometer needle in your car. To record the data, the anemometer must be read manually.

Cup anemometers, however, are far more common. The spinning cups drive either a DC generator or AC alternator. With the advent of pocket calculators, digital displays have become popular. (The data is the same, it just looks different.) Better-quality instruments typically use AC alternators and measure frequency. These anemometers are more accurate than those using DC generators.

Whatever system is used, either the sensor (the anemometer head) drives a meter that displays instantaneous wind speed or it drives a recorder. Unlike the displays of cheap wind speed meters that indicate only instantaneous wind speed, recorders store information for future retrieval. At one time all meteorological data was recorded on strip charts, but wind prospecting and the boom in electronics have revolutionized wind speed measurement.

Before we go any further, let's clear up a common misconception. Instantaneous wind speed indicators are useless for finding the average wind speed. They're fun to watch, but that's it. To be of value in a site survey you would have to check them every hour, 24 hours a day, for months on end. Instantaneous wind speed meters are helpful for developing a better understanding of the wind. But use them only to calibrate your wind sense, or to gauge the performance of a wind machine after you've installed it.

Strip-chart recorders are now obsolete. Their chief drawbacks were the tedious steps needed to analyze the data and their inaccuracy. If you can find a cheap one (less than $100), have plenty of time and patience, and can ensure accuracy, then a strip-chart recorder may be for you. An electronic odometer or accumulator is an overall better choice.

Similar to the odometer on the dashboard of your car, an accumulator counts the miles or kilometers of wind that pass the anemometer. To estimate average wind speed, simply divide the distance the odometer records by the elapsed time between readings. In North America this gives an average wind speed in miles per hour. For example, if 240 miles of wind have passed the anemometer within a 24-hour period, the average wind speed for that day is 10 mph.

Early electronic odometers required the observer to keep track of the elapsed time. Today most instruments do that for you, as well as much more. Zond Systems offers one of the least costly wind-run odometers on the market. For less than $400, Zond's Windrunner will measure and store wind data as well as give instantaneous wind speed. For slightly more (about $500), NRG Systems' Wind Challenger will collect, process, and store average wind speed, elapsed time, peak gust, and power density (in W/m^2). It will also record the amount of time the wind was calm (for sizing batteries in stand-alone systems), and how much time the winds were above the cut-in speed of typical wind turbines. The Challenger will record the data in either mph or m/s and will even give you instantaneous wind speed. An indoor version capable of reading signals from an anemometer up to 1000 feet (300 meters) away costs about $350.

Instruments like NRG's Challenger and Zond's Windrunner represent the bare minimum for professional surveying of the wind energy at a site. As Mick Sagrillo of Lake Michigan Wind & Sun emphasizes, you need more than merely the average wind speed. You need some measure of the distribution of wind speeds to accurately assess the potential of a site. The Challenger does that inexpensively by calculating power density and storing a running tally. To record the actual speed distribution requires a significant jump in sophistication and cost.

As the need for more sophisticated measurements has increased, accumulators have evolved into data loggers. In essence, data loggers consist of multiple accumulators that tally the data falling into each accumulator's domain. For example, each accumulator could represent a given wind speed range. Winds 0-1 mph would fall into the first accumulator, winds 2-3 mph would fall into the second, and so on. At the end of the observation period,

Figure 3-10. The serial data logger is used for electronically recording wind measurements. (NRG Systems)

the contents of each register can be used to plot the speed distribution. This distribution can then be used to calculate power density or can be compared with a wind turbine's power curve to project potential performance.

Most data loggers process some of the data as measured and record the results. These *smart* recorders significantly reduce the time and expense of analyzing the data later. By reducing the amount of data that must be stored, they can record data for longer periods. NRG builds a data logger ($1300) that records wind speed over one of four user-selectable averaging periods on a removable chip (see Figure 3-10). The data stored on the chip must then be analyzed on a personal computer. Second Wind's Nomad ($1450) performs a similar function, as does Zond Systems' Model E Windrunner ($1200). Unlike NRG's logger, Second Wind uses a removable random access memory (RAM) card. Like the chip used by NRG, the RAM card requires a computer to process the data. (Second Wind argues that a RAM card provides greater versatility.) Data is retrieved from Zond's Windrunner with a notebook computer.[6]

The degree of sophistication represented by a data logger is unwarranted for most residential and small business applications. Such sophistication can lead to a false notion that estimates based on the data are more reliable than those from simpler recorders. (The potential for error in estimating wind turbine performance is considerable, and it's not limited solely to wind measurements.) Likewise, wind direction is unnecessary for most siting studies. We perform a site survey as an aid in estimating potential wind energy. We don't really care what direction the wind comes from, just that it is there. Whether it is or is not will be measured by the anemometer.

Anemometer Towers

In the United States and Canada simple anemometer towers can be made from 16-gauge TV antenna masts. The 10-foot sections can be found in any building supply or hardware store. Using three sections, a tower 29 feet (about 10 meters) tall can be erected with two guy levels. At some stores a 5-foot section is available. This can be added to the tower to bring the total height to 34 feet—the limit to this type of tower (the sections are too weak to support anything taller). If you use this type of tower, guy it at two levels: one at the midpoint of the second section and one at the midpoint of

---

6. None of these data loggers record a true wind speed distribution. They approximate that function by logging a series of average speeds, often hourly. Subsequent computer processing of the data produces an approximate distribution of wind speeds during the sampling period.

the third section. Use four guy cables and anchors at each level. Four guy cables allow the tower to be raised by tipping it into place.

Unarco Rohn's telescoping masts or the equivalent are a better choice. Each bay is approximately 10 feet long. Masts with four and five bays are available. Taller towers can be assembled from a Rohn 25-G guyed tower. This is an expensive and labor-intensive approach, and can be justified only on commercial installations. If you go through this much trouble for a residential site survey you might as well install a small wind turbine instead of the anemometer. (The Rohn 25-G will support such small wind turbines as the Bergey 1500.)

Many homeowners try to cut corners by installing the anemometer on the roof of a building. Unfortunately, turbulence around the building lowers wind speed dramatically. Avoid this practice. Most of the time, mounting an anemometer on a building is impractical and a waste of time. Where do you attach the guy cables, for example? Will it be necessary to drill holes in the roof for anchors? If that doesn't stop you, imagine trying to erect a 30-foot mast on a steeply pitched roof; it's an accident waiting to happen.

From bitter experience, wind prospectors have learned there's absolutely no substitute for measuring the wind at the height where you plan to install your wind turbine. Since even at the best sites the wind turbine will be installed on a tower at least 80 feet (24 meters) tall, you'll need an anemometer mast of equivalent height. Fortunately, NRG Systems builds hinged masts designed specifically for this purpose (see Figure 3-11). They're not cheap—a 70-foot (20-meter) mast costs about $800—but they do their job well and can be erected by two people in a matter of hours. Their ease of use has made NRG's TallTowers™ the defacto standard for professional wind resource assessments worldwide. NRG also provides taller towers: 90, 110, and 140 feet in height, as well as metric versions reaching 30 and 40 meters.

One note of warning: Kevin Cousineau of Zond Systems urges extreme caution when erecting anemometer masts around overhead power lines. Cousineau also suggests securely grounding the anemometer tower—not the instruments—to guard against damage from electrical storms.

Survey Duration

"Yeah, you got a fine site here," said the dealer as he installed the anemometer George had ordered. "I bet you've got 12 mph."

Two days later the dealer returned. After examining the anemometer he

Figure 3-11. NRG Systems' lightweight, hinged anemometer mast simplifies wind measurements at remote sites.

said, "Just as I thought, an easy 12 mph average." The dealer then persuaded George to buy his wind machine.

A wet finger in the air on the first visit would have been just as accurate. Maybe the dealer didn't know how to measure wind speed. Then again, maybe he was a con artist. It's hard to tell. The site was indeed a good one, and the wind turbine (made by a reputable manufacturer) was installed in a workmanlike manner. The site could have had a 12-mph average wind speed, but the dealer or the buyer wouldn't have known that after 2 days of measurement.

How long is enough? That's another tough question. Average wind speeds can vary as much as 25 percent from year to year. But it's obviously impractical to gather 10 years of data from a site before you decide whether or not to install a wind machine. Battelle suggests gathering 1 year of data. Even so, your site's average speed will depend on how normal the year has been with respect to the long-term average. Was it a typical year, or was it windier? To answer that question, you must examine the wind data from the nearest airport or other long-term recording station and compare the test year's results with the station's historical average.

If you use an airport in the United States for this comparison you can order monthly summaries of local climatological data.[7] At the end of the year, NCC issues an annual summary. This summary includes the history of the anemometer, its location and height, the average daily wind speeds, and the average 3-hour speeds as well. The summary in comparison to the historical average will give you a good idea how typical the winds were during your year of measurements.

Try to establish a correlation between your site and the airport using the monthly summaries. If you're lucky, you may find that a full year of measurements isn't necessary. But 4 months are a minimum. Anything less is guesswork. If you're not going to gather a full year of data, make sure that you at least capture the wind season, typically winter and spring months for much of the United States, because the correlation may change with average speed.

Data Analysis

Making sense out of the data from a site survey is more akin to alchemy than science. Much is left to the judgment (or imagination) of the observer. You must determine whether the data is representative of the site and not

---

7. National Climatic Center, Federal Building, Asheville, NC 22801.

surrounding obstructions, whether the year is normal, and whether or not there is a direct relationship between the site and, for example, a nearby airport with long-term records.

1. Calculate the ratio between your site's average speed and the airport's.
2. Adjust the airport's historical average by this ratio.

Step one establishes whether your site is more or less windy than the airport. Step two normalizes the results for a typical year.

This approach assumes that a correlation exists between your site and the airport. There may not be one, particularly in rough or mountainous terrain. Try to use an airport in terrain similar to yours and with a similar exposure to the wind. The nearest airport may not always be the best choice.

Another method is to use linear regression analysis. This technique gives a measure of the ratio between the two sites, by testing the degree of correlation, and projects the site's average speed. Linear regression analysis is the same as graphically drawing the best fitting line between two sets of data. Pocket calculators with engineering or statistical functions make the job a cinch. Consult your calculator handbook for details.

You'll find that hourly or daily wind speeds are often too erratic to establish a correlation between the two sites. Average weekly speeds are more stable. They tend to smooth out the passage of weather systems and local diurnal variations.

Let's assume that you installed a wind-run odometer. Record the data from the odometer at least weekly, more often if you like. Note the date, time, and the count on the register.

The key to reliability is frequent inspection. Site visits can reveal a failure of the anemometer or the recorder. The more frequently you read the odometer, the less data you're likely to lose in the event of a malfunction.

You're not finished yet. What you're trying to derive is the power in the wind, and average speed is only part of the picture. You need to find some measure of the speed distribution as well. Once again, the best way to do this is to measure it. NRG's Wind Challenger and similar instruments provide a shortcut to this function by calculating power density internally. You don't need to touch a calculator. If you use simpler wind-run odometers, you'll have to assume that your site has a Rayleigh distribution and use the cube factor for the Rayleigh distribution of 1.9. With your wind speed and power estimates in hand, you can now estimate how much a typical wind turbine will generate using the techniques explained in the next chapter.

Another Option—Install a Small Wind Turbine

After all this you may have had enough. If you don't want to go through the trouble and expense of performing a site survey, there's one avenue left. "Install a small wind machine," says Mick Sagrillo, a remanufacturer and dealer of small wind turbines in Wisconsin. The idea sounds crazy at first. But it makes sense the more you look at it.

You can install a micro wind system for about $1500, not much more than the cost of NRG's TallTower™ and recording anemometer. Agreed, this isn't a low-cost way to test the wind, but it works. What you get is an operating wind system. You gain hands-on experience and you learn exactly what you want to know: how well a wind machine will perform at your site, that is, how much energy it will generate. Sagrillo points out that if the turbine doesn't work as well as you expected, it's far easier to resell a used wind turbine and tower than a used anemometer and mast. There are simply more people wanting small wind turbines than anemometers.

To take advantage of this strategy, it helps if you want to eventually use a larger machine. Installing the micro turbine enables you to gradually phase in your wind generation as you prove its worth. For best results you need to closely monitor the performance of the wind turbine just as you would record the wind speeds from an anemometer.

If you found, for example, that a 7-meter wind turbine would meet your needs best (it would fulfill your energy and cost-effectiveness goals) you could begin with a 0.5-kW micro turbine. Then, when you have successfully demonstrated that wind works at your site, you could graduate directly to the 7-meter, 10-kilowatt machine. If that's still too big a leap, you could install a 3-meter, 1.5-kilowatt turbine instead.

There are a couple of ways to do this. You could install a foundation and anchors suitable for the 7-meter machine and erect the smaller turbine on a light-duty tower. Or you could install the small turbine on a heavy-duty foundation and tower. In the first case, the tower and turbine would be traded in for a heavy-duty tower and the 7-meter machine. In the second scenario you would trade in only the small turbine for the bigger one. This won't work with all wind systems, as towers and turbines may not be interchangeable, but it gives you much more flexibility than confronting an all-or-nothing decision.

In the next chapter we'll put this wind data to work as we estimate how much energy you can expect from typical small wind machines.

4

Estimating Output—
How Much to Expect

Once we have defined the site where we want to put a wind machine and determined how much wind is available to it, we can proceed to the next step: estimating the amount of energy that typical wind machines can produce. With an estimate of the annual energy output (AEO) in hand we can then examine the economics of various sizes and brands of wind machines to find the one that offers the most for our money.

There are three methods that we'll use. The first is a "back-of-the-envelope" technique using the swept area of the wind turbine. With it you can quickly evaluate the potential output of any wind machine by finding the average speed or power and simply sizing up its rotor. If you do it often enough, the technique will become so familiar you will be able to do it in your head. The second method is more involved and requires a speed distribution for your site and a power curve for each wind turbine you would like to evaluate. The third approach uses manufacturers' published estimates for typical wind regimes.

As in the preceding sections, formulas are presented to describe precisely what we're doing (so you know where the numbers come from). They also aid in the explanation of important concepts. As before, where formulas are used a table will be included summarizing the results of calculations for a standard set of conditions.

Swept Area Method

Our first step is to find the power in the wind—the power density in W/m². Once we have determined power density (by whatever means necessary) we can easily estimate the potential power output from a wind machine. All we need is the area swept by the wind turbine's rotor.

Think about it for a moment. What captures the wind in a wind machine? Is it the tower, the transmission, the generator? No, of course not. It's the spinning rotor. Yet this concept is difficult for many to grasp, even those working in the wind industry. The tower is important, as we have learned, and so is the transmission and the generator. But they are not directly responsible for capturing the wind. Inevitably many look at the size of the generator first. Yet the generator tells you very little about the size of the wind machine. The rotor says it all.

Say we're considering the installation of Northern Power Systems' HR3 on the Great Plains. This wind machine uses a three-bladed rotor 5 meters (16 feet) in diameter. Since radius is half the diameter, the area for the HR3 becomes:

$$A = \pi R^2 = \pi(5/2)^2 = 19.63$$

The rotor disc sweeps nearly 20 square meters. Let's also assume that the turbine intercepts annual average winds at hub height of 250 W/m². From an earlier equation:

$$\text{Power} = \text{Power density} \times A,$$

where the power density is given in W/m² and *A* is the area swept by the rotor in square meters.

$$\text{Power} = 250 \text{ W/m}^2 \times 20 \text{ m}^2$$
$$= 5000 \text{ W}$$

The wind machine in our example intercepts 5000 watts of power or 5 kilowatts (1000 watts = 1 kilowatt).

Now you can see why the metric system is used to describe rotor size. Using meters makes the job easier because wind power is always given in units of W/m².

Power, unfortunately, is not what we are seeking. Energy is. When you pay your utility bill you are not paying for the power you used, you are paying for energy.[1] The amount of energy consumed is the product of power and time—how long the power was used. In the case of a wind machine, the

---

1. Commercial consumers do pay a "demand" charge for the instantaneous amount of power they use at any one time. Most small wind turbines are not paid for the power they deliver, only the energy.

energy it intercepts is a function of average power and how long it is available. In this example we're using the average annual wind power. There are 8760 hours in a year. Consequently, this wind turbine will intercept about 44,000 kWh of energy that pass through the rotor annually.

$$\text{Energy} = \text{Power x Time}$$
$$= 5 \text{ kW x } 8760 \text{ h/yr}$$
$$= 43,800 \text{ kWh/yr}$$

That's not what the wind machine will produce, because we can't capture all of it. If we did, the wind would come to a halt at the rotor and nothing further would happen. The maximum that we can capture at the rotor, the theoretical limit, was derived by a German scientist, Albert Betz. He theorized that a portion of the wind must keep moving through the rotor, and not all of it could be captured. The *Betz limit* is 59.3 percent of the energy available to the rotor.

In practice, wind turbine rotors convert much less than the Betz limit. Optimally designed rotors reach levels slightly above 40 percent. Usable energy is even less because energy is lost in transmissions, generators, and power conditioning (the equipment necessary to convert the energy into a form we can use). There are also losses due to rapid changes in wind speed and direction that are not accounted for in our simple formulas.

Well-designed drive trains operate consistently above 90 percent efficiency. The efficiency of generators, on the other hand, varies significantly depending on how they are loaded. When running at their rated output where they were designed to operate best, generator efficiency can also be above 90 percent. But wind turbines drive their generators at the rated output infrequently. Much of the time the generator is partially loaded and its efficiency suffers as a result. Power conditioning on some interconnected wind turbines can also contribute to significant losses.

Conventional wind turbines also miss some of the wind available. Unlike an anemometer, which measures gusts, a wind turbine does not respond to all gusts because of the turbine's inertia. The energy available in a gust as registered by an anemometer may not be used by the wind turbine. It may not "see" the gust, particularly if the speed of the gust goes above the turbine's operating limits. Yawing or changing the direction of the wind turbine as the wind changes direction causes a similar problem for conventional wind machines. Cup anemometers capture wind from all directions. But a conventional wind turbine takes time to change its position; thus it misses a portion of the wind recorded by the anemometer. Vertical-axis wind turbines, because they are omnidirectional, capture the wind from all directions like the anemometer.

When you put all this together, a well-designed wind turbine using a rotor performing at two-thirds of the Betz limit can deliver about 30 percent of the overall energy available. This is what you can get out of the wind and actually put to use.

Rotor		Transmission		Generator		Power Conditioning, Yawing, and Gusts		Overall Conversion Efficiency
40%	x	90%	x	90%	x	90%	=	29%

In practice wind turbines capture 12-30 percent of the energy in the wind, depending on the wind turbine and the wind regime. Wind turbines are designed for use in specific markets under specific wind conditions. Outside of these wind regimes the wind machines will perform less optimally. That's not to say they're less cost-effective, just that the overall conversion efficiency is lower. For example, wind turbines at high wind sites, such as those in California's mountain passes, will encounter far more periods of extremely high winds where the turbine either furls, turns itself off, or otherwise limits its production than at lower speed sites on the Great Plains. Typically, small wind turbines will convert 25-30 percent of the power in the wind at sites with average speeds below 12 mph (5.5 m/s) and less than 20 percent at windier sites. Medium-sized machines perform better at high wind sites than small turbines.

Assume Northern Power Systems' HR3 converts 20 percent of the energy in the wind to usable electricity. Therefore it will generate nearly 9000 kWh per year at our hypothetical site.

$$20\% \times 44,000 \text{ kWh/yr} = 8800 \text{ kWh/yr}$$

This is a reasonable amount of energy to expect from a 5-meter wind turbine at a modest site on the Great Plains with average wind speeds at hub height of about 13 mph (6 m/s).

When we look at the product literature describing a wind turbine, the swept area isn't always apparent, although most manufacturers now list it along with other measures of performance. What is always obvious, or clearly stated in the literature, is the rotor's size. For conventional wind machines the rotor's diameter is given. For vertical-axis wind turbines (those like the Darrieus or eggbeater that rotate about a vertical axis) the

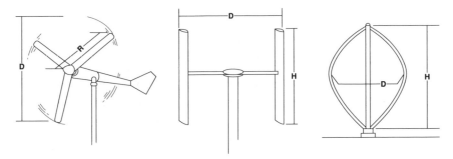

Figure 4-1. *Dimensions necessary for finding the swept area of three wind turbine configurations. D=diameter, H=height, R=radius.*

height of the blades and the rotor's diameter are usually both provided (see Figure 4-1).

If the swept area of a conventional wind machine isn't provided, you can calculate it by using the formula for the area of a circle; for an H-rotor use the formula for the area of a rectangle; and for Darrieus rotors use the formula approximating the area of an ellipse in Table 4-1.

To summarize, the steps necessary to estimate the AEO of any wind turbine are:

1. Find the wind power density P at the site and height where the wind machine will operate,
2. Find the area A swept by the wind turbine's rotor,
3. Assume a reasonable value for the conversion efficiency of the entire wind system,
4. Find the product of your results of steps 1, 2, 3 and the number of hours in a year.

Table 4-1

Formulas for Calculating Swept Area

Type of Wind Turbine	Formula for Swept Area
Conventional rotor	$A = \pi \times R^2$
H-rotor	$A = D \times H$
Darrieus rotor	$A = 0.85 \, D \times H$

where R is the radius of the rotor (1/2 the diameter), D is the diameter, and H is the height of the blades on a vertical-axis wind turbine.

Thus,

$$AEO = P/A \times A \times \% \text{ efficiency} \times 8760 \text{ h/yr.}$$

You can get the feel of this technique by working through another example. Assume you're planning to install a Bergey Excel at a site with an average wind speed of 5 m/s (about 11 mph) measured at about 10 meters (30 feet) above ground. The 7-meter turbine will be erected on a 100-foot tower. The wind speed distribution at the site is similar to that of the Rayleigh distribution, and wind speed increases with height according to the one-seventh power law. The turbine converts about 20 percent of the energy in the wind to electricity.

Solution: Wind Speed at 100 feet = 5 m/s x 1.2 = 6 m/s
Power Density at 6 m/s = 253 W/m²
Swept Area = π (3.5)² = 38.5 m²
AEO = 253 W/m² x 38.5 m² x 20% x 8760 h/yr
= 17,000 kWh/yr

If you have an aversion to such calculations, use Table 4-2. It summarizes the results for a range of average wind speeds in m/s at hub height for small wind turbines of various sizes. The same table is repeated in Appendix E for wind turbines up to 43 meters (141 feet) in diameter.

Table 4-2

Estimated Annual Energy Output

at Hub Height in Thousand kWh/yr *

Average Speed (m/s)	(mph)	Power Density (W/m²)	Total Effic.	Rotor Diameter, m (ft) 1 (3.3)	1.5 (4.9)	2 (6.6)	3 (9.8)	4 (13.1)	5 (16.4)	6 (19.7)	7 (23.0)
4.0	9.0	75	0.28	0.1	0.3	0.6	1.3	2.3	3.6	5.2	7.1
4.5	10.1	110	0.28	0.2	0.5	0.8	1.9	3.4	5.3	7.6	10
5.0	11.2	150	0.25	0.3	0.6	1.0	2.3	4.1	6.5	9.3	13
5.5	12.3	190	0.25	0.3	0.7	1.3	2.9	5.2	8.2	12	16
6.0	13.4	250	0.21	0.4	0.8	1.4	3.3	5.8	9.0	13	**18**
6.5	14.6	320	0.19	0.4	0.9	1.7	3.8	6.7	10	15	20
7.0	15.7	400	0.16	0.4	1.0	1.8	4.0	7.0	11	16	22
7.5	16.8	490	0.15	0.5	1.1	2.0	4.6	8.1	13	18	25
8	17.9	600	0.12	0.5	1.1	2.0	4.5	7.9	12	18	24
8.5	19.0	720	0.12	0.6	1.3	2.4	5.3	9.5	15	21	29
9	20.2	850	0.12	0.7	1.6	2.8	6.3	11	18	25	34

*Assumed effeciency based on published data.

Understanding Table 4-2

To estimate the potential generation from a 7-meter wind turbine installed on the Great Plains where the average annual wind speed at hub height is 6 m/s (about 13 mph) follow these steps. Find row 6.0 in the first column of Table 4-2. This is the wind speed in m/s. Then move along the row until it intersects with the last column, labeled 7. This is the wind turbine's rotor diameter in meters. The value where row 6.0 and column 7 intersect, 18, is the estimated generation in thousands of kilowatt-hours per year.

The overall conversion efficiency assumed in Table 4-2 is given in the column labeled "Total Effic." The assumed efficiencies have been derived from a survey of wind turbine manufacturers worldwide. The results are approximations and won't necessarily correspond to estimates by the manufacturer for the same conditions. According to Table 4-2, a 7-meter turbine under these conditions will produce approximately 18,000 kilowatt-hours per year, which compares well to that calculated for the same conditions.

Now let's try another example. In this case assume you don't know the average wind speed but you've identified the location on Battelle's wind power maps in the *Wind Energy Resource Atlas of the United States* (see Appendix I) and found it's a Class 3 site with a wind power density of 150 W/m². You want to know what you can expect if you install a Bergey 1500 on a 100-foot tower. The Bergey 1500 uses a rotor about 3 meters in diameter.

Battelle's wind power classes are based on wind speeds at 10 meters above the ground. They've assumed that the increase in wind power with height for most sites corresponds to the one-seventh power law. The 100-foot tower is equivalent to the 30-meter height used by Battelle in their assessments of the nation's wind resource. At a height of 100 feet the Bergey 1500 will intercept winds averaging about 6 m/s (13 mph).

Solution: Power Density at 10 m = 150 W/m²
Power Density at 30 m = 240 W/m²
Swept Area = $\pi (1.5)^2 = 7.1$ m²
AEO = 240 W/m² x 7.1 m² x 20% x 8,760 h/yr
    = 3,000 kWh/yr

Table 4-3 summarizes the results of similar calculations for a range of Battelle wind power classes. By using Table 4-3 you can reach approxi-

Table 4-3

Estimated Annual Energy Output

at 30 m (98 ft) Hub Height in Thousand kWh/yr

Class	Battelle Power Class at 10 m Power Density (W/m²)	Speed (m/s)	Wind Speed and Power at 30 m Hub Height Power Density (W/m²)	Speed (m/s)	Total Effic.	Rotor Diameter, m (ft) 1 (3.3)	1.5 (4.9)	2 (6.6)	3 (9.8)	4 (13.1)	5 (16.4)	6 (19.7)	7 (23)
1	50	3.5	80	4.1	0.28	0.2	0.3	0.6	1.4	2.5	3.9	5.6	7.6
2	100	4.4	160	5.1	0.25	0.3	0.6	1.1	2.5	4.4	6.9	9.9	13
3	150	5.0	240	5.9	0.21	0.3	0.8	1.4	3.1	5.6	8.7	12	17
4	200	5.5	320	6.5	0.19	0.4	0.9	1.7	3.8	6.7	10	15	21
5	250	6.0	400	7.0	0.16	0.4	1.0	1.8	4.0	7.1	11	16	22
6	300	6.3	480	7.4	0.15	0.5	1.1	2.0	4.5	7.9	12	18	24
7	400	7.0	640	8.2	0.14	0.6	1.4	2.5	5.5	9.9	15	22	30
	1000	9.5	1600	11.1	0.12	1	3	5	12	21	33	48	65

mately the same conclusion. In Table 4-3 a Class 3 site with a 3-meter turbine will generate 3100 kilowatt-hours per year. This table is repeated in Appendix E for wind turbines up to 43 meters in diameter.

The values in both Table 4-2 and Table 4-3 are for hub height. You must account for this when using the tables to estimate annual energy output. The preferred method, as used in the examples, is to adjust wind speed or power density for the height of the tower and find the AEO for the new speed or power.

Power Curve Method

Where you have access to the wind speed distribution for your site, or at least for the nearest long-term recording station, you can use the manufacturer's power curve to estimate the AEO. This is the technique used by meteorologists when determining the potential generation from a wind machine in a commercial wind power plant. Essentially you match the speed distribution with the power curve to find the number of hours per year the wind turbine will be generating at various power levels. The power curve for the Bergey 1500 used in the previous example is shown in Figure

4-2. We'll work through an example using this wind machine and a Rayleigh speed distribution for 5.5 m/s (12 mph) at an anemometer height of 30 feet.

First, here are some points of reference on the power curve. Note that *power* is shown on the vertical axis, and *wind speed* is shown on the horizontal axis. Thus, if the wind is blowing at a certain speed during any one instant, you can find the corresponding power produced by the wind turbine at that speed.

Newcomers to wind energy, in their zeal to estimate how much electricity they can generate with a particular wind turbine, sometimes confuse the power curve with graphs of AEO. For example, Mike Bergey of Bergey Windpower has found that some erroneously apply the *average wind speed* at their site to the *instantaneous wind speed* shown on the power curve. Unfortunately, this approach ignores the effect that the speed distribution has on a wind turbine's production. If you want to use the average wind speed method, you must use tables or charts of the AEO. If you want to use the *power curve* method you must use a wind speed distribution.

Here are the definitions of terms used with the power curve.

Start-up speed. The wind speed at which the rotor first begins to turn.

Cut-in speed. The wind speed at which the generator begins to produce power. In this example, the start-up and cut-in speeds are the same, 8 mph (3.6 m/s), because the Bergey uses a permanent-magnet alternator. Whenever the rotor is spinning, it's generating electricity. This isn't always the case. Like the Bergey 1500, the rotor on most medium-sized turbines is free-wheeling and begins to spin at about 8 mph. But nearly all medium-sized wind machines use induction generators. The rotor on these machines doesn't begin producing power until the generator's magnetic field is energized, usually at about 10 mph (5 m/s).

Rated speed and power. The wind speed at which the generator produces the advertised power. For the Bergey 1500, the generator produces 1500 watts at 28 mph (12.5 m/s). Though frequently used as a reference for the size of a wind machine, rated speed and power have little utility. Most wind machines will produce more than their rated power. At one time Northern Power Systems rated their 5-meter turbine at only 2 kilowatts, designating it the HR2, even though it consistently produced more than 3 kilowatts. Northern Power eventually changed the turbine's designation to correspond to the rating system used for engine generators it was designed to replace. The turbine thus became known as the HR3 due to its new rating.

Peak power. Simply the maximum power the generator is capable of producing. In this case, Bergey Windpower says the power of the 1500 is the same as its peak power.

Cut-out speed. The wind speed at which the generator stops producing power. This is accomplished by applying a brake or other mechanism for physically stopping the rotor. The Bergey 1500 does not use a brake and continues to produce power at all wind speeds. In contrast, most medium-sized wind turbines turn themselves off in winds above 55 mph (25 m/s) to reduce wear and tear, and to protect the turbine from damage.

Furling speed. The wind speed at which a wind machine using a tail vane begins to fold or furl toward the tail. The Bergey wind turbine does not use a brake to lock the rotor in place to protect itself in high winds. Instead, the wind machine folds toward the tail at a wind speed of 30 mph (about 13 m/s) to decrease the area of the rotor intercepting the wind.

To use the power curve in Figure 4-2, first find the power the turbine will produce at wind speeds from cut-in (4 m/s) through furling (20 m/s). We can neglect speeds above 20 m/s (45 mph), because little or no energy is captured when the turbine is furled (see Figure 4-3).

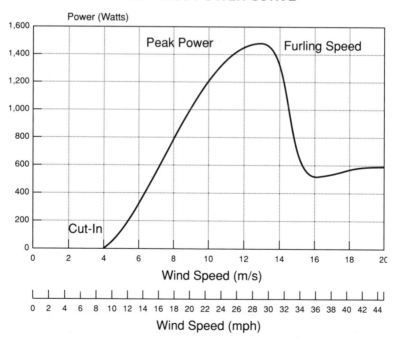

BERGEY 1500 POWER CURVE

Figure 4-2. Nomenclature for power curve of the Bergey 1500, a small wind machine 3 meters (10 feet) in diameter. (Bergey Windpower)

POWER CURVE

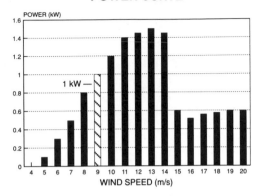

WIND SPEED FREQUENCY

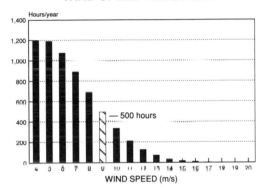

ANNUAL ENERGY OUTPUT

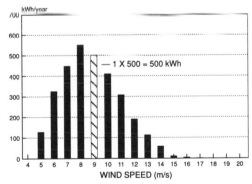

Figure 4-3. Power curve method of calculating annual energy output. At 9 m/s the Bergey 1500 will produce 1 kilowatt. At a site with a 5.5 m/s (12.3 mph) average wind speed and a Rayleigh frequency distribution, winds occur at this speed 500 hours per year. Winds in the 9 m/s bin contribute 500 kilowatt-hours of the wind machine's total annual generation.

Table 4-4

Estimating Annual Energy Output from Manufacturer's Power Curve for Bergey 1500

Average Annual Wind Speed = 5.5 m/s (12.3 mph), Rayleigh Distribution

Wind Speed Bin (m/s)	Instantaneous Power (kW)	Rayleigh Frequency Distribution	Hours/ Year	Energy (kWh/yr)
4	0	0.1371	1,201	0
5	0.1	0.1357	1,188	119
6	0.3	0.1224	1,072	322
7	0.5	0.1019	892	446
8	0.8	0.0789	691	553
9	1	**0.0571**	**500**	**500**
17	0.55	0.0005	4	2
18	0.575	0.0002	2	1
19	0.6	0.0001	1	0
20	0.6	.0000	0	0

Annual Energy Output (kWh/yr) =   3,025

In Table 4-4, the Bergey 1500 will produce 1 kilowatt at a wind speed of 9 m/s. Winds at this speed occur 5.71 percent of the time during the year. Because there are 8760 hours per year, winds at this speed will occur 500 hours per year (8760 h/yr x 0.0571), producing 500 kilowatt-hours. At 9 m/s:

$$1 \text{ kW} \times 500 \text{ h} = 500 \text{ kWh.}$$

Summing the energy produced by this turbine over the range of wind speeds represented by the Rayleigh distribution, we find that the Bergey 1500 will generate about 3000 kilowatt-hours per year at this site.

The values shown in the power curve are for the turbine's output at hub height. You must take into account the increase in power that accompanies an increase in height. To do so, determine the speed distribution at the height of the tower. This is a simple task when using a Rayleigh distribution. You merely adjust the average wind speed to the new height and calculate the new speed distribution. Tables of Rayleigh distributions for different average speeds are given in Appendix C. For example, assume the average speed at 30 feet is 5.5 m/s and increases to 6 m/s on a taller tower. Simply look up the 6 m/s column in Table C-1 in Appendix C. This

Table 4-5

Estimating Annual Energy Output from Manufacturer's Power Curve for Bergey 1500 by Adjusting Speed Distribution to New Height

Average Annual Wind Speed = 5.5 m/s (12.3 mph) at 30 ft Rayleigh Distribution

30 ft Wind Speed Bin (m/s)	100 ft Wind Speed Bin (m/s)	Instantaneous Power (kW)	Rayleigh Frequency Distribution	Hours/ Year	Energy (kWh/yr)
4	4.8	0.075	0.1371	1,201	90
5	6.0	0.3	0.1357	1,188	357
6	7.2	0.5	0.1224	1,072	536
7	8.4	0.9	0.1019	892	803
8	9.6	1.1	0.0789	691	760
9	10.8	1.3	0.0571	500	650
17	20.4	0.6	0.0005	4	3
18	21.6	0.6	0.0002	2	1
19	22.8	0.6	0.0001	1	0
20	24.0	0.6	.0000	0	0

Annual Energy Output (kWh/yr) =   4,261

column gives the speed distribution at an average wind speed of 6 m/s. Use Table C-2 the same way for speeds in mph.

It's more complicated when you're using a measured speed distribution instead of the hypothetical Rayleigh. You must adjust each speed bin in the measured distribution to the speed at the height of the turbine (see Table 4-5).

Understanding Table 4-5

Table 4-5 illustrates how to adjust the speed in a wind speed bin to a new height using the previous example. Assume that we measured the distribution at 30 feet and want to estimate the turbine's output on a 100-foot tower in a region where the one-seventh power law applies. To do so, increase each speed bin by 1.2 (the increase in speed with height). Thus, the 5 m/s speed bin becomes 6 m/s. Because the turbine produces 0.3 kilowatts at the higher speed instead of only 100 watts, the turbine generates more kilowatt-hours.

Flip back to Table 4-2 and note that overall conversion efficiency in the column labeled "Total Effic." drops off at higher average speeds. The same effect is produced by increasing wind speed with a taller tower. In Table 4-5 the Bergey 1500 generates about 1200 kilowatt-hours more per year than under conditions in Table 4-4. For this wind machine the total amount of energy extracted from the wind at the new height is greater, even though the turbine is slightly less efficient at doing so.

Power curves are usually presented as smooth lines. In the real world, however, the power output for a particular machine is a range of values at each wind speed, depending on whether the rotor was coasting from a previous gust at the time the measurement was made (in which case power would be greater than average) or was coming up to speed and the anemometer registered the gust but the rotor did not (power would be less than the norm). Power curves, like many other aspects of wind energy, are just approximations of what happens in a complex environment. They should not be taken literally.

Performance Ratings

The practice in the United States and the Netherlands has been to describe a wind machine's size by referring to its generator capacity in kilowatts at some rated wind speed. This power *rating* is then used extensively in product promotion. For a moment, though, let's take an excursion. Put yourself in Denmark—Roskilde, to be specific—at the Danish Test Center for Wind Turbines.

"Helge, that's a beaut, what size is it?"

"Ten meters."

"What? No, I meant, how big is it? My Danish isn't too good."

"It's a 10-meter, but there's a bigger 12-meter down the road. Would you like to see it?"

"We're not going another step until you tell me how big that is!"

"You Americans are so demanding."

"All right, one more time, how big a generator does it have?"

"Which one?" Helge asked quizzically.

"Boy, what a case of jet lag," the American said to himself.

"I want to know how big the generator is. You know, the thing that generates the electricity, the guts of the machine."

Helge's patience was beginning to wear thin. "It has two generators, the largest of which is 30 kilowatts, but it has a 10-meter rotor driving it."

"Phew, I thought I'd never get it out of you," said the exasperated Yankee.

"Tak (Thanks)," replied Helge.

Such an exchange has probably taken place more than once. We find ourselves in this predicament—identifying wind machines by their generator size—because many of the early pioneers in wind development came from the electric utility industry or were designing wind turbines for the utilities. In common parlance, we refer to power plants by the combined size of their generators. For example, an Eskimo village runs a 50-kilowatt diesel generator; it is the plume from a 500-megawatt (MW) coal-fired power plant that clouds the valley; or it was the 900-MW Unit 2 reactor that was damaged at Three Mile Island.

Utilities try to run their generators at a fairly constant output so the generators perform as efficiently as possible. As a result, a 500-MW generator normally produces 500 MW. Engineers understandably use this power rating when they talk to one another.

Wind machines are different because the wind is an intermittent resource. Fluctuations in wind speed cause generator output to vary as well. Seldom does a wind generator produce its rated output for any extended period of time. Moreover, optimum rotor and generator combinations depend on the wind regime. A wind turbine with a 10-meter rotor may, for instance, perform most efficiently (deliver the most energy) matched with a 10-kilowatt generator in regions of the country with low average wind speeds, but the same rotor may work better with a 25-kilowatt generator elsewhere.

Rotor diameter is a much more practical measure of size since it is the rotor and not the generator that captures the wind and converts it to a useful form. The generator comes later in the conversion process. Most Europeans, especially the Danes, refer to the size of their machines by rotor diameter.

Nothing says more about a wind turbine than rotor diameter. Nothing.

Fortunately, the rating practice common in the United States began falling out of favor during the early 1980s. Increased awareness of rotor diameter's importance led to the adoption of a hybrid designation using rotor diameter and generator capacity. Enertech's model 44-40, for example, used a rotor 44 feet (13 meters) in diameter to drive a 40-kilowatt generator. Similarly U.S. Windpower's 56-100 uses a rotor about 56 feet in diameter to drive a 100-kilowatt generator. Such designations are particularly helpful to wind farm developers for whom generator capacity is more critical.

The rated power system is not only confusing, it can also be misleading. First, there's no reference speed to compare one turbine to another: rated speeds range from 22 mph (10 m/s) to over 30 mph (15 m/s). Second, some manufacturers rate their machines at peak power output and others do not. Bergey Windpower, for example, rates their turbines at peak power. Northern Power Systems' HR3, on the other hand, will produce up to 3.2 kilowatt though it is rated at 3 kilowatts. Medium-sized machines, especially those using aerodynamic stall to regulate peak power, will often exceed their rated capacity, sometimes by up to 30 percent. It's not uncommon for a 65-kilowatt, stall-regulated turbine to produce 80 kilowatts.

A few, shall we say, less than reputable manufacturers have taken advantage of the emphasis on generator size by adding large generators to relatively small rotors. It's possible to slap a 6-foot plank on a 25-kilowatt generator and call it a 25-kilowatt wind turbine using this rating system. In one particularly notorious case Fayette Manufacturing, an American company, built a 10-meter turbine and saddled it with a 95-kilowatt generator. Most other manufacturers would have rated a turbine of this size 25-35 kilowatts. Years of results from California have proven that this machine performs no better than other turbines with rotors of similar size—when it works.

U.S. Windpower has even abandoned the hybrid designation in deference to the European system for its latest design. They now label their newest turbine 33M VS for its 33-meter, variable speed rotor. They do this in part because the machine will generate 300-400 kilowatts depending on the site.

Because of questionable power ratings by some manufacturers and general confusion by consumers as to what the numbers mean, the American Wind Energy Association (AWEA) has attempted to clarify machine designations and performance ratings by calling for a standard list of parameters. These include maximum power (not rated) and, most importantly, the AEO at various average speeds. AWEA hopes to eliminate the rated power at rated speed nomenclature with values that make more sense. It represents the wind industry's equivalent to the Environmental Protection Agency's mileage sticker on new cars. And like the mileage sticker, the AEO is based on assumed conditions. "Your performance may vary."

The AEO may be given as a table of production at several average wind speeds or as a graph. To standardize the estimates from one company to the next, AWEA specifies using the Rayleigh distribution to project the AEO. A heartening development has been the presentation of the AEO as a range of outputs (which include those produced by the Rayleigh distribution) rather

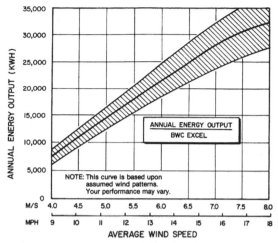

Figure 4-4. AEO curve for the Bergey Excel. Note that the output (in kilowatt-hours of annual generation) is presented as a range of values. Graphs such as this or tables of the annual energy output are the wind industry's equivalent to the EPA's mileage labels on new cars.

than as a single line. This range of values better reflects reality. Output estimates are not precise. They shouldn't be displayed that way.

If you find the calculations described in this chapter frustrating and you're not satisfied with Tables 4-2 and 4-3, use the AEO estimates provided by manufacturers. For most residential applications the estimated AEO is sufficiently accurate once the average wind speed is known with some confidence. Commercial applications, because far more money is at risk, demand more sophistication and require the power curve method.

To estimate the AEO at your site, simply find the average wind speed, adjust it for the height of your tower, and then look on the chart. Check the table or graph first for any footnotes. Most AEO values are presented for hub height, but some are not. Those that don't can catch you off guard.

Assume that you plan to install a Bergey Excel on a 100-foot tower. The average speed at hub height is 6 m/s (13 mph). From the AEO chart in Figure 4-4 you find that the Excel will generate 18,000-25,000 kilowatt-hours per year. The lower end of the range is about what we estimated the 7-meter turbine would produce at 6 m/s in Table 4-2. The difference lies in the conversion efficiency used in Table 4-2.

The AEO curve in Figure 4-4 is based on the manufacturer's estimate of the performance from this specific machine. Tables 4-2 and 4-3 average the efficiency for many different size machines and will seldom give exactly the same results as manufacturers' AEO estimates.

Efficiency

Now that you've estimated what a wind machine will produce at your site, you may conclude that it isn't enough. If that's the case, there are three ways to increase output:

1. Increase wind speed; that is, find a better site or use a taller tower.
2. Increase the swept area of the rotor by finding a bigger wind turbine.
3. Improve the wind turbine's conversion efficiency, that is, find another wind turbine.

Efficiency is placed last for a reason. Americans in particular have a disturbing fondness for the word "efficiency." Invariably people will say "Yeah, all those calculations are fine. But what's the most efficient turbine built today?" What they overlook is that energy output is more sensitive to speed (because of the cube law) and swept area (because of the square of the rotor's radius) than efficiency.

In the previous example of the 7-meter turbine, the difference in efficiency between the manufacturer's estimate and that assumed in Tables 4-2 and 4-3 is 6-11 percent. That difference can more than be made up by using a tower 30 feet (10 meters) taller or by lengthening the blades only 5 percent.

Wind turbines must first be reliable; second, they must be cost-effective. Efficiency is important, but it's not the sole criteria for judging the performance of a wind machine.

To gauge the potential of a wind machine, ignore the size of the generator or its purported efficiency and get right to what matters most: the AEO for your site. But if the AEO is not available, whether you're in Europe or the United States, nothing outside of the wind itself, no other single parameter, is more important in determining a wind machine's capability of producing energy than rotor diameter.

Investors have lost staggering amounts of money (not just homeowners but savvy executives who should have known better) because they didn't grasp this concept. Even after years of experience with operating wind machines, reams of technical documents and studies, there's always someone who has made a startling discovery of a way to beat the Betz limit. (Most of them have never heard of Betz, unfortunately.) These wonderful new devices are not only more efficient than any previous wind machine, but they will produce more energy than theoretically possible. That's not to say that it can't be done. The Betz limit, after all, is only a theory. But no one has done it yet, and plenty have said they could. And, as sure as the rain will fall, there'll be someone willing to part with their savings before punching a few numbers into a calculator.

5

Economics—

Does Wind Pay?

Will a wind machine pay for itself? Will it be a sound investment? Or, more simply, is it worth the trouble? The answers to these frequently asked questions are elusive. They depend on a number of speculative variables not subject to precise calculation, such as inflation, interest rates, and the desired rate of return. Nor is there just one straightforward way to look at the economics.

A better question may be, does it have to pay for itself? Certainly commercial-sized wind machines intended to produce bulk power in competition with conventional fuels must be cost-effective. But small wind turbines need not pay for themselves overnight or even within 10 years to prove beneficial. As long as they pay for themselves within their expected lifetimes, they're economic in the strict sense of the term.

Yet even this dictum may be too restrictive when the value of wind turbines is compared to other products. Consumers often buy items of equivalent cost that provide no monetary return whatsoever. But wind machines are not luxuries. Unlike a swimming pool or a rack of snowmobiles in the backyard, wind machines save or even earn their owners money. And they do more besides. They generate electricity cleanly. That has value as well, a value that's not currently incorporated in the price of utility-supplied power (though it might well be by the end of the decade).

All too often, consumers look only at the initial cost. They contrast this with what they are accustomed to paying their utility and throw up their hands in despair. There's no contest! "The wind may be free," they might be overheard saying, "but it sure costs a lot to catch it." They're right, wind turbines are expensive. But there are always two sides to every equation and the other side of the wind system equation is the revenue it earns or the money it saves.

Buying a wind machine is a lot like buying a house. You can always rent at a lower initial cost, but you will almost always spend more over the long term. Mike Bergey of Bergey Windpower likes to point out that you're paying for a wind machine whether you want to or not every time you mail your check to the local utility. Although, as with death and taxes, you may not be able to avoid dealing with the utility entirely, a wind machine can at least help you reduce those monthly payments. In the process, you assume responsibility for reducing some of the environmental impact from your energy consumption.

A wind machine need not meet your entire electrical load to be economic. Many mistakenly think that the wind system must eliminate their entire electric bill for it to be worthwhile. Sometimes this may be best. But a wind machine may be a good buy even when it provides only a small portion of your electricity. In fact, for many applications interconnected with the utility it's better to select a wind system that will produce less energy than currently used because utility buy-back rates are often less than retail rates. It doesn't make any sense to sell the utility energy for 3 cents per kilowatt-hour when they will turn around and sell it back to you for 8 cents per kilowatt-hour.

Cost of Energy and Payback

There are several methods for evaluating the economics of wind energy. Determining the cost of energy (the COE for those with a bent for acronyms) produced by a wind system is one approach popular with government agencies and academic institutions. The COE accounts for initial cost, maintenance, interest rates, and performance over the life of the wind system. This *life-cycle* method produces an estimate of cost, in cents per kilowatt-hour, for the wind system over its life span. The results can be easily compared with today's cost of energy from new conventional sources. As a result, the COE is a useful tool for planners and those charged with energy policy.

The COE has limited usefulness because it reveals only whether the wind system's generation will cost more or less than that from other sources. It doesn't tell you how much of a bargain—or cost—the wind system might be. The COE, consequently, cannot be used to judge whether or not your money would be more productive in an interest-bearing account at the local bank or in some other investment.

Finding the payback, or the time it will take for the investment to pay for itself, is an easy way to gauge an investment's worth. Simply divide the wind system's cost by its projected revenue. If the time to payback is less than the life of the wind machine, the turbine has paid for itself and makes sense. But there may be better ways to use the money. Payback is related to return on investment. A short payback offers higher returns than a long payback. To maximize the return (to get the most for your money) you want as short a payback as possible.

Nevertheless, simple techniques, such as payback, are quick but misleading. They don't account for effects that take place after payback occurs. Payback is well suited for low-cost items like storm doors and added insulation, but not for costly long-term investments.

Payback gives no indication of the future earnings a wind turbine will produce after it has paid for itself. Since wind generators are designed to last 20 years or more, much of their earnings take place in later years. Most people are overly concerned about payback. If the wind system doesn't pay for itself within 5 years, they quickly lose interest. They fail to realize that some wind machines will pay for themselves many times over even though they may have paybacks greater than 5 years. Wind systems are long-term investments and should be treated as such.

To get a better sense of how a wind system will perform financially, there's no better way than to estimate the cash flow from one year to the next. Constructing a cash flow table is the only way to examine the economics over the long term with any degree of realism. It's a technique commonly used by businesses when they're considering investments in new equipment.

Before we launch into our first example, let's examine the factors that affect a wind system's economics.

Economic Factors

The *installed cost* of a wind system is simply the cost of the wind turbine, tower, wiring, and installation less any state or federal tax credits. Where they exist, tax credits reduce an individual's tax liability instead of merely reducing taxable income as with tax deductions. In effect, tax credits reduce initial cost.

In the United States, Oklahoma offers a state tax credit, and the 1992

National Energy Policy Act provides a production tax credit of 1.5 cents per kilowatt-hour. The federal credit applies only to sales of electricity, not to savings. Thus you would have to dedicate the wind turbine's generation for sale to the utility, or to a third party, to qualify for the credit.

Rather than tax credits, Germany and the Netherlands give outright grants to individuals who install wind turbines for their own use. These grants vary, but can amount to 30 percent of the installed cost. For up-to-date information on incentive programs, including those that effectively reduce installed cost, contact your national trade association or your accountant.

Maintenance costs are expenses for servicing or repairing the wind system. They can be expressed in cents per kilowatt-hour or as a percentage of the initial cost. The vast amount of experience from California indicates that the cost of operating and maintaining medium-sized turbines is 1-2 cents per kilowatt-hour, somewhat less than that for coal and nuclear plants.

The maintenance cost for small wind turbines is less well defined. Advanced small wind turbines using integrated assemblies are designed to be nearly maintenance-free. OnSite Energy operated a Bergey Excel for more than 8 years without any maintenance other than replacement of the leading edge tape on each blade. After monitoring another Excel for 3 ½ years, Wisconsin Power & Light found that the owner paid only $100 in maintenance expenses during the entire period for a cost of less than 0.3 cents per kilowatt-hour. However, Wisconsin Power & Light found that the cost of maintaining other brands sold during the early 1980s was prohibitive. Nevertheless, there's a consensus within the wind industry that maintenance will cost about 2 percent of the total wind system cost annually.

The cost of *financing* the purchase of a wind system can add significantly to overall costs. You immediately become aware of financing costs if you have considered using a loan to buy your wind system. It's much like building an addition on your house with a loan from the bank. You pay for the use of the bank's money. You can't avoid financing costs simply by paying with cash. The cash invested in the wind system could have been earning interest at the bank. For homeowners and farmers the installation of a wind turbine can be financed by increasing the mortgage on the property instead of taking out a short-term loan. The financing or interest cost of the loan will then reflect the current mortgage rate.

There are some hidden costs as well. A wind system could potentially increase property taxes, though in some areas they are exempt. Insurance is another often overlooked cost. Not only should there be insurance on the wind system itself but there should be insurance for any accidents due to the wind machine. Since most wind systems require an extended period to pay for themselves, an owner would be foolhardy to operate an uninsured machine. An unexpected event could wipe out an expensive wind turbine before the investment had been recouped. Wind systems are also a potential hazard. Personal injuries and property damage can occur in a multitude of ways: the tower can fall over, or (the most likely form of accident) someone can fall off the tower. Many insurance companies have little or no experience with small wind systems, and their coverage varies from a simple inclusion under a homeowner's existing policy at no cost to a policy written specifically for the wind turbine with an attendant premium.

Despite the absence of tax credits in most states, *taxes* still have a profound effect on wind turbine economics, largely on the cost side of the ledger. For homeowners, the costs of financing a wind system through a home mortgage are tax deductible. Because the interest costs are high in the early years of a loan, this deduction dramatically improves the attractiveness of small wind turbines to homeowners over what's apparent from a simple payback calculation. The tax bonus is even more important for businesses because they can deduct the cost of maintenance and other expenses.

The *tax bracket* of the investor determines the value of the deductions. Those in higher tax brackets save more by reducing their tax burdens than those who pay a lower tax rate. If you're in the 30 percent tax bracket, you save 30 cents for every dollar in tax deductions. Similarly, you receive more than 1 dollar in value when you offset 1 dollar's worth of electricity with a wind turbine because such savings are currently not taxed. (No one taxes you for using less electricity.) Someone in the 30 percent tax bracket receives 70 cents after taxes for every dollar earned. But if they offset one dollar's worth of electricity they receive $1.43 in value after taxes. For those in lesser tax brackets the benefits are not as great, but they're still considerable.

Now for the income side of the account. The income derived from the wind system is a product of the annual energy output and the value of the electricity. Like costs, the value of the electricity generated depends on a number of factors, the most important of which is the retail rate for electricity. If we assume that whatever energy the wind machine generates will displace electricity that otherwise would be bought from the utility, the value of this energy (per kilowatt-hour) equals the utility's retail rate.

For a homeowner, the cost of electricity per kilowatt-hour is easy enough to estimate. Simply pull out the last 12 months of your electric bills, sum them, and divide by the total amount of electricity consumed to get an average cost in cents per kilowatt-hour. For businesses it may not be so clear.

Commercial customers pay not only for the energy they use, they also pay a demand charge. This charge is based on the maximum power drawn during the billing period in relation to their total energy consumption. This compensates the utility for maintaining generators on line to provide power at the demand of the customer. A business installing a wind generator lowers its total consumption while hardly affecting its peak demand. Thus the wind system could actually increase the demand charge while lowering costs for the energy consumed. This isn't a problem with all utilities, but farmers and businesses need to watch for it.

Once the value of the wind-generated energy in cents per kilowatt-hour has been found, the annual value of the electricity produced can be calculated. Multiply the retail rate in cents per kilowatt-hour by the amount of energy (kilowatt-hours) produced yearly. The result, in dollars per year, is the gross proceeds from offsetting the consumption of utility-supplied electricity.

One of wind energy's chief advantage over generating electricity by conventional means is that the fuel (the wind) is free. The bulk of the cost for a wind system occurs all at once. Once paid for, the energy produced costs little over the remaining life of the wind system. Conventional generation, on the other hand, consumes nonrenewable fuels whose costs continue to escalate. Thus, our analysis would be incomplete if we didn't account for the rising price of electricity. Like other aspects of energy production, the rate at which utility prices will rise over time is hotly debated.

Utility rates escalated sharply during the 1970s and early 1980s. These price hikes were due in part to the rapid rise in the cost of oil caused by the two oil embargoes, and in part by completion of expensive new power plants. Oil-dependent utilities were hit the hardest, and their rates jumped dramatically. Energy costs at other utilities eventually also rose because the prices for coal and gas often track price increases in oil. Further, many utilities in the 1970s committed themselves to massive construction programs to meet expected growth in electricity consumption. These plants, both coal-fired and nuclear-powered, were completed during the early 1980s and were enormously expensive, particularly the nuclear reactors.

After these price shocks, electricity prices stabilized. Rate increases were much less severe during the late 1980s. In real terms, after accounting for inflation, the rates at some utilities even declined. But by the early 1990s electric utility rates began to rise again even though the world price of oil—

the benchmark of energy costs—remained low. Analysts expect rates to rise gradually though the 1990s as utilities begin to incorporate the costs of stricter pollution abatement into the price of electricity. And, as the war in Kuwait illustrated, future conflicts in the Middle East could once again cause shortages and higher prices. The international oil market remains as volatile as ever.

The most important aspect of utility rate escalation is its relationship to inflation. We can safely say that utility rates will rise during the 1990s. How much they will rise relative to inflation is altogether more difficult to define. But it is important. Critics of utilities or those hawking energy saving devices tend to overemphasize the effect over time of rising utility rates. They fail to mention that inflation reduces the effect of rising rates in real terms. While utility rates may be rising 10 percent per year, an inflation rate of 5 percent erodes the potential value of each kilowatt-hour. In general most analysts expect utility rates to rise faster than the inflation rate.

Residential Economics

Let's put all this together and construct a cash-flow table of costs and income over the life of a typical wind turbine. Assume that you plan to install a 7-meter wind turbine on a 100-foot tower costing about $20,000. You've identified the site on Battelle's wind power maps and found that it has a Class 3 wind resource. After measuring the wind at the site you confirm the average wind speed at hub height is about 6 m/s (13 mph). The manufacturer estimates that the turbine will generate 18,000-25,000 kilowatt-hour per year under these conditions. For the purpose of estimating the turbine's cost-effectiveness you assume an AEO of 20,000 kilowatt-hours. You expect to use all the electricity in your home at a retail rate of 10 cents per kilowatt-hour, and no tax credits or other incentives apply. The 10 year payback doesn't appear particularly attractive so you decide to find the net revenues over the 20-year life of the machine.

From your observation of the local utility you expect rates to continue rising 10 percent per year for the life of the wind system. This represents a real price increase of only 5 percent because inflation has been hovering around 5 percent per year. You have convinced your banker to take out a second mortgage on your house at 10 percent interest for 20 years to pay for the wind system. Your accountant says you're in the 30 percent tax bracket. After explaining that future dollars are worth less than today's dollars due to inflation, your accountant agrees to calculate the net present value for your results. Table 5-1 summarizes your year-by-year tally of net revenues.

Table 5-1

Economics of Small Wind Turbines for Home Use

Total Cost	$20,000		Rate Escalation	0.1
AEO (kWh/yr)	20,000		Mortgage Cost	0.1
Rate ($/kWh)	0.10		Inflation Rate	0.05
O&M, Insurance	0.02		Tax Bracket	0.3
Simple Payback	10			

Year	Gross Savings	O&M	After-Tax Mortgage Cost*	Net Revenue (Loss)
1	$2,000	($400)	($1,749)	(149)
2	$2,200	($420)	($1,760)	20
3	$2,420	($441)	($1,771)	208
17	$9,190	($873)	($2,126)	6,191
18	$10,109	($917)	($2,174)	7,018
19	$11,120	($963)	($2,227)	7,930
20	$12,232	($1,011)	($2,285)	8,936
				$62,000
Net present value of savings				$30,000

* Assumes interest payment is deductible against federal taxes.

Understanding Table 5-1

Gross savings derive from offsetting the purchase of electricity from the utility with that from the wind turbine at the utility's retail rate. O&M is the cost of operations and maintenance. The after-tax mortgage cost is the net cost of payments on the wind turbine's debt. This includes the value of a tax deduction for interest payments on the debt. For example, during the first year, payment toward the principal is $349 and toward interest is $2000 (not shown). The value of an interest deduction of $2000 to someone in the 30 percent tax bracket is $600. Thus, the after-tax cost of the mortgage ($2349-$600) is $1749. The value of the interest deduction declines over time as the mortgage principal is paid. The net present value discounts future earnings for the effect of inflation.

You find that the turbine will not only pay for itself but will earn an additional $62,000 in future dollars. Your accountant says that's equivalent to $30,000 in today's money. But you're still not convinced and decide to use the same method to determine what benefits, if any, you could get by putting the money in the bank and using the interest to pay your utility bill. To your dismay the bank will pay only 7 percent on your account, and to top it off your accountant advises you that interest from a savings account is taxable income. Table 5-2 summarizes your results.

To your surprise, the wind system looks more attractive all the time. If you invested the $20,000 and continued to pay the utility bill as before, you find that by the twentieth year you're in the hole the equivalent of $29,000 in today's dollars.

If the wind system were to earn $30,000 over its life while you avoided paying $29,000 in net expenses, you could conclude that the overall benefit

Table 5-2

Purchasing Electricity from a Savings Account

Deposit	$20,000	Rate Escalation	0.1
Use (kWh/yr)	20,000	Inflation	0.05
Rate ($/kWh)	0.10	Tax Bracket	0.3
Interest	0.07		

Year	Interest * Less Taxes	Purchased Power	Net Revenue (Loss)
1	$980	($2,000)	($1,020)
2	$980	($2,200)	($1,220)
3	$980	($2,420)	($1,440)
17	$980	($9,190)	($8,210)
18	$980	($10,109)	($9,129)
19	$980	($11,120)	($10,140)
20	$980	($12,232)	($11,252)
			($95,000)
Net present value			($49,000)
Plus deposit			$20,000
Total net present value			(29,000)

*Assumes interest is taxable.

Understanding Table 5-2

Interest earnings are taxable in the United States. A person in the 30 percent tax bracket will pay $420 on $1400 in interest from $20,000 invested at 7 percent. Thus, the value of the interest earnings ($1400-$420) is only $980. The cost of purchased power is simply the product of the per kilowatt-hour cost of electricity and the total amount consumed.

---

approaches a staggering $60,000. And that's after the wind turbine had paid for itself.

The results look so encouraging you now wonder whether a smaller turbine would work equally well while not tying up so much of your money. So you consider a 3-meter turbine costing about $5000 under similar conditions. The manufacturer estimates that the turbine will generate

Table 5-3

Economics of a 3-meter Wind Turbine for Home Use

Total Cost	$5,000		Rate Escalation	0.1
AEO (kWh/yr)	3,000		Mortgage Cost	0.1
Rate ($/kWh)	0.10		Inflation Rate	0.05
O&M, Insurance	0.02		Tax Bracket	0.3
Simple Payback	17			

Year	Gross Savings	O&M	After-Tax Mortgage Cost*	Net Revenue (Loss)
1	$300	($100)	($437)	(237)
2	$330	($105)	($440)	(215)
3	$363	($110)	($443)	(190)
17	$1,378	($218)	($531)	629
18	$1,516	($229)	($543)	744
19	$1,668	($241)	($557)	871
20	$1,835	($253)	($571)	1,011

				$4,000
Net present value of savings				$1,400

* Assumes interest payment is deductible against federal taxes.

3000 kilowatt-hours per year at your site. Tables 5-3 and 5-4 summarize your findings.

The 3-meter turbine will pay for itself over the life of the system, earning $1400 and saving $1000 that would have been spent purchasing electricity. You conclude that the small machine is far less attractive for this application than the larger version, which you decide to install.

Farm and Business Economics

Let's try another example using the 7-meter turbine under the same conditions except that we'll install it on a farm. In this application the maintenance costs are a deductible business expense. This reduces the farm's after-tax costs. However, the farm's purchase of electricity is also a deductible expense, effectively reducing the after-tax value of the electricity offset by the wind turbine. Once again no special incentives apply, but the wind system qualifies for a 5-year depreciation deduction as a business investment.

Table 5-4

Purchasing Electricity from a $5000 Savings Account

Deposit	$5,000	Rate Escalation	0.1
Use (kWh/yr)	3,000	Inflation	0.05
Rate ($/kWh)	0.10	Tax Bracket	0.3
Interest	0.07		

Year	Interest* Less Taxes	Purchased Power	Net Revenue (Loss)
1	$245	($300)	($55)
2	$245	($330)	($85)
3	$245	($363)	($118)
17	$245	($1,378)	($1,133)
18	$245	($1,516)	($1,271)
19	$245	($1,668)	($1,423)
20	$245	($1,835)	($1,590)
		($17,000)	($12,000)
		($9,000)	
Net present value			($6,000)
Plus deposit			$5,000
Total net present value of savings			($1,000)

* Assumes interest is taxable.

Depreciation deductions offer substantial tax benefits to users of capital-intensive equipment such as wind systems. A business in a 30 percent tax bracket saves 30 cents for every dollar in deductions. Because of the tax rate, for every dollar deducted from gross income the business no longer owes the government 30 cents. At the end of 5 years you have obtained a piece of capital equipment for 30 percent less than its original cost; see Tables 5-5 and 5-6.

> The conditions at your site may vary from those used here. Always consult your accountant or financial adviser before buying a wind system. They may be able to offer suggestions on how to maximize your wind investment.

Table 5-5

Economics of a 7-meter Wind Turbine for Business Use

Total Cost	$20,000		Rate Escalation	0.1
AEO (kWh/yr)	20,000		Mortgage Cost	0.1
Rate ($/kWh)	0.10		Inflation Rate	0.05
O&M, Ins.	0.02		Tax Bracket	0.3
Simple Payback	10			

Year	Gross Savings	O&M	Net After-Tax O&M *	After-Tax Loan Cost *	After-Tax Benefit of Depreciation *	Net Revenue (Loss)
1	$2,000	($400)	($280)	($1,749)	$1,200	1,171
2	$2,200	($420)	($294)	($1,760)	$1,200	1,346
3	$2,420	($441)	($309)	($1,771)	$1,200	1,540
4	$2,662	($463)	($324)	($1,784)	$1,200	1,754
5	$2,928	($486)	($340)	($1,798)	$1,200	1,990
6	$3,221	($511)	($357)	($1,813)		1,051
17	$9,190	($873)	($611)	($2,126)		6,453
18	$10,109	($917)	($642)	($2,174)		7,293
19	$11,120	($963)	($674)	($2,227)		8,219
20	$12,232	($1,011)	($708)	($2,285)		9,239
						$72,000
Net present value of savings						$38,000

* Assumes interest payment, O&M expense, and depreciation are deductible.

Understanding Table 5-5

For a business in the United States, maintenance and other costs are expenses deductible from gross earnings. Thus, the true or net costs of operations and maintenance (O&M) after taxes is $400-$120=$280 in the first year. Wind turbines also typically qualify for a 5-year depreciation deduction. For a $20,000 wind turbine, each year a farmer can deduct $4000 in depreciation from the farm's gross income. The net benefit at a 30 percent tax rate is $1200.

Table 5-6

Purchasing Electricity from a $20,000 Savings Account for Business Use

Deposit	$20,000	Rate Escalation	0.1
Use (kWh/yr)	20,000	Inflation	0.05
Rate ($/kWh)	0.10	Tax Bracket	0.3
Interest	0.07		

Year	Interest Less Taxes *	Purchased Power *	Net After-Tax Purchased Power Cost	Net Revenue (Loss)
1	$980	($2,000)	($1,400)	($420)
2	$980	($2,200)	($1,540)	($560)
3	$980	($2,420)	($1,694)	($714)
17	$980	($9,190)	($6,433)	($5,453)
18	$980	($10,109)	($7,076)	($6,096)
19	$980	($11,120)	($7,784)	($6,804)
20	$980	($12,232)	($8,562)	($7,582)
				($61,000)
Net present value of savings				($31,000)
Plus deposit				$20,000
				($11,000)

* Assumes interest is taxable, and purchased power is a deductible expense.

Understanding Table 5-6

The cost of purchased power is the same as with the example for home use. Unlike the case of a homeowner, however, businesses can deduct the cost of purchased power as an expense. Thus the after-tax net cost for a business in the 30 percent tax bracket is $2000-$600=$1400 during the first year.

Net earnings in today's dollars are $49,000 after the turbine has paid for itself. Thus, the 7-meter turbine is less attractive financially in this application than for a homeowner, primarily because the farm's electricity

Table 5-7
Economics of a 25-meter Wind Turbine for Business Use

Total Cost	$250,000		Rate Escalation	0.1
AEO (kWh/yr)	300,000		Mortgage Cost	0.1
Rate ($/kWh)	0.10		Inflation Rate	0.05
O&M, Insurance	0.02		Tax Bracket	0.3
Simple Payback	8			

Year	Gross Savings	O&M	Net After-Tax O&M *	After-Tax Loan Cost *	After-Tax Benefit of Depreciation *	Net Revenue (Loss)
1	$30,000	($5,000)	($3,500)	($21,865)	$15,000	19,635
2	$33,000	($5,250)	($3,675)	($21,996)	$15,000	22,329
3	$36,300	($5,513)	($3,859)	($22,140)	$15,000	25,301
4	$39,930	($5,788)	($4,052)	($22,298)	$15,000	28,580
5	$43,923	($6,078)	($4,254)	($22,473)	$15,000	32,196
6	$48,315	($6,381)	($4,467)	($22,664)		21,184
17	$137,849	($10,914)	($7,640)	($26,572)		103,637
18	$151,634	($11,460)	($8,022)	($27,174)		116,438
19	$166,798	($12,033)	($8,423)	($27,836)		130,538
20	$183,477	($12,635)	($8,844)	($28,564)		146,069
						$1,191,000
Net present value of savings						$623,000

* Assumes interest payment, O&M expense, and depreciation are deductible.

consumption is a deductible expense. As this example shows, the deductibility of energy expenses significantly limits the use of renewable energy in the United States.

Now let's examine a medium-sized wind turbine under the same conditions. Assume you're considering a wind turbine 25 meters (80 feet) in diameter costing about $250,000. Using the AEO tables in Appendix E for Battelle's wind power Class 3 you find that this machine could generate 300,000 kilowatt-hours per year. At windier sites, such as those on the west coast of Denmark or in California's mountain passes, wind turbines of this size can easily generate 500,000 kilowatt-hours per year.

Though this size wind turbine is used cooperatively by residential consumers in Denmark, it's much bigger than that used by most homeowners in the United States. Typically a machine of this size is used to off-load the

Table 5-8

Purchasing Electricity from a $250,000 Savings Account for Business Use

Deposit	$250,000	Rate Escalation	0.1
Use (kWh/yr)	300,000	Inflation	0.05
Rate ($/kWh)	0.10	Tax Bracket	0.3
Interest	0.07		

Year	Interest Less Taxes *	Purchased Power	Net After-Tax Purchased Power Cost *	Net Revenue (Loss)
1	$12,250	($30,000)	($21,000)	($8,750)
2	$12,250	($33,000)	($23,100)	($10,850)
3	$12,250	($36,300)	($25,410)	($13,160)
17	$12,250	($137,849)	($96,494)	($84,244)
18	$12,250	($151,634)	($106,144)	($93,894)
19	$12,250	($166,798)	($116,758)	($104,508)
20	$12,250	($183,477)	($128,434)	($116,184)
				($958,000)
Net present value of savings				($492,000)
Plus deposit				$250,000
				($242,000)

* Assumes interest is taxable, and purchased power is a deductible expense.

electrical consumption of a business or farm or is used in wind power plants. Tables 5-7 and 5-8 summarize the economics for this system.

Although the medium-sized turbine is more costly than the smaller turbine, it's also more cost-effective. The 25-meter machine will not only pay for itself but will earn more than $600,000 over its life span while saving $240,000 for a total benefit of at least $840,000. The medium-sized machine looks so attractive that you now consider installing a dozen of them in your own wind farm.

Before you do, there are several refinements needed in your spreadsheet. First, wind turbines in the massed arrays found in wind plants deliver less generation than comparable machines sited singly because upwind turbines interfere with those downwind. Depending on the configuration, wind machines in arrays produce 10-15 percent less than individual turbines. Second, medium-sized wind machines are more complex than integrated small wind turbines. Periodically the generators and transmissions must be rebuilt in machines of this size. Their brakes must be replaced regularly as well. The costs for these services may be covered by the amount set aside for maintenance. But they may not be. The maintenance of the larger, more complex wind turbines introduces another element of risk into the equation.

Risks

Analysis of the economics of a small wind system is fraught with assumptions about the future. The assumptions you use may or may not reflect conditions over the 20-year life of a wind system. No one knows with certainty what the future will bring. There's a degree of risk associated with every investment. Consequently, there's no simple answer to the question, "Is it a good deal?"

When buying a wind system, you're betting that utility costs will increase faster than inflation. That certainly was true during the 1970s and early 1980s. But utility costs lagged behind inflation during much of the 1980s, and in some places actually declined. The preceding examples assume that the value of electricity increases 5 percent relative to inflation. Interest rates and utility rate escalation have a profound effect on a wind turbine's economics because they compound year after year. Slight changes in their relationship could change the attractiveness of wind energy as an investment.

Maintenance costs are equally important. Small wind systems are generally more expensive than bigger machines relative to the amount of energy they produce. For small wind turbines, low maintenance costs can be a deciding factor at marginal sites. In the foregoing examples we assumed that maintenance cost 2 percent of the installed cost per year. At low wind sites, maintenance can consume much of a small wind turbine's revenue. Cutting annual maintenance costs by using an integrated, advanced, small wind turbine that requires little or no maintenance can measurably improve the overall economics at such sites.

Implicit in these payback calculations has been the assumption that all the energy generated was used on site, offsetting energy bought from the utility. This may not be the case. Some of the energy may be generated at night when there is little or no need for it. The excess energy then will be sold back to the utility, and often at a price far lower than the purchase or retail rate. This reduces the wind system's overall savings. Where there's a demand for heat or hot water, such as at a dairy, the surplus generation can be dumped into a thermal storage system instead of selling it back to the utility at a reduced rate.

If you're going to use some of the wind machine's excess generation for heating, first convert the surplus kilowatt-hours of electricity to gallons of oil using the conversion factors in Appendix A. For example, let's assume you plan to use the entire output from the 7-meter wind machine of the previous examples to heat your house instead of burning oil.

The energy content in 0.034 gallons of fuel oil is equivalent to that of 1 kilowatt-hour of electricity. This isn't the amount of oil saved. For every gallon of oil the furnace burns, 40 percent goes up the chimney; only 60 percent is put to work heating your house. You need to take this into account because a gallon of oil generated by the wind machine has more heat value than a gallon of oil burned. Putting it together:

20,000 kWh x (0.034 gal/kWh) x (1/60%) = 1000 gallons of oil.

For heating oil at $1 per gallon, the value of the wind turbine's output in the first year is $1000, about one-half that used in the example for electricity at 10 cents per kilowatt-hour. Electricity is frequently more costly than oil or natural gas for home heating. This is the principal reason why so much of the work with wind energy is devoted to generating electricity.

Cost-Effectiveness

Many people confuse the cost-effectiveness of wind turbines with efficiency. The reason for installing wind machines is to generate low-cost energy. You could have the most inefficient wind machine every built, the kind that brings a tear to an engineer's eye, but if it works and is cheap enough, it could be more cost-effective than a modern engineering marvel costing much more. If you deliver lower cost energy with an inefficient wind machine than with an efficient one, so be it.

There are several measures of cost-effectiveness in use: cost per kilowatt of installed capacity ($/kW), cost per rotor swept area ($/m²), and cost per kilowatt-hour of energy generated ($/kWh). The most frequently used measure of cost-effectiveness is cost per kilowatt. However, it's about as meaningful as the power rating in describing the size of a wind machine. This measure came into use the same way—utility engineers were accustomed to using it. Like power ratings, the cost per kilowatt works well for power plants that run at constant output. But for wind machines, the cost per kilowatt just confuses matters.

During the early 1980s a few manufacturers took advantage of this situation. Because consumers were using cost per kilowatt to compare one wind machine to another, these manufacturers began offering products with higher generator ratings relative to the size of the rotor than was the norm. Their products were not any better, nor would they produce more energy than their competitors, but their cost per kilowatt was lower. One company, Fayette Manufacturing, rated their wind turbine nearly three times greater than wind machines of comparable size. Even though their machine cost more than others, it appeared more cost-effective because of its greater generator rating. They sold 1600 wind turbines this way.

A truer measure of a wind machine's size is the area swept by its rotor, and a more useful measure of cost-effectiveness is the cost per swept area ($/m²). The limitation on using cost per swept area is the assumption that all wind machines are equally efficient at converting the energy in the wind into electricity. They are not. The cost per swept area is just a shortcut for

Table 5-9

Relative Cost per kWh for Examples

Size Turbine (m)	Cost ($)	AEO (kWh)*	Cost/kWh/yr
3	$5,000	3,000	1.7
7	$20,000	20,000	1.0
25	$250,000	300,000	0.8

* AEO at Class 3 site, 100-ft (30-m) tower

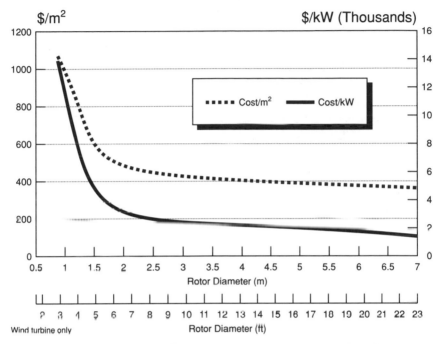

SMALL WIND TURBINE COST
RELATIVE TO SIZE

(Wind Turbine Only)

Figure 5-1. Relative cost of small wind turbines. The relative cost of small wind machines generally declines with increasing size.

what counts most: the relative cost per kilowatt-hour generated at your specific site.

This cost per kilowatt-hour isn't the same as the cost per kilowatt-hour you pay for electricity from the utility. They're two different animals. The cost per kilowatt-hour measure should be used only for comparing one wind machine to another. It's not appropriate for comparing a wind machine to other forms of energy because it doesn't account for all the costs and benefits from the wind turbine over its entire life cycle. It's merely a measure for comparison shopping—nothing more.

Table 5-9 confirms a common trend that wind turbines become more cost-effective as size increases (see Figure 5-1). This relationship holds true for wind turbines up to about 40 meters (130 feet) in diameter. It may seem arcane at first, but it's a technique used by professionals to compare the cost-effectiveness of various wind turbine designs in different wind regimes. Cost-effective turbines at windy sites produce lower costs per kilowatt-hour per year. More costly wind turbines or those at less windy sites produce higher values. For utility applications, such as in California wind power plants, wind turbines must generate more than twice their cost in U.S. dollars each year. Small wind turbines often produce values greater than 1 U.S. dollar per kilowatt-hour per year.

The cost-effectiveness of any wind system is more sensitive to initial cost and the AEO than to any other factors. That's why proper siting is so critical. For the wind turbines used in the examples, moving from a Class 2 to a windier Class 3 site increases energy output by 25-30 percent. This produces an equivalent drop in the cost per kilowatt-hour, making the wind systems that much more attractive. Where a windier site is not possible, nearly the same result can be accomplished by installing the turbines on towers 160 feet (50 meters) tall because of the increase in wind speed with height.

6

Evaluating the Technology—

What Works and What Doesn't

"Hey, that's a funny lookin' windmill you got there. What is it?"

"A VAWT."

"Don't get smart with me, son. I asked you a simple question."

"Actually, it's an articulating, straight-bladed VAWT."

"Can't you speak in English? I don't work for the government, you know."

"Some call it a giromill."

"Well, that's better. Why didn't you say so in the first place? For a moment there I thought you were speaking in tongues."

As wind technology has grown, so has its vocabulary. At times it may seem as if the wind industry does speak in tongues. Nearly every conceivable wind turbine configuration has been tried at least once—most only once. Designs have run the gamut from the familiar farm windmill to contraptions such as the giromill. Despite the plethora of imaginative designs developed during the 1970s, only a few approaches have since proved successful.

During the past decade small wind machines designed for residential or remote uses where simplicity is required, have evolved into highly integrated designs with few moving parts. These advanced small wind turbines typically use a rotor with three blades that spin about a horizontal axis upwind of the tower. Most of these designs drive a permanent-magnet alternator and have demonstrated exceptional reliability with little or no maintenance. Similarly, most of today's medium-sized wind turbines, like those used in California wind plants, share many characteristics. Most use three blades and drive induction generators. And until the early 1990s, nearly all successful designs used simple, fixed-pitch blades.

In this chapter we'll look at where the technology stands today, and why designs such as these have become commonplace. We'll also look at the important difference between wind machines that use drag to drive their rotors and those that use lift, why modern wind machines use only two or three blades, what materials are used to make these blades and the advantages of each, the kinds of controls used to protect the wind turbine, and the types of transmissions and generators now being used.

Orientation

There are two great classes of wind turbines, horizontal and vertical axis machines (see Figure 6-1). Conventional wind turbines, like the Dutch windmill found throughout northern Europe and the American farm wind mill, spin about a horizontal axis. As the name implies, a vertical axis wind turbine (VAWT) spins about a vertical axis much like a top or a toy gyroscope.

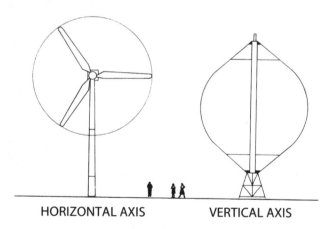

HORIZONTAL AXIS   VERTICAL AXIS

Figure 6-1. Horizontal- and vertical-axis wind turbines. Although the Darrieus or eggbeater turbine on the right (FloWind 19-meter model) spins about a vertical axis, it's equally efficient at harnessing the energy in the wind as the conventional wind turbine on the left (WindMaster 23.5-meter model). Both wind machines are typical of the medium-sized wind turbines used in wind power plants, and each is capable of producing about 200 kilowatts. (Pacific Gas & Electric Co.)

Horizontal Axis

Because the wind changes direction, all horizontal axis wind machines have some means for keeping the rotor into the wind. Consequently, either the entire wind machine and its tower, or the top of the wind machine where the rotor is attached must change its position relative to the wind.

Traditionally the rotors of horizontal axis wind machines have been placed upwind of their towers and there was some device for keeping the rotor into the wind. On the Dutch windmill, for example, the miller had to constantly monitor the wind. When the wind changed direction the miller laboriously pushed a long tail pole or turned a crank on the milling platform that moved the windmill's massive rotor back into the wind. Later versions liberated millers from their labor by using fan tails that mechanically turned the rotor toward the wind. On smaller wind machines, such as the farm windmill, the task is much easier and a simple tail vane will do. The tail vane keeps the rotor pointed into the wind regardless of changes in wind direction.

Both tail vanes and fan tails use forces in the wind itself to orient the rotor upwind of the tower (see Figure 6-2). They passively change the orientation, or *yaw*, of the wind turbine with respect to changes in wind

Figure 6-2. Upwind rotor. Many small wind machines, such as this HR1, use a tail vane to orient the rotor into the wind. (Northern Power Systems)

direction without the use of human or electrical power. Without a tail vane or a fan tail, upwind turbines won't automatically stay into the wind. (There are some modern exceptions.) Downwind rotors don't need tail vanes or fan tails. Instead the blades are swept slightly downwind, giving the spinning rotor the shape of a shallow cone with its apex at the tower. This *coning* of the blades causes the rotor to inherently orient itself downwind.

Upwind and Downwind

Downwind machines are certainly sleeker than small upwind machines with their tail vanes (see Figure 6-3). Some believe this gives downwind machines a more modern look. They do have one clear advantage. Downwind machines eliminate the cost of tail vanes. But they pay a price, say proponents of upwind turbines. Downwind machines occasionally get caught upwind when winds are light and variable. Some downwind turbines

Figure 6-3. During the 1970s and early 1980s several manufacturers built wind machines with rotors downwind of the tower. Although several thousand are still in operation, most wind machines built today use rotors upwind of the tower. Shown here in the Altamont Pass is a cluster of U.S. Windpower's 56-100, a downwind machine nominally 56 feet (17.5 meters) in diameter. The rotor drives a 100-kilowatt generator.

like the Enertech and Storm Master (both no longer manufactured) occasionally *hunt* the wind after a strong gust subsides. Such a turbine may walk completely around the tower as it searches for the wind. Critics of downwind machines add that the tower creates a *shadow* that disrupts the air flow over the blades as they pass behind the tower. This decreases performance, they charge, increases wear, and emits a characteristic sound some people find annoying.

West Texas State University's Vaughn Nelson asserts that upwind machines suffer a similar, but less severe, performance penalty. The wind piles up in front of a tower much like the small zone of turbulence just upstream from a stone in a swiftly flowing brook. The stone creates a much bigger zone of disturbance or wake downstream.

One significant disadvantage of passive downwind machines is yaw control in a high wind. A common method for protecting upwind turbines in high winds is to orient the rotor 90 degrees to the wind. On upwind machines with a tail vane the rotor can be furled (turned) and the rotor will soon move out of the wind. Upwind turbines with active yaw controls mechanically swing the turbine out of the wind. Unless a downwind machine has an active yaw drive, the rotor will always stay downwind of the tower during high winds. Other mechanisms for controlling the rotor in high winds must be used.

Although tail vanes are simple and effective devices for passively controlling yaw, they are limited to wind machines less than 10 meters (33 feet) in diameter. Above this size the tail vane becomes too unwieldy, and electric motors or fan tails are necessary to mechanically orient the rotor into the wind. (The wind machines in Figure 2-4 use motors to yaw the turbine into the wind.)

Vertical Axis

The principal advantage of modern vertical-axis wind machines over their conventional counterparts is that VAWTs are omnidirectional—they accept the wind from any direction. This simplifies their design and eliminates the problem imposed by gyroscopic forces on the rotor of conventional machines as the turbines yaw into the wind. The vertical axis of rotation also permits mounting the generator and gear at ground level (see Figure 6-4). This is a feature you'll learn to appreciate if you ever have to service a conventional wind turbine 80 feet above the ground in a howling blizzard.

Vertical-axis turbines, like their conventional brethren, can be divided

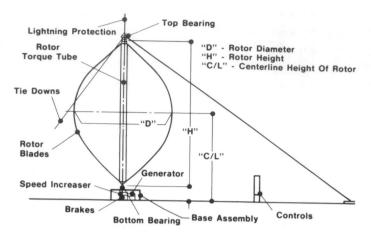

Figure 6-4. Detail of Darrieus rotor with nomenclature.

into two major groups: those that use aerodynamic drag to extract power from the wind (for example, the cup anemometer) and those that use lift from an airfoil. We can further subdivide those VAWTs using airfoils into those with straight blades and those with curved blades. The simplest configuration uses two or more straight blades attached to the ends of a horizontal cross-arm. This gives the rotor the shape of a large H with the blades acting as the uprights of the H. Unfortunately, this configuration permits centrifugal forces to induce severe bending stresses in the blades at their point of attachment.

During the 1920s French inventor D.G.M. Darrieus patented a wind machine that cleverly dealt with this limitation. Instead of using straight blades he attached curved blades to the rotor. When the turbine was operating the curved blades would take the form of a spinning rope held at both ends. This *troposkein* shape directs centrifugal forces through the blade's length toward the points of attachment, thus creating tension in the blades rather than bending. Because materials are stronger in tension than in bending, the blades could be lighter for the same overall strength and operate at higher speeds than straight blades. Although the phi Φ or *eggbeater* configuration is the most common, Darrieus conceived several other versions including Delta, Diamond, and Y. All have been tried at one time or another (see Figure 6-5).

Darrieus' concept eventually faded into obscurity. Canada's National Research Council reinvented the idea in the mid-1960s, and subsequently

DARRIEUS OR VAWT CONFIGURATIONS

"H" DELTA DIAMOND "Y" PHI Ø

Figure 6-5. Darrieus configurations. There are several other Darrieus configurations besides the common "eggbeater" design.

Canadian wind development has focused solely on Darrieus turbines. Sandia National Laboratories in the United States has pursued the technology for nearly two decades. Several firms have attempted to commercialize Sandia's work, and 5 percent of the wind turbines in California are of the Darrieus design. However, no one currently manufactures Darrieus turbines, and outside of the programs in Canada and at Sandia, work on the technology has practically ceased.

Darrieus turbines were never reliably self-starting. Their fixed-pitch blades can't drive the rotor up to operating speed from a standstill unless the blades are parked in just the right position relative to the wind. This isn't necessarily a serious limitation. But to provide self-starting capability several researchers in this country and abroad reverted to the H-rotor configuration. By using straight blades they can vary blade pitch as the blades orbit around the rotor's axis. (The blades on Darrieus rotors are attached rigidly to the torque tube or central shaft.) Technocrats identify this kind of wind machine as an articulating, straight-bladed, vertical-axis wind turbine. It's also known as a giromill or cycloturbine (see Figure 6-6).

The H-rotor has one important advantage over the Darrieus design. It captures more wind. The intercept area of an H-rotor is a rectangle. For the same size wind machine—that is, where the height and diameter are the same—the H-rotor will sweep more area than an ellipse does.

Lift and Drag

Regardless of whether a wind machine rotates about a vertical or horizontal axis, it depends on either of two aerodynamic principles to derive power from the wind: drag or lift. Drag devices are simple wind machines that use flat, curved, or cup-shaped blades to turn the rotor. Both cup anemometers and panemones are representative of drag devices. In each the wind merely

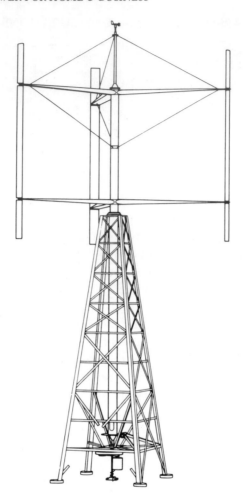

Figure 6-6. *Giromill or cycloturbine. Like all vertical-axis wind machines, this turbine could transmit mechanical power to ground level via a long shaft. The wind vane at the top of the rotor orients the blades with respect to the wind.*

pushes on the cup or blade, forcing the rotor to spin (see Figure 6-7). Lift devices, in contrast, use airfoils to propel the rotor.

Drag devices characteristically produce high starting torques. Because much of the rotor's swept area is covered with blades, there's a lot of surface for the wind to push against. Drag devices are ideal for pumping water in low volumes. But their inherent drawbacks limit their use for generating

Figure 6-7. Panemone. Simple drag device used in ancient Persia for grinding grain. The vertically mounted blades were made by fastening bundles of reeds onto a wooden frame. The surrounding wall guides the prevailing wind onto the retreating blades. (Sandia)

electricity. At best only one-third of the power in the wind can be captured by such machines. In comparison, the maximum possible for a lift device is 59 percent (the Betz limit). Drag devices also require more materials than comparable wind machines using lift.

Experimenters have tried numerous approaches to improving the performance of drag devices. Backyard tinkerers often turn their attention first to drag devices because they're easier to understand and construct. The wind pushes on a big wide blade and it moves. What could be simpler? (Lift devices are more complicated. We'll see why in a moment.)

Although the cup anemometer is the most widely used drag device, the farm windmill and the Savonius rotor are the two most successful. Both deliver slightly better aerodynamic performance than true drag devices. Early farm windmills, for example, used flat wooden slats for blades. In 1888 Aermotor introduced its *mathematical* windmill, which substituted sheet-metal blades for those of wood (see Figure 6-8). Aermotor stamped a

Figure 6-8. Multiblade farm windmill. The curved metal blades resulted from the pioneering work of Thomas Perry, who conducted 5000 experiments on various "wheel" (rotor) designs. The Aermotor embodied all the principles Perry learned and was almost twice as efficient as the wood wheels then commonly in use. It also included "back gearing," which allowed the wheel to make several revolutions for each stroke. Since then, Perry's design has been widely copied.

broad curve into the metal blade to trap more air. It did, and in doing so directed the air to flow over the backside of the following blade. This cascade effect heightened the difference in pressure from one side of the blade to the other, improving Aermotor's performance over that of its rivals. Unfortunately, the "new and improved" farm windmill—all now use Aermotor's technique—still extracts only 15 percent of the power in the wind.

In 1924 Finnish inventor Sigurd Savonius developed an S-shaped vertical-axis wind machine. Principally it's a drag device, but Savonius improved its performance by recirculating some of the air flow between the two halves of the rotor (see Figure 6-9). Air striking one blade is directed through the separation between the two halves of the S and onto the other blade. Researchers have measured conversion efficiencies of almost 30 percent under optimum conditions, considerably more than that extracted by other drag devices. In practice, however, S-rotors, like the farm windmill, extract less than 15 percent of the power in the wind. Because of this limitation Savonius rotors have never been commercially successful. Today they're found only in experimenters' garages.

There are other hybrids as well, wind machines that don't fall neatly into either category. Anton Flettner built several such devices. Using the Magnus effect as a means of propulsion on two upright spinning cylinders, he sailed the Atlantic for New York in 1925. The following year he built a horizontal-axis wind turbine 65 feet in diameter in which he used four spinning cylinders to drive the rotor.

SAVONIUS ROTOR

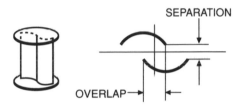

Figure 6-9. Savonius rotor. Another hybrid device where performance exceeds that of a simple wind machine dependent on drag alone. To achieve optimum performance, the two blades must be separated to permit some recirculation of flow. Because of its simplicity, a Savonius rotor is often the first choice of do it yourselfers.

The Magnus effect is the lift or thrust produced when air moves over the surface of a spinning object (see Figure 6-10). It's what produces the curved flight in the curve ball. The pitcher imparts spin to the ball as it leaves his hand. Air rushing over the spinning ball forces the ball off its normal straight-line course. In 1933 J. Madaras constructed a 90-foot cylinder 28 feet in diameter at Burlington, New Jersey, in hopes of using this phenomenon to drive cars around a track. Like similar attempts to harness the Magnus effect, it was abandoned because the spinning cylinders were material-intensive. The same results could be obtained from true airfoils at less expense.

Instead of paddles or cups, lift devices use airfoils like those in the wing of an airplane to power the turbine. The limited number of blades on lift devices contrasts markedly with the multiple blades of drag devices. It seems mysterious that a wind machine with only a few blades can operate more

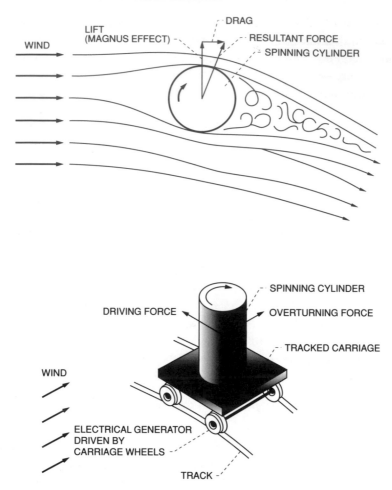

Figure 6-10. *A hybrid device that uses the lift created by a spinning cylinder demonstrates the Magnus effect. Spinning cylinders have been used to drive several kinds of wind machines, including a flat car on a track.*

efficiently than one with a large number of blades. But a modern wind turbine, because it uses lift, can capture the same amount of power with a smaller rotor using fewer blades than a multiblade drag device. On the basis of blade surface area, wind turbines using lift can extract 100 times more power from the wind than a drag device. That's why today's wind machines look so different from the farm windmill. Modern wind turbines, those using lift, do much more with less.

Why is this? Why do some wind machines use multiple blades like the farm windmill where others use only a few? Why do some blades taper from the root to the tip where others taper from the tip to the root? To understand the answers to these questions we need to briefly delve further into wind turbine aerodynamics.

Aerodynamics

It's intriguing that a sailboat can travel faster than the wind, and even more so when we learn that the boat sails faster across the wind than when the wind is pushing from behind. Mariners discovered this fact intuitively centuries ago. Today we explain the paradox by speaking in terms of lift and drag. The blade of a modern wind turbine is much like the sail of a sailboat—lift propels them both.

To begin, let's look at the factors affecting the lift from an airfoil as found in a wind turbine blade. Air flowing over the blade causes both lift and drag. When you're driving down the highway and you stick your hand outside the window, lift from the air flowing over your hand (a crude airfoil shape) literally lifts your hand toward the roof. Drag pulls your hand toward the rear of the car, dragging it with the wind. The sum of these two forces on a wind turbine blade generates a thrust that pulls the blade on its journey through the air, much like it pulls a sailboat through the water. This thrust is greatest when the blade is slicing through the wind, or the sailboat is sailing across the wind.

Engineers rate airfoil performance by the ratio of lift to drag. Designers want a high lift-to-drag ratio for best performance. The lift-to-drag ratio is determined by the blade's angle of attack—the blade's angle with respect to the *apparent* wind. The lift-to-drag ratio increases with increasing angles of attack until a point is reached where the air flow over the blade becomes turbulent. Lift then deteriorates rapidly, the ratio declines, and the airfoil *stalls*. The angle of attack at which this occurs varies from one airfoil to another.

Stall is a deadly condition in flight. Airplanes literally fall out of the sky when stall occurs, when there's no longer enough lift to support them. It's one of the leading cause of light plane accidents. In a wind machine stall can be put to good use. We'll see why in a moment. But first consider the angle of attack: it's a function of the blade's angle to the plane of rotation—its pitch—and the apparent wind (see Figure 6-11). For now assume the pitch is fixed, which it is on most wind machines.

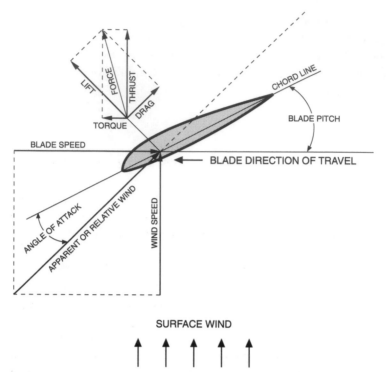

Figure 6-11. Airfoil performance is gauged by the ratio of lift to drag. Lift is determined by the angle of attack. The pitch of the blade, the speed of the blade through the air, and the speed of the wind control the angle of attack and, consequently, lift.

Apparent Wind and the Angle of Attack

The apparent wind is the wind "seen" by the blade. It is a combination of the blade's own motion and the wind across the ground. If you recall some of your high school physics you'll note that both the magnitude and direction of the apparent wind depend on the magnitude and direction of its components. For example, if both were equal in speed (say the wind speed was 10 mph and the speed of the blade was also 10 mph) and if they were acting at right angles to each other, the apparent wind would be acting at 10 mph at a 45-degree angle between the two (see Figure 6-11).

If wind speed increases while the blade's speed through the air remains constant, the position of the apparent wind swings toward the wind direction because it has become more influential. As the apparent wind changes position it also changes the angle of attack. Reverse the process and see what happens. Blade speed now becomes stronger and causes the apparent wind to shift toward the direction of the blade's motion, decreasing the angle of attack. Designers must decide how best to deal with this relationship for each airfoil, because there's an optimum angle of attack, a point where the lift-to-drag ratio is optimum and performance reaches a maximum.

For a fixed-pitch blade to maintain an optimum angle of attack, blade speed must increase in proportion to wind speed. As wind speed increases the rotor must spin faster. Another way to say it is that the tip-speed ratio, the relationship between blade speed and wind speed, must remain constant to maximize aerodynamic performance. Most small wind turbines operate this way because it's not only more efficient but also simpler.

Most medium-sized wind machines, those above 10 meters in diameter, don't maintain the optimum tip-speed ratio because of the type of generator they use (more on this in the section on generators), and because they take advantage of the airfoil's decreasing performance in high winds. Designers are willing to sacrifice some performance for the simplicity of fixed-pitch rotors driving constant speed generators. As wind speed increases the airfoil begins to stall and performance declines. Stall is desirable because it reduces the rotor's power in high winds, making it easier for designers to build protective controls to keep the rotor from destroying itself.

The amount of thrust driving the rotor is not only a function of the lift-to-drag ratio and the blade's angle of attack, but also the area of the blade and its speed through the air. To increase the load an airplane can lift, you either increase the size of the wings, increase the plane's speed, or do both.

Taper and Twist

For the sake of simplification we've been looking at a blade as if the conditions it sees were constant along its entire length. That may be true for airplanes, but not so for wind turbines. Even when the pitch of the blade is fixed and rotor speed constant, the speed through the air of a point on the blade changes with its distance from the hub. The speed is higher at the tip than near the hub because it has more distance to cover in the same amount of time.

Because blade speed increases with distance from the hub, the apparent wind varies as well. The apparent wind increases in strength, and its position shifts toward the plane of rotation as you move out along the blade toward the tip. If the blade designer wants to maintain the angle of attack (to optimize performance) at the same time blade speed is increasing, the pitch of the blade must decrease toward the tip. As a result, wind turbine blades are twisted from root to tip (see Figure 6-12). The greatest pitch appears at the root and the least at the tip. Glance up at the next wind turbine you see and note that the tip of the blade is almost parallel with its direction of travel for this reason.

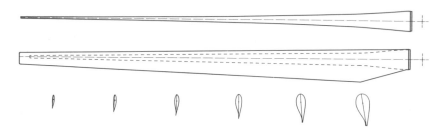

Figure 6-12. Taper and twist. Rotor blades on conventional wind machines often taper from the root where the blade attaches to the hub, to the tip. This saber-like shape minimizes solidity. Depending on the size and its construction, the blade may also be twisted to optimize performance along its entire length. (Vestas DWT)

Solidity

Wind machine designers long ago learned that blade area (the number of blades as well as their length and width) governs the amount of torque, or turning force, a rotor can produce. The more blades the rotor has for the wind to act on, the more torque it will produce. Greater solidity, the ratio of blade area to the area swept by the rotor, generates greater torque. If you were designing a wind machine for pumping water on the semi-arid Great Plains, you'd want a rotor that provides high torque at low wind speeds. This would assure you that the windmill would be able to lift the pump's piston (and the water with it) during late summer when the need is greatest but the winds are lightest. During the rest of the year you wouldn't really care how well the rotor performed because there would be less demand for water and more than enough wind to pump it.

There's no better example of a high-solidity rotor fitting this description than the American farm windmill. It uses multiple "sails" that taper from the tip to the root so that nearly the whole rotor disc is covered by blades (80 percent solidity). It was designed to deliver high torque at low wind speeds, and it does its job remarkably well.

Southeastern Pennsylvania is renowned for the tidy, prosperous farms of the Pennsylvania Dutch. These descendants of German settlers live a simple life and husband their land. They shun power equipment, farm with horse-drawn teams, and use buggies to get about. Water-pumping windmills are a familiar sight on the Pennsylvania Dutch landscape.

Many in the eastern United States who have seen the Pennsylvania Dutch conclude that the water-pumping windmill must obviously be a good idea if the thrifty Dutch are still using them. These people fail to realize that the Pennsylvania Dutch abhor modern conveniences such as electricity. The farm windmill is their only source of water. Cost is no object. The farm windmill was never designed to deliver water economically. It was designed to deliver water, period. It only had to compete with lifting water by hand. When you're dying of thirst you don't care how much the windmill costs or how much power it produces just so long as it pumps water.

The demands on electricity- or power-generating wind machines are different. We don't need power from wind turbines. Our lives are not dependent on them like those of the Pennsylvania Dutch. For us there are many other sources of power. the utility, photovoltaics, a stand-alone generator. We want power from a source and are willing to pay for it only when it's a better buy—when it's cheaper than from competing sources. To compete with these other sources the wind machine must be designed to extract power from the wind in the most efficient and least costly manner possible. The farm windmill isn't the way to do it.

The multiblade farm windmill looks like it would capture more wind than a modern machine with two or three slender blades. Intuitively, we feel that the rotor should have more blades to capture more wind. If this were true, however, consider what would happen if we carried this belief to its logical extreme. The optimum rotor would cover the entire swept area with blades, in effect producing a solid disc. No air would pass through. The wind would pile up in front of the rotor and flow around rather than through it. The wind speed behind the rotor would be zero. Instead of capturing more wind, we wouldn't capture any. There must be some air moving through the disc and it must retain enough kinetic energy so it can keep moving to make way for the air behind.

We must strike a balance between a rotor that completely stops the wind and one that allows the wind to pass through unimpeded, between the amount of wind striking the rotor and the amount flowing through. Albert Betz demonstrated mathematically that this optimum is reached when the rotor reduces wind speed by one-third.

When the wind flies into the wide vanes of the farm windmill it is deflected slightly as it moves downstream. The combined effect from the wind moving across the blades causes the wind leaving the rotor to spin in a spiral like a corkscrew. To maximize the amount of work the wind can perform (to approach the theoretical limits as near as we can), designers must minimize this deflection and the spiralling of the wind stream in the turbine's wake.

This spiralling effect is greatest, and so is the amount of power lost, in rotors producing high torque. High-solidity rotors such as the farm windmill produce plenty of torque, but they also lose more energy in the wake than do lower solidity rotors delivering less torque. But, you may ask, if we were to lower the rotor's torque wouldn't we be lowering the power it can produce even if it's going to be more efficient at producing it? Yes, if we kept everything else the same. We don't. Power is a product of torque and rotor speed. To deliver the same amount of power we need to increase rotor speed. Decreasing torque while at the same time increasing rotor speed will improve the rotor's efficiency.

Torque is the product of a force acting on a lever. In our case the lever is the blade of a wind turbine. The lever is longest at the tip and shortest near the hub. To reduce the torque produced by the blade, the force acting on it must progressively decrease as the lever arm increases in length or as we proceed along the blade to the tip.

Lift provides the force needed to produce torque. We learned previously that lift is a function of blade area and speed. As we move toward the tip, blade speed increases and so does lift. We now have two factors reinforcing each other: blade speed increases toward the tip, as does the length of the lever through which it will act. Thus, to decrease torque we need to decrease blade area, or solidity, more at the tip than near the hub. The blade, as a result, tapers from the root to the tip. This also explains why so few blades are used. Designers want to keep solidity as low as possible.

Blade speed is a function of rotor diameter and rotor speed. Both are described by a single term, the ratio between the speed of the blade through the air at the tip and the wind speed: the tip-speed ratio. Tip speed increases as either rotor speed increases or the length of the blade increases. Large-diameter rotors have higher tip-speed ratios than do smaller rotors spinning at the same speed. A rotor spinning at a faster rate than another of the same size also has a higher tip-speed ratio. For optimum performance, solidity should decrease as tip-speed ratio increases. Drag devices operate at tip-speed ratios of one or less compared to lift devices, which operate at tip-speed ratios of five or more.

To summarize, lift devices are capable of extracting more power from the wind with less material than drag devices. For lift devices to perform optimally they must operate at low torque but at a high tip-speed ratio. To achieve high tip-speed ratios and low torque the rotor must have low solidity. This is why modern wind turbines operate at high speeds: the rotor is more efficient. It is not, as has been said by others, because electric generators operate at high speeds. If this were the case you could always use a transmission to increase the speed to the generator.

Wind turbines need only one slender blade to efficiently capture the energy in the wind. The giant German conglomerate Messerschmitt-Bolkow-Blohm built just such a series of one-bladed wind turbines. But there are other, equally important design criteria besides efficiency. Two blades are often used for reasons of static balance. (The one-bladed MBB turbines use a counterweight.) Many modern wind turbines use three blades because they give greater dynamic stability than either two blades or one.

Two-bladed turbines suffer a dynamic imbalance when the wind machine changes direction. When the blades are vertical little force is needed to yaw or to swing the turbine around the tower. But when the blades are horizontal much more force is required because of the rotor's inertia. (The blades like to stay just where they are.) When the rotor is spinning and the wind turbine changes direction, the wind machine and tower are subjected to these oscillating forces twice per revolution: once each time the rotor is horizontal. This causes the rotor to yaw unevenly. On larger machines this effect is dampened with shock absorbers or by allowing the rotor to teeter. Three blades minimize this dynamic problem and are preferred on small machines where yaw dampening or teetering hubs would be too costly.

Self-Starting

Low-solidity rotors have one drawback. They may not be self-starting. Remember that the apparent wind flowing over the blade is partly due to the blade's motion. When the rotor is stopped, the lift on the blades from the wind alone may not be enough to start the rotor moving. One solution for rotors using fixed-pitch blades is to spin the rotor up to a speed where it can drive itself. This is a common practice for Darrieus turbines. There are also conventional wind machines that require motoring the rotor up to speed. But most designers are willing to sacrifice a slight amount of performance to gain a self-starting capability.

Bergey Windpower cleverly surmounted this self-starting problem by using a torsionally flexible blade and a pitch weight. This design allows the blades to be set at a high angle of attack for start up conditions. As rotor speed increases, a weight attached to an outboard portion of the blade twists the blade, changing its pitch progressively toward the optimum running position. This pitch weight isn't used to control overspeed but solely to improve the rotor's performance over a full range of wind speeds.

Wind machines that are self-starting begin turning in winds of 8-10 mph (4-5 m/s). Don't be misled by glib talk promoting a "new" wind turbine that runs in low winds. Anybody can design a wind machine to turn in light winds. But why bother? There's no energy in winds at low speeds. The rotor may spin but it won't do much else.

Darrieus turbines are typically not self-starting, though it is now known that Darrieus turbines can self-start under the right wind conditions. These conditions—though infrequent—do occur. When the Darrieus rotor is at a standstill, only the wind across the ground acts on the blade. Because the pitch of the rotor is fixed, the blades stall and nothing happens. Normally, the rotor must be motored up to speed. But on July 6, 1978, all that changed and a new corollary was added to Murphy's Law: "wind turbines that won't self-start, will." Canadian researchers were testing a 230-kW experimental Darrieus rotor on Magdalen Island in the Gulf of St. Lawrence. While repairs were underway the brake was released. Because it was thought the turbine could not start itself, it was left unattended overnight. During the night the wind picked up. By the next morning the rotor was spinning out of control. Eventually the rotor spun off the tower and corkscrewed itself into the ground.

Darrieus rotors have also been plagued by a misperception that they're less efficient than conventional wind machines. This results because one-half of the time the blades are traveling either with the wind or against it. Even so the blades on a Darrieus turbine produce lift for most of their orbit around the turbine's axis. Even when the blades are moving downwind (retreating) they see an apparent wind due to their own motion sufficient to create lift. Only when the blades are parallel to the wind does lift fall to zero.

Sandia Laboratories have found that under ideal conditions Darrieus turbines can extract more than 40 percent of the power in the wind. In other words, their performance is similar to that of conventional wind turbines, not any better but certainly no worse.

Developers of the Musgrove H-rotor have also found that a vertical-axis rotor with fixed-pitch blades can be made to start itself. (The Musgrove design, named after its English inventor Peter Musgrove, uses a novel means of protecting the rotor in high winds as explained in the section on controls.) The designers discovered that decreasing the blade's aspect ratio—its height to its width—by shortening and widening the blades created more lift while the rotor was at rest. The stubbier blades could start the rotor without robbing from performance at operating speeds.

An H-rotor with articulating blades, such as a giromill or cycloturbine, is also self-starting (see Figure 6-6). The pitch of each blade is set according to a predetermined schedule and the position of the blade relative to the wind. The blade's angle of attack is optimized at each position of its orbit around the rotor's axis. Controlling blade pitch with respect to the wind gives the rotor a reliable self-starting capability not found in the Darrieus rotor. It should also deliver better performance because lift can be maximized regardless of whether the blade is advancing into, across, or with the wind. Giromills, however, have never lived up to expectations. They're also material-intensive.

Blades

Blades are one of the most critical, and visible, components of a wind turbine's rotor. Blades can be made from almost any material—and have been. Wood has always been popular. Early farm windmills used wooden slats, and windchargers of the 1930s used wood almost exclusively. Wood is still the material of choice for many small wind machines. It's strong, readily available, easy to work with, relatively inexpensive, and has good fatigue characteristics. "Wood flexes for a living," explains Mick Sagrillo, who runs a small turbine repair shop in Wisconsin. "It works well in high-fatigue applications."

Wood blades are built either from single planks of Sitka spruce or from wood laminates. The blades are then machined into the desired shape and coated with a tough weather-resistant finish. Then the manufacturer covers the leading edge with fiberglass tape to protect the blades from wind erosion and hail damage. This tape is the same as that used on the leading edges of helicopter blades. It's resistant to ultraviolet light and abrasion.

Few new to wind energy appreciate the wind's erosive force. If you need to be convinced of this, pay a visit to the Texas Panhandle or the Tehachapi Pass during the spring wind season. But don't forget to take your goggles. Sand and blowing grit scour anything in their path. This airborne sandpaper has deeply etched the galvanizing on the windward side of towers in the Tehachapi Pass. In areas prone to blowing sand, wooden wind turbine blades have had their leading edges eroded away after only 2 years of use.

Though solid wood planks work well for small machines up to 5 meters in diameter, some manufacturers prefer laminated wood. Designers prefer laminated wood (used in the construction of a butcher's block) in larger machines because they can better control the blade's strength and stiffness, as well as limit shrinkage and warpage. In the laminating process slabs of wood are bonded together with a resin. The resulting block can then be carved into a blade. By varying the types of wood, the direction of their grains, and the resin, a material can be produced that is stronger than any one part alone and stronger than a single plank of the same size. Laminated wood blades have been used on small wind turbines of all sizes.

Thinner slices of wood are also used to produce veneers. Layer upon layer of razor-thin slices are sandwiched together with a resin and molded into the airfoil shape. The process is widely used to build the hulls of sailboats and has been adapted successfully for wind turbine blades both in the United States and Europe. Wood-composite blades fabricated by Michigan's Gougeon Brothers have earned a reputation for strength and reliability in wind turbines up to 43 meters (142 feet) in diameter.

In the late nineteenth century galvanized steel began replacing the wooden blades on the farm windmill, and steel has been used ever since. Steel is strong and well understood. That's why it was chosen by Boeing engineers for the blades on the 300-foot (91-meter) diameter Mod-2, and the 320-foot (98-meter) diameter Mod-5B. It's nothing fancy, just plain structural steel—the same steel used in bridges. Because steel is so heavy, the hub, drive train, and tower must be more massive than on a wind machine with a lighter-weight rotor.

Aluminum is lighter and, for its weight, stronger. It's used extensively in the aircraft industry for this reason. We can fabricate aluminum blades with the same techniques used to build the wings of airplanes: form a rib and then stretch the aluminum skin over it. The blades on NASA's early Mod-OA were built this way. On smaller machines a simpler method can be used by stamping a curve into the leading edge, folding the sheet metal over the spars, and then riveting it in place.

Aluminum can also be extruded, eliminating all other fabrication steps. It was once thought that blades could be mass-produced this way, extruding blades in the same way we manufacture drain spouts and window moldings by squeezing a hot piece of aluminum through a die. Alcoa and a Canadian company spent considerable money developing extruded aluminum blades for Darrieus turbines. They believed that the Darrieus rotor was well suited for aluminum extrusions because the forces on the blades are in tension.

The blades endure less stress in the Darrieus rotor than they would in a conventional wind machine. They can also use a blade of a constant width such as produced by extrusion. Aluminum, unfortunately, has two weaknesses. It's expensive and it's subject to metal fatigue.

Ever take a piece of wire and break it by flexing it back and forth a few times? That's metal fatigue, and it works the same way in the wing of an airplane or the blades of a wind turbine. Aluminum is a good material when used within its limits. On wind turbines aluminum hasn't been successful. Most of the problems Darrieus turbines in California have encountered are due to metal fatigue. Operators of these Darrieus turbines may eventually replace the aluminum blades with fiberglass. The only successful manufacturer of extruded aluminum blades used them in home light plants during the 1940s when Wincharger switched from wooden blades to extruded aluminum. Some Winchargers can still be found with their blades intact. No major manufacturer builds wind turbines today with metal blades.

Another drawback to metal blades, whether steel or aluminum, is television and radio interference. Metal reflects television signals, and this can cause "ghost" images on nearby TV sets. This has proven to be far less of a problem than first thought, even among the existing wind turbines using metal blades including the 500 Darrieus turbines still operating in California.

Fiberglass (glass-reinforced polyester, or GRP to Europeans) has grown increasingly popular (see Figure 6-13). Like wood, fiberglass is strong, relatively inexpensive, and has good fatigue characteristics. It also lends itself to a variety of designs and manufacturing processes. Fiberglass can be pultruded, for example. Instead of pushing the material through a die, as in extrusion, fiberglass cloth (like the cloth used in fiberglass auto body kits) is pulled through a vat of resin and then through a die. Pultrusion produces the side rails for fiberglass ladders and other consumer products. The pultruded blades on Bergey Windpower's turbines can be easily identified by their constant width and thickness. Pultrusion gives Bergey's single-surface airfoil a strength and torsional flexibility not found in other constructions.

For pleasure boaters fiberglass has become the material of choice. In fact, the techniques used to build fiberglass boats have been successfully adapted by Danish, Dutch, and American companies to build wind turbine blades. These manufacturers place layer after layer of fiberglass cloth in half-shell molds of the blades. As they add each additional layer, they coat the cloth with a polyester or epoxy resin. When the shells are complete they

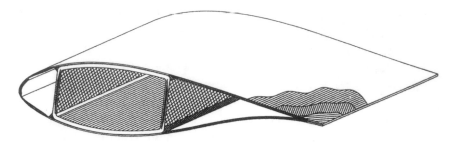

Figure 6-13. Blade cross section. Construction of a fiberglass blade found on many medium-sized wind machines. The central section is the spar, which provides the blade's principal structural support. (Vestas DWT)

literally glue them together to form the complete blade. Nearly all medium-sized European wind turbine blades are made with this technique.

Filament winding is another process where fiberglass strands are pulled through a vat of resin and wound around a mandrel. The mandrel can be a simple shape like a tube, or a more complex shape like that of an airfoil. Originally developed for spinning missile cases, filament winding delivers high strength and flexibility. Though some blades have been made entirely from filament winding, the process is often used only to produce the blade's main structural spar. The blade is then assembled in a mold with a smooth fiberglass shell using the boat-building technique.

Hubs

Like the spokes in a bicycle wheel, the blades become part of the rotor when attached to a hub. The hub holds everything together and transmits the motion of the blades into torque, or turning force. Three aspects of the hub are important: how the blades are attached, whether the pitch is fixed or variable, and whether or not this attachment is flexible.

All conventional wind turbines today use blades cantilevered from the hub, that is, they're supported only at the hub just as the wing of a modern airplane is attached only at the fuselage (see Figure 6-14). During the late 1970s and early 1980s some European designs used struts and stays to brace the blades, following the pattern of the famous Danish wind machine at Gedser. Struts increase the drag on the rotor, but they reduce bending on the root of the blade where it attaches to the hub. Consequently, the spar, the main structural support of the blade, and its attachment to the hub need not be as massive as on a cantilevered blade. Struts and stays work fine on upwind machines as long as the turbine stays upwind. They tend to fail when the turbine inadvertently swings downwind. Early Danish designs were susceptible to this weakness, which led to an industry-wide abandonment of struts and stays for bracing the rotor.

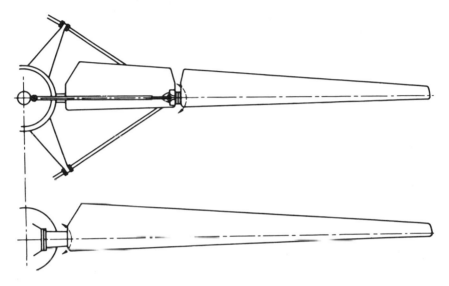

Figure 6-14. Blade attachment. Blades can be attached to the hub with stays, or cantilevered (attached at only one point). Early Danish wind machines used rotors braced with stays. All contemporary designs use cantilevered blades. (Danish Ministry of Energy.)

Most hubs are rigid: they don't allow the blades to flap back and forth in gusty winds. The blades may change pitch by turning about their long axis, but they don't change from the plane of rotation. In nearly all wind turbines currently on the market, the blades are bolted directly to a rigid hub. Most don't change pitch. During the late 1980s and early 1990s several manufacturers of medium-sized wind turbines for commercial applications, 25 meters in diameter and larger, reintroduced pitchable blades to control the rotor in high winds.

On some wind machines a rotor made up of two blades may teeter or rock about the hub. The rotor, as a unit, swings in and out of the plane of rotation like a teeter-totter. This teetering action relieves forces on the blade during gusty winds, when the turbine yaws in response to changing wind direction, and when the blade passes through the tower's wake. Though engineers have long stressed its advantages and simplicity, no wind turbines using the technique have proven commercially successful.

Following the hub, the remainder of the drive train consists of the main shaft to which the rotor is attached, the transmission (where used), and the generator (see Figure 6-15).

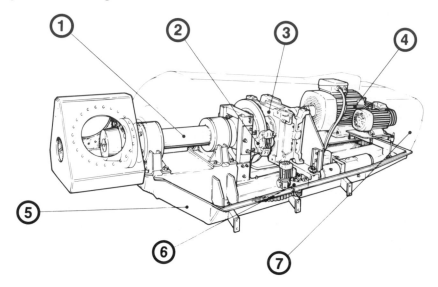

Figure 6-15. Drive train for typical medium-sized wind turbine. (1) Main shaft. Note that the bearings supporting the main shaft are independent of the transmission. On some wind machines the bearings in the transmission housing support the rotor. (2) Disc brake. Note position on the main shaft. On some wind machines the brake is located on the output side of the transmission. (3) Transmission. (4) Induction generator. Many wind machines, such as this one, commonly use two generators, or they use dual windings on a single generator. (5) Bedplate (frame or strongback). (6) Yaw or slewing drive for pointing the turbine into the wind. (7) Nacelle cover. (Vestas DWT)

Transmissions

There are three ways to transfer power from the rotor to the generator: direct drive, mechanical transmissions, and hydraulic transmissions. The simplest method is to drive the generator directly with the rotor. This eliminates the need for a transmission and reduces the complexity of the drive train. Direct drive also offers slightly higher conversion efficiencies because no power is lost going through a gearbox. Direct drive, though, requires a specially designed slow-speed generator that may be larger and may demand greater amounts of expensive materials than a conventional generator driven at higher speeds via a transmission. The most successful of the pre-REA windchargers, the Jacobs home light plant, used direct drive. Although the industry flirted with gear-driven machines during the 1970s and early 1980s, most small wind turbines today, especially those featured in this book, use direct drive.

The chief competitor of the old Jacobs generator, Wincharger, took the transmission approach. They used one large helical gear on the main shaft of the rotor to drive a small gear on the generator. During the 1970s Sencenbaugh Wind Electric's Model 1000 used a similar approach. The transmission increased the 350-rpm speed of the wind turbine to 1100 rpm at the generator. Even with this 3:1 gear ratio a low-speed alternator was still necessary. Most generators need 1800 rpm to operate properly; others are designed for 1200 rpm operation.

On small machines such as the Sencenbaugh or old Wincharger it's also possible to use belts and chains instead of gearing. Most home-built machines use belts and pulleys because they're cheap and readily available. Cogged belts, for example, were used by the defunct manufacturer Aeropower. In practice, belts and chains have proven unreliable. Today no one uses either belts or chains as the principal means to transmit power to the generator on commercial wind machines of any size.

As wind turbine models grow in size the need for a transmission becomes more pressing because the speed of the main shaft decreases. For small machines, transmissions with only one or two stages of parallel shafts may suffice. But with medium-sized wind turbines more stages may be necessary, or designers may even opt for planetary or epicyclic gear boxes.

Inventors often suggest using hydraulic transmissions because they can more easily be matched to the torque characteristics of a wind turbine rotor than a mechanical transmission. In principle they should also be simpler. These advantages are offset, however, by greater inefficiencies. The only large-scale test of hydraulic transmissions, the Bendix-Schachle turbine once owned by Southern California Edison, ended in ignominious failure. No wind turbine using a hydraulic transmission has ever been either a technical or a commercial success.

Generators

First and foremost, generators are not perpetual motion machines. They transfer power, not create it. Power must be delivered to a generator before you can get power out of it. (In our case the prime mover, as it's called, is the rotor.) Nor are generators 100 percent efficient at transferring this power. The rotor will deliver more power to the generator than the generator produces as electricity. This leads us to a fundamental principle about the size of wind turbines. The size of a generator indicates only how much power the generator is capable of producing if the wind turbine's rotor is big enough, and if there's enough wind to drive the generator at the right speed. Thus, we once again confront the fact that a wind turbine's size is primarily governed by the size of the rotor.

The generator converts the mechanical power of the spinning wind turbine into electricity. In its simplest form a generator is nothing more than a coil of wire spinning in a magnetic field. Consequently, whether generating direct current (DC) or alternating current (AC), a generator must have:

1. Coils of wire in which the electricity is generated and through which it flows.
2. A magnetic field.
3. Relative motion between the coils of wire and the magnetic field.

By varying each of these conditions you can design a generator of any size for any application.

Power in an electrical circuit is the product of current and voltage. In a generator the armature is the coil of wire where output voltage is generated and through which current flows to the load (see Figure 6-16). The portion of the generator where the magnetic field is produced is the field. Relative motion between the two is obtained by either spinning the armature within the field or spinning the field within the armature. As you would expect, the stationary part of the generator is the stator; the spinning part is the rotor.

The power produced by a generator depends on the size and length of the wires used in the armature, the strength of the magnetic field, and the rate of motion between them. Increase any one, and you increase the potential power of the generator. The size of the wire in the armature determines the maximum current that can be drawn from the generator before it overheats, melts its insulation, shorts out, and otherwise destroys itself. The

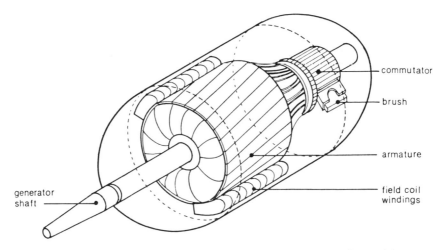

Figure 6-16. DC generator. Sketch of direct-drive generator on Jacobs windchargers. Power is drawn off the spinning armature through brushes. Some of this power is used to energize the field coils.

heavier the wire, the more current it can carry. As long as the wind turbine's rotor continues to provide greater and greater amounts of power as wind speed increases, the generator will continue to produce more current until the generator overheats. To prevent such occurrences, generators have a mechanism for limiting current to a safe maximum.

Generators are rated in terms of the maximum current they can supply at a specified voltage and (for AC generators) at a specific frequency. This rating is given on the name-plate in amps and volts (and frequency where appropriate), as kilowatts and volts, or as kilovolt-amperes (kVA). The generator may be rated for the current it can supply continuously or the current it can supply for only a short period. If generator size is of concern to you, always check which rating is being used. Reputable manufacturers rate their generators for continuous rather than intermittent duty.

Let's turn to voltage, the other half of the power mix. Generated voltage depends on the rate at which magnetic lines of force are crossed by the wire loops in the armature. Designers alter voltage by changing the magnetic field, by changing the rate of motion between them, or both.

The generator's field is provided by magnets. With electromagnets, some power is used to "excite" or "energize" the field around the armature. The strength of this field is a function of the length of wire (the number of coils) in the field windings and the current flowing through them. For example, double the length of wire in the windings and you double the strength of the field, doubling generated voltage.

Many of the windchargers built during the 1930s produced 32 volts. Resistance losses are high when transmitting low-voltage power. Because of this, most reconditioned windchargers were rewound for 110 volts. The old wire was stripped off the generator and replaced with more turns of thinner wire. Less current could be drawn through the smaller wire than before, but the increased length of wire produced a stronger field, increasing the voltage. Generating capability was not affected, power from the generator remained the same, but the balance between the voltage and current changed: voltage increased and current decreased by an equivalent amount.

Permanent magnets can also provide the field. They don't require power for excitation, because they're inherently, that is, permanently, magnetic. The principal means for increasing field strength with permanent magnets is to use magnets with greater magnetic density.

The voltage can also be increased by adding more or larger field coils, by adding more permanent magnets, or by increasing the speed at which the armature windings pass through the field. This can be accomplished by increasing the diameter and length of the generator so there's room for more magnets, or by spinning the rotor faster.

Yes, all this does have some bearing on the design of wind-driven generators. To get a feel for how, let's examine two popular pre-REA wind machines. Both Jacobs and Wincharger used about the same size rotor (14 feet, or 4 meters); thus the power available to the generator and the speed of the rotors were roughly equivalent. Yet Jacobs chose to use a direct-drive generator whereas Wincharger chose a transmission.

To produce the same power and voltage as Wincharger without a transmission, Jacobs had to design a generator that would operate at lower shaft speeds. Jacobs did so by increasing both the diameter and the length of its generator. This allowed the use of more field coils (six to Wincharger's four). The coils were also longer.

The Jacobs generator's greater diameter also increased the speed at the periphery of the armature where it passed the field coils. Doubling the diameter doubles the rate at which the armature cuts through the field. The effect is the same as that from a 2:1 transmission that doubles the speed of the generator.

All in all, the Jacobs generator was considerably larger and used much more copper and iron than did Wincharger's to do the same job. But the Jacobs generator could do that job at a slower speed. Jacobs chose a slow-speed generator for long bearing life and simplicity, believing these advantages offset the greater cost.

Barry Commoner's adage, "There's no such thing as a free lunch," puts it succinctly. Whether it's the design of generators or any other wind machine component, there's always a trade-off. You gain something only by giving up something else. You hope that what you gain is more valuable than what you've lost. It's as true today as it was during the 1930s. Small wind turbine designers who stress long life and low maintenance choose lower generator speeds. The price they pay is increased costs.

Manufacturers of small wind turbines intended for remote sites in harsh environments may opt, as Jacobs did, for building slow-speed generators tailored to their wind turbine. That's just what most of today's manufacturers, such as Bergey Windpower, Northern Power Systems, and others, have done. They build specially designed, direct drive, slow-speed alternators.

The trade-offs are also apparent in medium-sized wind turbines. During the early 1980s many American-designed wind turbines that operated at high speeds were installed. These machines were not only noisy, they were also trouble prone. Danish designs operating at much more modest speeds eventually won more than half the California market. Like Jacobs before them, the rugged Danish designs opted for lower speeds to reduce wear and tear. The Danish turbines typically drove a six-pole generator at 1200 rpm, while their American competitors used four-pole generators running at 1800 rpm. Today none of the early U.S. designs are still being built.

Alternators

Pre-REA windchargers produced DC by spinning the armature within the field. Power was drawn off the rotating armature through brushes. During the 1960s the auto industry began replacing DC generators with alternators. Alternators offer several advantages over DC generators. For a given output, alternators cost less than generators, and an alternator's slip rings last much longer than the brushes in a generator. Slip rings are more durable because they don't carry the alternator's current output as brushes do in a generator.

The battle between alternators and generators is far from over. Some die-hards, such as Mick Sagrillo at Lake Michigan Wind & Sun, believe that DC generators still offer promise. Sagrillo, who rebuilds DC generators, argues that special-purpose generators, such as the Jacobs home light plant, use oversized brushes to ensure long life. These brushes don't wear out as quickly as many imagine, he says. Further, Sagrillo maintains that a generator gives better high-end performance than an alternator. Still, Sagrillo concedes that alternators now dominate the market.

In today's alternator the field, rather than the armature, revolves. Power is drawn off the stator from fixed terminals. Excitation of the alternator's field is provided through slip rings on the rotor, but only enough power passes through the slip rings to excite the field (a small percentage of the alternator's output). There are no brushes and no commutators to wear from the passage of high current. There's no arcing at the brushes. There are no slip rings—no moving contacts—in a permanent-magnet alternator because the field is permanently excited.

In a conventional alternator the field revolves inside the stator. But Bergey Windpower, Marlec Engineering, and SOMA Power all spin the permanent-magnet field outside the stator. They attach the blades of their rotors to the magnet ring that spins around the armature. Power is drawn off from inside the generator. This arrangement, which eliminates the need for slip rings, has become common among small wind turbines.

As the name implies, alternators generate AC. As the rotor spins, current rises and falls like waves on the ocean (electrons in the armature are first jostled in one direction, then alter course and are jostled the other way). The alternator's frequency is the rate at which current rises and falls; it's given in cycles per second or hertz. The speed of the rotor and the number of poles determine the alternator's frequency. Drive the alternator faster and frequency increases; slow the rotor and frequency decreases. This explains why most small wind turbines generate variable-frequency AC. When wind speed rises, the turbine spins faster, increasing frequency (as well as voltage and current). When the wind subsides, frequency decreases.

In a simple alternator the four poles are wired together in series as a single circuit producing single-phase AC. When three groups of poles are arranged symmetrically around the stator, the alternator produces three-phase AC, each phase one-third out of sync with the next. Most alternators used in wind systems produce three-phase AC. Three-phase alternators do more with less. The designer can more efficiently pack poles within the generator. Power is determined by the rate at which lines of force are cut by the armature. Thus we can increase power by increasing the number of poles to take up all the available space within the generator.

If you've ever spun the shaft of a toy generator in your hand, you remember how it felt when the rotor would stick slightly as the coils in the armature aligned with the magnets in the field. As the coils passed by the magnets, the shaft would turn more easily. This same effect, cogging, occurs in large generators and motors. Cogging is of interest in wind machines because it can retard the startup of the wind turbine in light winds when the poles are aligned. Increasing the number of poles by arranging them in three phases reduces cogging, enabling the turbine to start more easily in light winds.

Variable- or Constant-Speed Operation

Wind machines driving electrical generators operate in either of two ways: at variable speed, or at constant speed. In the first case, the speed of the wind turbine varies with the speed of the wind. In the second, the speed of the wind turbine remains relatively constant as wind speed fluctuates.

In all small wind turbines built today, the speed of the rotor varies with wind speed. This simplifies the turbine's controls while improving aerodynamic performance. When such wind machines drive an alternator, both the voltage and frequency vary with wind speed. The electricity they produce isn't compatible with the constant-voltage, constant-frequency AC produced by the utility. If you used the output from these wind turbines directly, your clocks would gain and lose time, and your lights would brighten and dim as wind speeds fluctuated. Eventually you'd burn up every motor in the house. Unless you have a use for this low-grade electricity (heating, pumping water, and so on), the output from these wind machines must be treated or conditioned first, even if it's simply for charging batteries.

Because batteries can't use AC, the alternator's output must be converted to DC. As in your car alternator, diodes—electrical check valves that permit the current to flow in only one direction—rectify the AC output to DC, which is then used for battery charging.

To produce utility-grade electricity, either the alternator's AC or rectified DC can be treated with a synchronous inverter to produce constant-voltage 60-cycle AC like that from the utility (see Figure 6-17). Most of these inverters, though not all, are line commutated. They must be interconnected with the utility to operate. The utility's AC provides a signal that triggers electronic switches within the inverter, which transfers the variable quality electricity at just the right time to produce 60-cycle AC at the proper voltage. No utility power is consumed in the process. It's merely used as a signal to coordinate the switching.

UTILITY COMPATIBLE WIND MACHINES

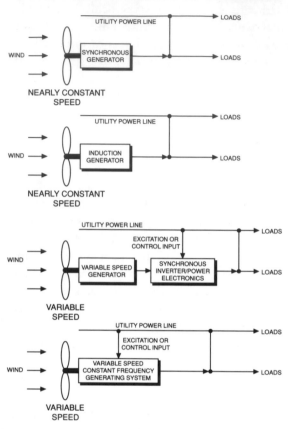

Figure 6-17. Techniques for generating utility-compatible electricity. Most medium-sized wind turbines use induction generators. Most small (and now some medium-sized wind turbines as well) use power electronics with a variable speed generator. Very few wind turbines have ever been designed to drive true synchronous generators.

Though some manufacturers of medium-sized wind machines are now building variable-speed turbines, most models operate the rotor at or near constant speed. The latter produce utility-compatible power directly. They don't require power conditioning, because they produce electricity synchronized directly with that from the utility. There are two ways to do this.

One method is to drive an alternator at a constant speed so voltage and frequency remain constant. Synchronous alternators are used primarily in the utility industry. Power to the generator is controlled precisely by limiting steam to the turbines. This keeps the generator spinning at just the right speed to maintain synchronization with the utility's other generators. On a wind machine it's much more difficult to control the turbine's speed because the source of power, the wind, constantly varies. Complex and fast-reacting mechanisms for changing the pitch of the blades are required to maintain a constant generator speed. To date only a few giant experimental wind machines have been able to justify the cost of synchronous alternators. No commercial wind turbines of any size use this technique today.

Induction Generators

Another approach uses induction generators. These have two advantages over alternators: they're cheap, and they can supply synchronous power without sophisticated controls. Induction generators are simply induction motors (like the motor in your refrigerator) in disguise.

An induction motor becomes a generator when driven above its synchronous speed. Plug an induction motor into an outlet and the motor will turn at 1800 rpm, consuming power. Leave it plugged in, but now drive the motor at 1800 rpm. The motor will no longer consume power from the outlet. You're now supplying it. Spin the rotor just a little faster, say at 1820 rpm, and it won't be consuming electricity, it will be generating it. As you try to spin the motor faster it gets harder to turn. The utility consumes the additional power as you produce it, without rotor speed appreciably increasing.

In a wind turbine which drives an induction generator, when wind speed increases, the load on the generator automatically increases, as more torque (power) is delivered by the rotor. This continues until the generator reaches its limit and either breaks away from the grip of the utility or overheats and catches fire. Technically, induction generators are not true constant-speed or synchronous machines. As the load increases, the generator speed slips by 2-5 percent, or 36-90 rpm on an 1800-rpm generator.

Induction generators have proven extremely popular for wind turbines because they're readily available in a range of sizes, and interconnection with the utility is straightforward. Literally, plug it in and go. Early promotions for the defunct manufacturer Enertech showed its wind machines being plugged into a wall socket. Interconnection is a little more sophisticated today, but the principle remains the same. The wind machine is wired to a dedicated circuit in your service panel or directly with the utility. Utilities are much more comfortable with induction generators than with synchronous inverters because they understand them better. Synchronous inverters, for all their benefits, still remain a mystery to many utilities.

When looking at a wind machine's generator there's no need to be dazzled by the technology employed. Your primary concern is what kind of power it produces. If you want utility-compatible power, then you can't use a wind machine with an alternator that doesn't also include a synchronous inverter.

The inverter shouldn't be something slapped together just for the occasion. Experience with inverters has shown that they must be carefully tailored to the generator. The inverter not only produces 60-cycle AC at the correct voltage but it also performs another important function. It loads and unloads the generator as more or less wind is available. When the inverter and generator are improperly matched, the wind machine will not perform optimally. The rotor may require higher winds than necessary to start, or it may never reach the tip-speed ratio where it performs most efficiently. At the other extreme the inverter may not load the generator sufficiently to extract all the power available.

Likewise, if you want a wind machine for charging batteries at a remote hunting cabin, you won't be able to use an induction generator. They only work when interconnected with the utility.

You will need to look at what's available. Wind machines larger than 10 meters in diameter use induction generators almost exclusively. If you want a slow-speed, permanent-magnet alternator on a wind machine larger than 10 meters (33 feet) in diameter, you're simply out of luck. Mid-sized wind turbines typically are much more complex than their smaller counterparts. Nearly all use transmissions, and many use two generators.

Dual Generators

As mentioned elsewhere, don't be swayed by the size of the generator alone. It's only an indication of how much power the generator is capable of producing, not how much it will generate. Ask Danish manufacturers what size generator they have in their machine and they'll look at you quizzically and ask, "Which one?" Danish wind machines often use two induction generators, one for low winds and another, much larger, generator for higher winds.

Induction generators operate inefficiently at partial loads. For a wind machine with a generator designed to reach its rated output in a 25-35 mph (11-15 m/s) wind the generator would operate at partial load most of the time. Rather than use only one generator, Danish designers bring a smaller one on line first so that it operates at nearly full load in low to moderate winds. As wind speed increases they drop the smaller generator while energizing the larger or main generator. Thus, both generators operate more efficiently than either alone, and overall performance of the wind machine is improved.

The two generators may be in tandem and driven by the same shaft, or they can be side by side with the small generator being driven by belts from the main generator. Usually both generators are spun at the same time and are not brought on line mechanically but by energizing the field electrically. In some designs, the generator is wired in two stages: during light winds the first stage uses only a portion of the generator's capacity, and in higher winds the second stage uses the generator's full potential.

The use of dual generators permits most Danish turbines to operate at two speeds. This enables them to operate the generator and the rotor at a higher efficiency. Though they are not true variable-speed machines and can't take full advantage of the optimum tip-speed ratio, these turbines can bracket the optimum range. This is particularly useful in low winds where efficiency is most crucial.

Rotor Controls

The rotor is the single most critical element of any wind turbine. It's what confronts the elements and harnesses the wind. Because the blades of the rotor must be relatively large and operate at relatively high speed to capture the energy in the wind, they're the most prone to catastrophic failure. How a wind turbine controls the forces acting on the rotor, particularly in high winds, is of the utmost importance to the long-term, reliable functioning of a wind machine.

The simplest and most foolproof method for controlling the rotor is to decrease the area of the rotor intercepting the wind as wind speeds exceed the turbine's operating range. As frontal area decreases, less wind acts on the blades. This reduces the rotor's torque, power, and speed. The thrust on the blades (the force trying to break the blades off the hub) and the thrust on the tower (the force trying to knock the tower over) are also reduced. This method of rotor control permits the use of lighter-weight and less expensive towers than on wind machines where the rotor remains facing into the wind under all conditions.

Halladay's umbrella mills exemplified the concept (see Figure 6-18). These nineteenth-century water-pumping wind machines automatically opened their segmented rotor into a hollow cylinder in high winds, letting the wind pass through unimpeded. Each segment was composed of several blades mounted on a shaft, allowing the segment to swing into and out of the wind. When the segments are closed, Halladay's windmill looked like any other water-pumping windmill from the period. But in high winds thrust on the segments would force them to flip open. This action was

Figure 6-18. Halladay's rosette or umbrella mill. Segments of the rotor furl in high winds by swinging out of the wind's path. The rotor also passively orients itself downwind of the tower.

balanced by counterweights so the farmer could adjust the speed at which the windmill would open and close.

Horizontal Furling

Later developers, such as the Reverend Leonard Wheeler, chose to use the same control concept (changing the area of the rotor intercepting the wind) but in a different manner. Rather than swing segments of the rotor parallel to the wind, Wheeler thought it simpler to swing the entire rotor out of the wind. He couldn't do this with the downwind rotor used by Halladay. Instead, he used a rotor upwind of the tower; a tail vane kept the rotor pointed into the wind (see Figure 6-19).

The tail vane and rotor (wind wheel to oldtimers) were hinged to permit the rotor to swing sideways toward the tail. As the rotor furled toward the tail, the rotor disc took the shape of a narrower and narrower ellipse, gradually decreasing the area exposed to the wind. The mechanism for ex-

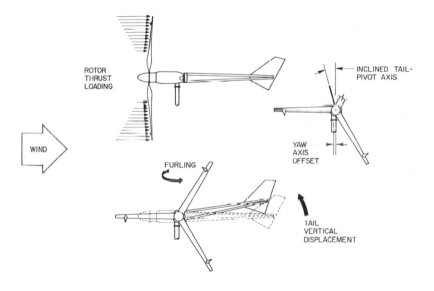

Figure 6-19. *Horizontal furling. This wind machine furls in high winds by swinging the rotor toward the tail. Note that the yaw tube (the tube connecting the wind machine to the tower) pierces the nacelle off-center. The rotor axis is offset from the yaw axis, causing the rotor to fold toward the tail in strong winds. Also note that the tail vane pivots about an inclined axis. The weights near the blade tips twist the flexible fiberglass blades to improve performance in low winds. (Bergey Windpower)*

ecuting this was the pilot vane. The pilot vane extended just beyond and parallel to the rotor disc. Unlike the tail vane, the pilot vane was fixed in position relative to the rotor. Wind striking the pilot vane pushed the rotor toward the tail and out of the wind. In the folded position the rotor and pilot vane were parallel to the wind like the segments of the Halladay rotor. The thrust on the pilot vane was counterbalanced with weights. By adjusting the weights the farmer could determine the wind speed at which the rotor would begin to furl.

As the American farm windmill evolved, the pilot vane went the way of hand cranks on cars. Offsetting the axis of the rotor slightly from the axis about which the wind machine yaws or pivots around the top of the tower produced the same results: self-furling in high winds. When the rotor axis is offset from the yaw axis, the wind's thrust on the rotor creates a force acting on a small moment arm (lever) represented by the distance between the two axes. The wind's thrust turns the rotor out of the wind. The tail vane is hinged, allowing the rotor to furl.

On contemporary farm windmills there are no weights and levers to counteract the furling thrust. Instead they use springs. By adjusting the tension in the spring, the farmer controls the wind speed at which the rotor furls. To see this for yourself, find an operating farm windmill and watch it in high winds. It will constantly fold toward the tail and reopen without any intervention.

Millions of machines using Wheeler's approach to overspeed control have been put into operation around the world. It's what you might call a proven concept. And if it worked reliably for all those machines for all those years, it should still work today. It does.

Nearly all small wind turbines today use furling of one form or another. Bergey Windpower's turbines have operated unattended in wind speeds above 120 mph (54 m/s) as have products by other manufacturers that also use furling.

The Bergey series of small wind turbines carries simplicity even further than the farm windmill. Rather than using springs to control furling, Bergey designs use gravity for returning the rotor to its running position. The hinge pin for the tail vane on the Bergey machines is skewed a few degrees from the vertical. When the rotor furls in high winds it lifts the tail vane slightly. As the wind subsides, the weight of the tail pulls itself down into position, forcing the rotor to swing back into the wind.

Wind machines using this approach, like the water-pumping windmills before them, can be controlled manually by furling the rotor with a winch and cable. The rotor doesn't come to a complete stop when furled. It will continue to spin but at low speeds and power.

Vertical Furling

During the 1930s Parris-Dunn built a windcharger that used a variation on the furling theme. Rather than turning the rotor parallel to the tail vane, they chose to tip the rotor up out of the wind. In high winds the turbine would take on the appearance of a helicopter (see Figure 6-20). As the winds subsided, the rotor would rock back toward the horizontal. Northern Power Systems' High Reliability series, SOMA, and Wind Baron's NEO have all adopted this technology. It's a simple strategy that

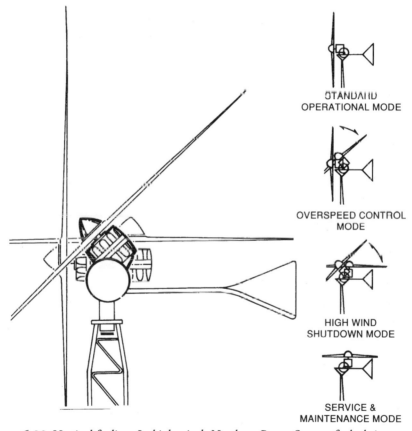

STANDARD
OPERATIONAL MODE

OVERSPEED CONTROL
MODE

HIGH WIND
SHUTDOWN MODE

SERVICE &
MAINTENANCE MODE

Figure 6-20. Vertical furling. In high winds Northern Power Systems furls their HR3 model by tilting the rotor skyward, following the example of a 1930s-era windcharger. A shock absorber dampens the rate at which the rotor returns to the running position. (Northern Power Systems)

works well. Northern Power Systems' HR3 model has survived winds in excess of 176 mph (79 m/s) using this approach.

Like the original Parris-Dunn, Northern Power and Wind Baron use a spring to control the wind speed at which the rotor begins to furl. The wind speed at which the rotor begins to pitch back is governed by tension in the spring. The SOMA turbine accomplishes the same effect with a sliding weight. The principal difference among the three manufacturers is the means they use for dampening the action of the rotor and generator as it rocks back and forth. Gusty winds can cause the rotor to tip up, then quickly rock forward, dropping the rotor and generator onto the wind machine's frame, severely jarring the blades and the rotor's main shaft. Both Northern Power and SOMA use a shock absorber that dampens the return of the rotor to the running position. Wind Baron's NEO follows the Parris-Dunn example and simply uses a rubber pad to cushion the blow.

Dr. Peter Musgrove's contribution to wind technology was designing a way to reduce the rotor's intercept area on vertical axis turbines (see Figure 6-21). As mentioned before, the H-rotor configuration offers several advantages over a conventional Darrieus turbine. Its weakness is the tremendous forces trying to bend the blades at the juncture between them and the cross-arm. These bending forces can be reduced and the speed of the rotor controlled by hinging the blades. In the Musgrove turbine the blades are hinged to the cross-arm in such a manner that the portion of the blade above the cross-arm is not equal to the portion below. As the rotor spins, centrifugal force throws the heavier portion of the blade away from the vertical, varying the geometry of the rotor. The wind and rotor speed at which this occurs is

VARIABLE GEOMETRY VAWT

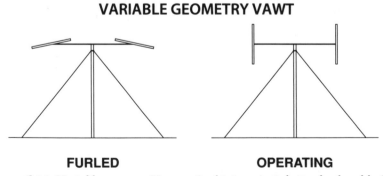

FURLED   **OPERATING**

Figure 6-21. Variable-geometry H-rotor. In this ingenious design developed by Dr. Peter Musgrove of Reading University, the straight blades of the rotor are hinged so that they tilt toward the horizontal at high rotor speeds.

determined by the weight of the blade and the tension in a spring restraining the blades in the upright position. At high wind speeds the blades approach the horizontal, reducing the intercept area. Though there are some experimental versions in England, no wind turbine using Musgrove's "variable geometry" has been commercially successful.

Changing Blade Pitch

When most people first consider the problem of controlling a rotor in high winds, they think immediately of changing blade pitch. This probably results from our exposure to propeller-driven airplanes. Indeed, a wind turbine rotor can be controlled much like the blades on the propeller of a commuter plane. Like changing intercept area, changing blade pitch affects the power available to the rotor. By increasing or decreasing blade pitch we can control the amount of lift that the blade produces.

There are two directions in which the blades can be pitched: pitching them toward stall, or pitching them toward feather. Blade pitch is usually set a few degrees into the wind. If the blade is nearly parallel with its direction of travel (perpendicular to the wind) it stalls. The blade is feathered, on the other hand, when it is at right angles to its direction of travel (90-degrees pitch), or parallel to the wind. To feather a blade it must be turned farther than when stalling the blade, causing the pitch mechanism to act through a much greater distance.

Stall destroys the blades' lift, limiting the power and speed of the rotor, but it does nothing to reduce the thrust on the rotor or the tower. Though it is simpler to build a mechanism for stalling the blade than it is to build a feathering governor, the technique is less reliable. On upwind machines, thrust on the blades bends them toward the tower. Designs dependent on blade stall as the sole means of overspeed protection have a poor survival record. The blades have a nasty habit of striking the tower. Downwind turbines using stall regulation have had fewer problems because the blades are forced to cone farther downwind and away from the tower. Still, they too have had an overall poor reliability record.

Where changing blade pitch is the primary means of control on an upwind rotor in high winds, the blade should rotate toward full feather. By so doing the drag on the blade is reduced to one-fifth of that on a blade flatwise to the wind.

Governors for pitching the blades appear in a variety of forms. During the 1930s, the Jacobs brothers popularized the flyball governor (see Figure 6-22). Above normal rotor speeds the weights would feather all three blades simultaneously via a mechanical linkage to three weights. (It's important that all blades change pitch at the same rate. If they don't, the rotor will become unbalanced, causing severe vibrations.) This massive 100-pound governor protected the 14-foot rotor reliably when carefully adjusted. On later models, Jacobs also marketed a clever version called the Allied (after a windcharger on which it first appeared) or blade-actuated governor.

Why use weights when you don't have to? The blade-actuated governor uses the weight of the blades themselves to change pitch (see Figure 6-23). Unlike the blades on the flyball governor, the blades not only turn on a shaft in the hub, they also slide along the shaft. Each blade is connected to the hub through a knuckle and springs. The knuckle in turn is attached to a triangular spider. As the rotor spins, the blades are thrown away from the hub, causing them to slide along the blade shaft. When they do, the blades pull on the spider, which rotates the blades toward feather. The springs

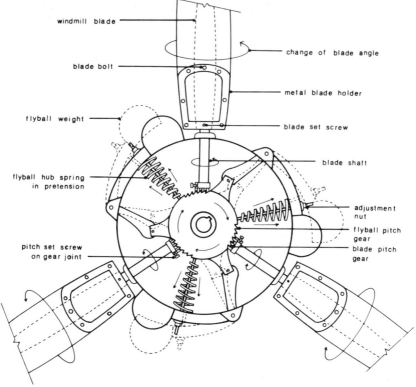

Figure 6-22. *Jacobs flyball governor. Centrifugal force throws the weights away from the governor, changing the pitch of the blades via a mechanical linkage.*

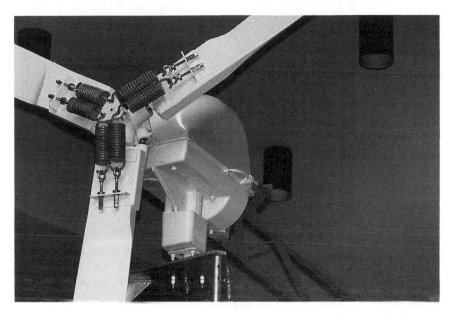

Figure 6-23. Blade-actuated governor. Many of the small wind turbines built during the 1970s used this design, patterned after later versions of the Jacobs windcharger. The force acting on the blades in high winds causes them to collectively change pitch.

govern the rotor speed at which this occurs. Like the flyball governor, the blade-actuated governor works reliably when properly adjusted and built to the highest material standards.

In the late 1970s Marcellus Jacobs, the sole surviving founder of the original Jacobs Wind Electric Company, reentered the wind business. (The original firm ceased activity during the 1950s.) Along with his son Paul, Marcellus began manufacturing wind turbines patterned after his earlier models. His company briefly built wind turbines 7-8 meters (21-26 feet) in diameter. Jacobs' redesigned machine didn't depend solely on blade feathering to control rotor speed, since the blade-actuated governor was inadequate for a machine of this size. The new Jacobs turbine was also self-furling. The governor feathered the blades to limit power output to the alternator; overspeed protection was provided by furling the rotor toward the tail.

On small wind turbines mechanical governors, whether or not blade actuated, have proven too costly and unreliable. They are also too maintenance-intensive for the modern wind turbine market. Only the French manufacturer Aerowatt still uses pitch weights to govern the rotor (see Figure 6-24). Aerowatts are used only in France or in French overseas territories. None of the advanced small wind turbines on the market today use this technology.

Despite this trend there are hundreds of small windchargers from the 1930s and from the Jacobs' revival in the 1970s operating in the United States that still rely on mechanical governors. With proper maintenance and a supply of spare parts, these machines will last for several more decades. Owners of these turbines argue that mechanical governors provide better power output in high winds than does furling. In winds above the rated speed, power output drops sharply on machines that furl, while small wind turbines using mechanical governors are able to maintain near constant peak power.

You can change blade pitch without using a mechanical governor. In the bearingless rotor concept the blades are attached to the hub with a torsionally flexible spar. At high speeds the blades twist the spar toward zero pitch, stalling the rotor. Weights attached to the blades are sometimes used to provide the necessary force. There are no moving parts in the hub: no

Figure 6-24. Pitch weights. The French Aerowatt is the only contemporary wind turbine that uses weights to change blade pitch in high winds.

bearings, knuckles, or sliding shafts. Several have attempted to market this technology on medium-sized wind turbines. All have failed because rotor dynamics are far more complex than first envisioned.

Carter Wind Systems' 10-meter turbine best represented this control strategy. The filament-wound, fiberglass spar permitted the blade to twist torsionally. During high winds small weights inside the blades would rotate the blade toward stall in one movement. The flexible spar also permitted the blade to cone progressively downwind of the tower in high winds like the fronds on a palm tree during a hurricane. This design, though elegantly simple, was too unreliable and is no longer manufactured.

Several medium-sized wind turbines currently on the market successfully use pitch control to limit power in high winds. Most of these machines are greater than 25 meters in diameter, so the higher costs and complexity of pitch control can be justified.

Aerodynamic Stall

Almost all wind machines without pitch control use aerodynamic stall to some extent for limiting power from the rotor. This is particularly true of medium-sized wind turbines, which use fixed-pitch rotors to drive induction generators. In winds above the rated speed, the tip-speed ratio for these turbines declines because the speed of the rotor remains constant. The angle of attack increases with increasing wind speed for wind turbines operating at constant speed, lowering the performance of the blades below the optimum.

Designers seldom rely on blade stall as the sole means of overspeed protection. Stall is most effective on induction wind machines with fixed-pitch rotors. Induction generators, however, are dependent on the utility for controlling the load. During a power outage the generator immediately loses this load. The rotor, no longer restrained to run at a constant speed, immediately accelerates. Stall now becomes ineffectual for regulating power until a new equilibrium is reached. Unfortunately, this occurs at extremely high rotor speeds.

On an upwind machine with a tail vane the rotor can be prevented from destroying itself by furling the rotor out of the wind. Since tail vanes are limited to small turbines, medium-sized upwind machines and all fixed-pitch downwind machines must use a different strategy. Brakes are the most popular.

Once brakes have been selected as the means to limit rotor speed during a loss-of-load emergency, they're also frequently used during normal operation. In a typical fixed-pitch wind machine the brake is applied at the cut-out speed to stop the rotor. Wind turbines using this approach require strong blades should the brake fail and the rotor accelerate to destructive speeds.

Consider the case of a small downwind turbine driving an induction generator that was braked to a halt at the cut-out speed. When the manufacturer, Enertech, first introduced the machine, it stressed that the rotor was stall-regulated and that it could operate safely above the cut-out speed without the brake. The rotor was braked, asserted Enertech's vice president, only to minimize wear on the drive train at high wind speeds. The amount of energy in the wind at these higher speeds, he said, did not warrant the cost of capturing it. Mother Nature soon gave Enertech ample opportunities to prove its mettle. The brake failed on several occasions. Rotors went into overspeed, and several Enertechs destroyed themselves. Stall alone wasn't enough to protect the rotor. Enertech later added tip brakes for such emergencies.

Mechanical Brakes

Brakes can be placed on either the main (slow-speed) shaft or on the high-speed shaft. Brakes on the high-speed shaft are the most common because the brakes can be smaller and less expensive for the needed braking torque than those on the main shaft. When on the high-speed shaft, the brakes can be found between the transmission and the generator or on the tail end of the generator. In either arrangement braking torque places heavy loads on the transmission and couplings between the transmission and generator. Moreover, should the transmission or high-speed shaft fail, the brake can no longer stop the rotor.

In general, brakes on fixed-pitch machines should be located on the main shaft where they provide direct control over the rotor. (There's always a greater likelihood of a transmission failure than a failure of the main shaft.) But the lower shaft speeds require more braking pressure and greater braking area. As a result, the brakes are larger and more costly than those on the high-speed shaft.

Brakes can be applied mechanically, electrically, or hydraulically. Most operate in a fail-safe manner. In other words it takes power to release the brake. The brake automatically engages when the wind machine loses power. Springs provide the force in a mechanical brake; batteries, the electricity in an electrical brake; and a reservoir, the pressure in a hydraulic brake.

The problem with brakes of any kind is that they fail—not often, it's true, but once is enough. Brake pads require replacement or adjustment periodically. After extensive use the calipers have to travel farther to reach the disc. If the brakes are spring applied, pressure from the springs decreases and so does braking torque as travel increases. In one 12-meter model built briefly during the late 1970s the brake just didn't supply enough torque to stop the rotor, and there were no backup devices to protect the turbine. Several machines ran to destruction, but some were brought under control when the designer, Terry Mehrkam, climbed the tower, wedged a lever into the brake, and manually forced the pads against the disc. This dangerous practice eventually cost Mehrkam his life.

Experience has taught wind turbine designers that wherever a brake is used to control the rotor, there must be an aerodynamic means to limit rotor speed should the brake fail. There are three common choices for aerodynamic overspeed protection on wind machines without tail vanes and pitch controls: tip brakes, spoilers, and pitchable blade tips. These devices are found frequently on medium-sized wind turbines using fixed-pitch rotors to drive induction generators.

Aerodynamic Brakes

Tip brakes are plates attached to the end of each blade (see Figure 6-25). They're activated by centrifugal force once the rotor reaches excessive speed. When deployed, they slap or drag at the wind. They're simple, effective, and they have saved many a fixed-pitch rotor from destruction. Tip

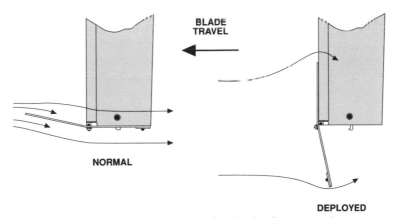

Figure 6-25. Enertech popularized the use of tip brakes for overspeed protection on downwind, induction wind machines. Tip brakes are often noisy, and rob power.

brakes, however, have been likened to keeping your foot on the accelerator at the same time you're stepping on the brake. They keep the rotor from reaching destructive speeds but do nothing to reduce the lift of the blade or the thrust on the wind turbine and tower. Tip brakes are also noisy and reduce the performance of the rotor under operating conditions by increasing drag at the tip where blade speed is greatest.

Most of the power in the wind is captured by the outer third of the rotor. Consequently, it's not necessary to change the pitch of the entire blade to limit the rotor's power and speed. The performance of the blade in this region can be reduced by using spoilers or movable blade tips. In one Danish design if there's a loss of load or the brake fails, centrifugal force activates spoilers along the length of each blade. The spoilers pop out of the blade and change the shape of the airfoil, destroying its effectiveness, reducing power and rotor speed.

Many stall-regulated Danish wind turbines use pitchable blade tips. Medium-sized turbines 15-25 meters in diameter use passive controls to activate the blade tips. At higher than normal rotor speeds the tips are thrown away from the rotor by centrifugal force, causing them to slide along a grooved shaft. As they move along the shaft the tips pitch toward feather. This action decreases lift where it's greatest while dramatically increasing drag. Both spoilers and pitchable blade tips have proven highly successful, though refinements have been necessary. For example, frequently only one or two of the blade tips would activate at the same time, rather than all three.

Both Boeing's 2.5 MW Mod-2 and its 3.2 MW Mod-5B used a similar approach (see Figure 6-26). But instead of using passive controls, Boeing hydraulically drove the tips toward the feathered position. Boeing actively positioned the blade tips to regulate power. Today most fixed-pitch turbines greater than 25 meters (80 feet) in diameter actively regulate the pitch of the blade tips to control power, much like Boeing did.

Putting It All Together

Let's look at two types of wind machines and how their manufacturers put all the pieces together. We'll also look at how they operate under normal and emergency conditions. The first group are the advanced small wind turbines designed specifically for high-reliability, low-maintenance applications. The second group are medium-sized wind turbines like those found in California wind plants and throughout Denmark.

Figure 6-26. Small Enertech wind turbine (1.8 kilowatts) in foreground used tip brakes for overspeed protection. The Boeing Mod-2 (2500 kilowatts) in the background controlled the pitch of the blade tip to limit power during normal operation and was used to stop the rotor. Neither turbine is currently manufactured.

Small Turbines

All small wind turbines today are specially designed for simplicity, ruggedness, and low maintenance. Mike and Karl Bergey quote Antoine de Saint Exupéry to express Bergey Windpower's design philosophy, "Perfection is achieved not when there is nothing more to add but when there is nothing more to take away."

Today's advanced small wind turbines, such as those in Table 6-1, typically employ an upwind rotor and are passively directed into the wind by tail vanes (see Figures 6-27, 6-28, and 6-29). All but Northern Power Systems' HR3 drive permanent-magnet alternators directly without the aid of transmissions. All are well suited for stand-alone, battery-charging applications. Through use of a synchronous inverter, some, such as the Bergey Excel, can also be used by homes, farms, and businesses with existing utility service.

These turbines all operate at variable speed. From startup through furling, the rotor rpm increases with increasing wind speed. Similarly, voltage and frequency increase with wind speed. In winds above the rated speed, the blades begin stalling and performance of the rotor decreases. Above the furling speed, the rotor begins to swing toward the tail vane. Power drops dramatically as furling proceeds. (This drop in power is characteristic of self-furling machines. Small, pitch-regulated turbines typically limit peak power.) When high winds subside, the turbine returns to its operating position automatically.

Blade stall and furling are the only means for limiting the rotor's speed and power during both normal and emergency conditions. There are no brakes either to stop the rotor in high winds or to park the rotor for servicing. Nor are there brakes to prevent the wind machine from yawing about the top of the tower in response to changes in wind direction.

The designers of these machines stressed simplicity and ruggedness over greater control. Bergey Windpower, SOMA Power, and Marlec Engineering go one step farther and integrate the hub and rotor housing into one assembly. Because they furl the rotor horizontally toward the tail in their designs, Bergey goes so far as to combine the mainframe and stator assembly into one unit. All are designed for little or no maintenance, with good reason. At windy sites it's not uncommon for a wind machine to be in operation for two-thirds of the time, or about 6000 hours per year. At that rate, a wind machine would operate as many hours in the first 4 months of the year as an automobile driven 100,000 miles at 50 mph. Over a 30-year lifetime

Table 6-1
Characteristics of Selected Small Wind Turbines

Manufacturer	Model	Rotor Dia. (m)	Swept Area (m²)	Rated Power (kW)	Rated Wind Speed (m/s)	No. of Blades and Blade Material	Means of Control*
Marlec	Rutland 500	0.51	0.20	0.02	10.0	6 nylon	nc
LVM Products	Aero4gen F	0.87	0.59	0.07	10.3	6 nylon	nc
Marlec	Rutland 913	0.91	0.65	0.09	10.0	6 nylon	nc
Ampair	100	0.92	0.66	0.05	10.0	6 poly	nc
Hamilton Ferris	Windpower 100	1.10	0.89	0.10	11.2	2 GFRP	v
Southwest Windpower	Air 303	1.10	1.02	0.30	12.5	3 CFRP	nc
Southwest Windpower	Windseeker 503	1.50	1.81	0.50	12.5	3 wood	v
Hamilton Ferris	Windpower 200	1.50	1.83	0.20	11.2	2 wood	ab
LVM Products	Aero8gen F	1.60	1.89	0.22	10.3	3 wood lam.	h
Wind Baron	NEO Plus 5	1.60	1.94	0.78	12.0	3 wood	v
Atlantis	WB 20H	2.00	3.14	0.60	11.0	4 GFRP	v
J. Bornay	Inclin 250	2.00	3.14	0.25	11.0	2 CFRE	v
World Power Tech.	600	2.10	3.46	0.60	11.0	2 wood	v
Bergey Windpower	850	2.40	4.52	0.85	12.5	3 GFRP	h
SoWiCo	AeroCraft 1000	2.40	4.52	1.00	9.5	3 GFRP	p
LMW	LMW 1000	2.50	4.91	1.00	12.0	3 GFRP	p
Proven Wind Turbines	WT600	2.60	5.31	0.60	10.0	3 Poly	p
Giacobone	Eolux	2.70	5.73	0.60	12.0	3 GFRP	v
SOMA	1000	2.70	5.73	1.00	10.0	2 GFRP	v
Survivor Energy Sys.	S-3000	2.70	5.73	0.50	9.0	3 GFRP	v
World Power Tech.	H1500	2.70	5.73	1.50	12.5	2 CFRE	v
World Power Tech.	Whisper 1000	2.70	5.73	1.00	11.0	2 wood	v
J. Bornay	Inclin 1000	2.90	6.42	1.00	12.0	2 CFRE	v
LMW	LMW 1500	3.00	7.07	1.40	12.0	3 GFRP	h
Bergey Windpower	1500	3.10	7.31	1.50	12.5	3 GFRP	h
Proven Wind Turbines	WT2200	3.40	9.08	2.20	12.0	3 poly	p
Westwind	Standard	3.60	10.20	2.50	14.0	3 GFRP	h
World Power Tech.	Whisper 3000	4.50	15.90	4.50	12.5	2 GFRP	v
J. Bornay	Inclin 2500	4.70	17.35	2.50	12.5	2 CFRE	v
Northern Power Sys.	HR3	5.00	19.60	3.00	12.5	3 wood lam.	v
Vergnet	GEV 5.5	5.00	19.60	5.00	13.0	2 wood lam.	ps
Bergey Windpower	Excel	7.00	38.50	10.00	12.1	3 GFRP	h
Vergnet	GEV 7	7.00	38.50	10.00	11.5	2 wood lam.	ps
Westwind	Standard	7.00	38.50	10.00	13.5	3 GFRP	h
Wind Turbine Ind.	23-10	7.00	38.60	10.00	11.6	3 wood lam.	p,h

* nc=no control, h=horizontal furling, v=vertical furling, ab=air brake, p=pitch to feather, ps=pitch to stall, GFRP=glass reinforced polyester or fiberglass, wood lam.=wood laminate, CFRP=carbon fiber reinforced polyester, CFRE=carbon fiber reinforced epoxy, poly=glass fiber reinforced polypropylene **Note:** An expanded version of this table is available on diskette from Real Goods (1-800-919-2400). The expanded table lists nearly 100 wind turbines from 0.5 meters to 13 meters in diameter.

Figure 6-27. The Bergey 1500 typifies today's integrated small wind machines: upwind, direct drive, and self-furling. The wind machine shown here is furling during high winds. Note the junction box between the guy bracket and the top of the tower. This box simplifies wiring the wind turbine to the cables running down the tower.

Figure 6-28. The Aero3gen is one of the smallest wind machines on the market. This machine is 0.9 meters (2.8 feet) in diameter and rated at 50 watts. (L.V.M. Ltd.)

Figure 6-29. Battery charger. Wind Baron's NEO 500 is a direct-drive, self-furling, micro turbine, suitable for small battery-charging applications. The Neo is 1.5 meters (5 feet) in diameter and produces 500 watts. (Wind Baron Corp.)

a wind machine 3 meters in diameter will accumulate 4 billion revolutions, or the equivalent of a car driven 9 million miles.

In programs where performance has been monitored, wind machines built by both Northern Power Systems and Bergey Windpower have chalked up an impressive record of reliability. Wind turbines built by these manufacturers have been available for operation 98-100 percent of the time they were in service. After a decade of development these designs have proven more dependable in remote power systems than the conventional engine generators they were originally designed to supplement.

Though extremely reliable, the initial cost of turbines using the integrated design philosophy becomes too prohibitive in wind machines above 10 meters in diameter (about 25 kW). At this size and above, a different approach becomes necessary.

Medium-Sized Turbines

In contrast to these integrated small machines, the wind turbines found in California and Denmark use off-the-shelf induction generators, transmissions, brakes, yaw drives, electrical sensors, and controls. The numerous components on a medium-sized wind turbine require regular maintenance, a function simplified by clustering the turbines together in one location, as in California wind plants, or within easy reach of the manufacturer, as in Denmark (see Figure 6-30).

This complexity and the resulting need for maintenance thwarted many American manufacturers who attempted to market similar designs for dispersed applications in the United States. These machines, best represented by Enertech's designs, used a fixed-pitch rotor downwind of the tower to drive an induction generator and used a brake to stop the rotor under normal and emergency conditions. If the brake failed, Enertech relied on tip brakes to protect the rotor from self-destruction.

Most of today's medium-sized turbines use fixed-pitch blades bolted rigidly to the hub. Like small wind turbines, most of these machines are self-starting (though some early models, like the Enertech, were motored up to their operating speed). From startup to the cut-in wind speed, the rotor speed varies with wind speed until it reaches the speed at which the generator can be synchronized with the utility.

From cut-in to rated wind speed, the rotor continues turning at the same speed while delivering more and more power. The wind machine may from time to time switch from one generator to another as wind speed varies, changing rotor speed like shifting gears in a car. But overall the turbine operates at constant speed. (The rotor speed does vary slightly, but this is imperceptible to the untrained observer.)

Above the rated speed the blades begin to stall, dumping excess power. Above the turbine's cut-out speed, or with any abnormal occurrence such as excessive vibration, the brake is applied, bringing the rotor to a stop. In the typical Danish design, the turbine then yaws 90 degrees out of the wind to a parked position. When wind speeds fall back below the cut-out speed, the

Figure 6-30. Danish wind power plant. Vestas wind turbines at Taendpibe-Velling Maersk, one of Europe's largest wind plants. This 100-unit project includes both stall-regulated and pitch-regulated wind machines 17-27 meters in diameter on the west coast of the Jutland Peninsula.

turbine yaws back into the wind and releases its brake. Soon after, the rotor accelerates until it reaches its operating speed.

On American downwind designs, such as Enertech's, there's no yaw control. The rotor stays downwind even when the rotor is parked.

The brake on both types of machines is normally "on" or engaged. The brake can be released only when there is electrical power to the wind machine. Thus, if there's a loss of power the brake automatically returns to its engaged position, stopping the rotor. If the brake is unable to prevent the rotor from unsafe speeds, the backup aerodynamic controls are deployed. These then slow the rotor to a safe speed. Once deployed, these aerodynamic controls must typically be reset manually to ensure at least a cursory inspection of the turbine.

There are important variations in the manner in which some medium-sized turbines operate. Unlike most such turbines, U.S. Windpower's model 56-100 employs a variable-pitch rotor downwind of the tower. More than 4000 of this 18-meter, 100-kilowatt turbine have been installed, the majority in the Altamont Pass. Even though it uses a variable-pitch rotor, the 56-100, and other turbines like it, maintains a constant pitch from cut-in to the turbine's rated output. Only in winds above its rated capacity does the rotor begin to pitch its blades. Thus the turbine uses pitch only to control peak power, and to protect itself from overspeed by feathering the

Table 6-2

Characteristics of Selected
*Medium-Sized Wind Turbines Suitable for Home and Business**

Manufacturer	Model	Rotor Dia. (m)	Swept Area (m²)	Rated Power (kW)	No. of Blades	Speed of Rotor	Rotor Control	Overspeed Control
Vergnet	GEV 10.25	10.0	79	25	2	c	s	variable pitch
Enercon	E-12	12.0	113	30	3	v	p	variable pitch
Sudwind	N1230	12.5	123	30	3	c	s	pitchable
Jacobs Energie	Aeroman	14.8	172	33	2	c	p	variable pitch
Enercon	E-18	18.0	254	80	3	v	s	pitchable tips
Lagerwey	LW 18/80	18.0	254	80	2	v	p	variable pitch
Micon	M300	19.5	299	55	3	c	s	pitchable tips
Ecotecnia	20/150	20.0	314	150	3	c	s	pitchable tips

* A more complete listing can be found in Appendix H;
  c=constant speed; v=variable speed; s=stall regulated; p=variable pitch.

blades. Other manufacturers use a similar strategy in larger upwind turbines, those 25-40 meters in diameter.

The introduction of variable-speed, medium-sized turbines, as shown in Table 6-2, is potentially another important innovation. Lagerwey, Enercon, and U.S. Windpower build variable-speed machines that, they argue, will produce about 15 percent more energy than conventional constant-speed turbines. All three depend on sophisticated electronics to deliver utility-grade power.

Wind turbines, however, are only part of a wind system. Towers, the subject of the next chapter, are also an essential element.

 7

Towers

Towers are as integral to the performance of the wind system as the wind turbine itself. Without the proper tower your wind machine isn't much more than an expensive lawn ornament and could even become a hazard to all in the vicinity.

Towers, as a rule, are one of the few wind system components where you have some choice. Unlike the selection of blades, transmission, and generator, which have been determined by the manufacturer, you have a variety of towers to choose from—at least for wind machines up to 7 meters in diameter. With small wind machines you can select the kind of tower as well as the height that best suits your site, temperament, and budget. At a minimum the many options can be confusing; at worst they could lead to a mismatch between the wind machine and the tower.

When considering tower options it's imperative to keep in mind that the tower must be strong enough to withstand the thrust on the wind turbine (the force trying to knock the wind turbine off the tower) and the thrust on the tower (the force trying to knock the tower over). And unless it's a hinged tower, the tower must support not only the weight of the wind turbine but also the weight of the people who will service it.

Choosing the right tower for a small wind machine depends on what designs are available, your site, and what you can afford. Foremost among these is whether a tower is available in the height desired.

Height

As the wind industry has matured—and wind system users as well—selecting a tower of the proper height has become increasingly important. In the early

1970s anything that would get the wind machine off the ground was acceptable.

Towers for wind machines used on the Great Plains during the 1930s were never very tall. The flat terrain and the few obstructions present didn't call for towers taller than 60 feet. Even so, by the late 1940s Wincharger was installing guyed towers 85 and 105 feet in height, and Parris-Dunn was advising its customers that "the higher the tower the greater the power."

As the technology has matured we've learned that economic power generation and good performance are only obtained on a tall tower. We have known for some time that wind speed and power increase with height, but it didn't begin to sink in until wind systems started to be installed in numbers across the United States. We gained far more experience with power-robbing turbulence and what it can do to a wind machine's performance than we ever needed. As a result, recommended tower heights have gradually increased.

Manufacturers prefer taller towers because they want their products to perform well and want to minimize turbulence-induced service and warranty claims. Taller towers also allow more flexibility in siting. If buildings and trees are present—and they usually are—a tall tower can redeem an otherwise unusable site. A minimum by today's standards is 80 feet (24 meters). And when trees are nearby, 100-120 feet (30-35 meters) is the norm.

The height requirements for micro wind turbines are somewhat different. Micro turbines, those less than 3 meters (10 feet) in diameter, are often used in low-power applications (like weekend cabins) where maximum generation isn't necessary. Because of their relatively low cost they're often used with inexpensive towers. These towers are generally not suited for heights above 60 feet (20 meters). It doesn't make a lot of sense to install them on taller towers that cost three or four times the cost of the turbine unless you plan to eventually install a bigger machine.

Strength

Next to the heights available, the most important factor is the ability of a tower to withstand the forces acting on it in high winds. Towers are rated by the thrust load they can endure without buckling. Standards in the United States call on manufacturers to design their wind systems to withstand 120-mph (54-m/s) winds without damage. The thrust on the tower at this wind speed depends on the rotor diameter of the wind turbine and its mode of operation under such conditions.

Two wind turbines of the same size may require entirely different towers because of differing approaches to protecting the rotor in high winds. Those that furl the rotor substantially reduce thrust loads on the tower in comparison to those that feather the blades. For wind turbines that furl the rotor, thrust reaches a maximum at the furling speed, and remains fairly constant thereafter. Thrust continues to increase with increasing wind speed on small turbines with mechanical governors.

For small wind turbines on tall towers, the drag on the tower in high winds adds significantly to the thrust loads the tower must withstand. In contrast, the rotor presents far more frontal area to the wind, proportionally, on a medium-sized wind machine than on a small turbine and so thrust on the rotor dominates.

All towers flex to some degree in response to the thrust on the rotor. One dealer discovered this the hard way. After he finished wiring his newly installed 7-meter turbine to the service panel he was eager to see it in operation (an affliction that wind pioneer Jack Park diagnoses as "fire-'em-up-itis"). The wind was strong, blowing near the rated speed of his Jacobs wind turbine. To ensure that all was well and to get a bird's-eye view of his new investment, he unwisely climbed up the 100-foot, heavy-duty truss tower. Stopping just below the rotor, he decided to check the operation of the feathering governor by unloading the generator and letting the rotor speed increase. When an assistant disconnected the wind turbine, he suddenly found himself hanging on with all his might as the blades feathered and the tower sprang several feet back into the wind like a giant whip. He was lucky. If he hadn't been strapped to the tower and kept his wits, he could easily have been killed. (This example violates one of the fundamental safety rules of working around wind turbines. Never, ever climb the tower when the rotor is spinning. For more on safety see Chapter 16.)

All towers sway to some extent. Slender pole towers are far more flexible than truss towers and visibly deflect in strong winds. Deflection isn't a problem unless the turbine and tower are mismatched. If the tower or the blades flex too much and at the wrong time, the blades could strike the tower. This dynamic interaction between the wind machine and tower is a major concern of manufacturers.

As the rotor and tower deflect in the wind they begin to oscillate like the swaying spans of a rickety suspension bridge. Should the turbine and tower begin to sway in harmony, the oscillations could gradually increase in magnitude until they destroy the wind machine, tower, or both.

The reported accounts of tower failures in wind system applications involve nearly equal numbers of truss towers and the apparently less secure guyed towers. In one widely discussed case, a truss tower failed when the bolts holding two 20-foot sections together sheared. The tower manufacturer asserted that the tower was overloaded by the dynamic interaction of the turbine and the tower. Witnesses noted that the tower was vibrating wildly prior to the accident. The turbine manufacturer, Jacobs Wind Electric, countered that the failure was due to "bad steel" in the bolts. Whether bad steel or not, a wind system vibrating in resonance can exert tremendous force on the tower, creating loads well beyond its design limits.

It's this dynamic interaction between the wind machine and the tower that leads some manufacturers to restrict the type of tower for their wind machines. The pairing becomes increasingly important as size increases.

Tower Types

Towers fall into two categories: free standing and guyed. Free-standing towers, or self-supporting towers as they are also known, are just that—free standing. They depend on a massive foundation to prevent the tower from toppling over in high winds, and must be strong enough internally to withstand the forces trying to bend the tower to the ground. Guyed towers, in contrast, employ several far-flung anchors and connecting cables to achieve the same ends. Free-standing towers are more expensive than guyed towers but take up less space.

Free-Standing Towers

There are two types of free-standing towers (see Figure 7-1). The most common is the lattice or truss tower, so called because it resembles the lattice work of an arbor. The Eiffel Tower is the best-known example of a truss tower. The tubular or pole tower is another form of free-standing tower, but one that encompasses several different varieties. Truss towers are typically more rigid than pole towers.

Towers can be designed to withstand any load. But as the size of the wind machine increases, so do the weight and cost of the tower supporting it. The same is true as the tower increases in height. The components become heavier, harder to move, and more costly to ship.

In the United States truss towers are assembled from a series of 20-foot sections. For small wind machines, the sections may be preassembled and welded together prior to delivery. For larger machines, the tower is shipped "knocked-down" or in parts and must be assembled on the site.

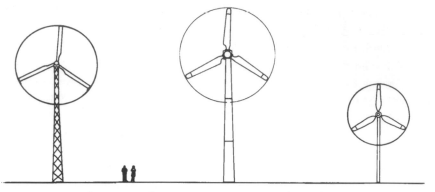

Figure 7-1. *Left: Free-standing truss tower (13-meter, Enertech E-44). Center: Free-standing tubular tower commonly used on medium-sized European wind machines (15-meter Micon). Right: Guyed pole tower without guys (10-meter Fayette). (Pacific Gas & Electric Co.)*

Installation of truss towers usually requires a crane. The tower is assembled on the ground, then hoisted into place and bolted to the foundation. Some enterprising individuals assemble the tower piece by piece in the upright position with the aid of a gin pole, or portable davit, mounted on the tower. This is time-consuming and risky. It isn't an approach for amateurs.

Another method is to hinge the tower at its base. The tower is bolted together on the ground, the wind turbine attached, and the whole assembly tipped into place with a gin pole and winch, or a small crane.

Pole Towers

The diameter and shape distinguish a pole tower from the related tubular tower. As turbines increase in size, or the method of overspeed control changes from that of self-furling to stall regulated, tower proportions change. Slender, free-standing pole towers are used for small wind turbines. Squat tubular towers are used with medium-sized wind machines like those found in California wind plants.

Unlike pole towers, which are primarily used as light standards, tubular towers are built specifically for wind turbines. The large volume of medium-sized wind turbines used in California and Europe has enabled manufacturers to design and build towers just for this market. The volume of small wind turbines installed on pole towers remains too low to justify mass production of these towers solely for this application.

Pole towers are made from tapered steel tube, steel pipe, wood, concrete, or even fiberglass. Though most are made from steel, pole towers of pre-stressed concrete have been used by some manufacturers. Several wind turbines built by Grumman Aerospace were installed on concrete towers during the mid-1970s with no reported failures. And the German manufacturer Enercon has built a reputation around the sound-deadening qualities of spun concrete towers. Enercon installs wind turbines up to 33 meters (110 feet) in diameter on concrete towers throughout northern Germany.

Unfortunately, pole towers are available only in limited sizes and strengths. For example, the selection of wood and concrete poles suitable for wind machines is limited by the length of pole that can be shipped conveniently. Like truss towers, pole towers are difficult to handle without heavy equipment. Because wood and concrete poles are shipped in one piece, they are the most difficult to handle. Largely for this reason they're seldom used in the United States.

Tapered tube differs from steel pipe by its relatively larger diameter and tapered sections. These towers are commonly used as light standards. Though both tapered steel tube and steel pipe can be guyed to give the tower added strength, steel pipe is more commonly used in guyed applications for small wind turbines.

Installation of pole towers usually requires a crane. A pole tower, though, can be hinged at the base and tipped into place with a gin pole. When upright, the tower is then bolted to the foundation. Some European manufacturers erect hinged tubular towers with powerful hydraulic jacks.

Many consider free-standing pole or tube towers more aesthetically pleasing than truss towers. This may be true in foreground views, but it isn't always the case. Surprisingly, at a distance the tubular towers used in some California wind power plants are more visible than nearby lattice towers. The lattice towers tend to blend into the background at distances greater than 2 miles (3 kilometers).

Pole and tubular towers are significantly more expensive than guyed towers, but only modestly more costly than truss towers. However, pole towers require a more substantial foundation than truss towers, which spread the overturning force over a wider base. If you're sensitive to cost, consider a guyed tower.

Guyed Towers

Guyed towers, by far, are the most common choice for small wind machines (see Figure 7-2). They offer a good compromise between strength,

Figure 7-2. *Guyed lattice tower. In the United States modular lattice mast is produced in 10-foot and 20-foot sections that can readily be assembled by hand.*

cost, ease of installation, and appearance. Unfortunately they take up more space than free-standing towers, and they suffer from the unfounded fear that a guy cable will fail and the tower will come crashing down. This rarely happens. Guy cables seldom fail, and when they do, often the tower will remain standing.

Guyed towers include a mast, guy cables, and earth anchors. The mast itself may be made from steel lattice or pipe. In the United States most guyed towers for wind machines up to 7 meters in diameter use masts of welded lattice made from steel tube and rod. These masts are popular because they're mass-produced for the telecommunications industry, making them available for other uses at modest cost. They're also produced in a convenient range of sizes from lightweight sections designed for radio antennas to heavy sections for mountaintop microwave dishes. Tower height is practically unlimited. A guyed tower using a lattice mast can be assembled by bolting sections together in the upright position a section at a time with a tower-mounted gin pole, or the entire mast can be assembled on the ground and set in place with a crane.

Masts of steel pipe and tube are also popular (see Figure 7-3). Masts of the desired height are assembled from several sections bolted or slipped together. Guyed towers using masts of steel pipe or steel tube are usually assembled on the ground and tipped into place with a crane or gin pole.

Guyed towers for small wind turbines typically use extra-high-strength stranded-steel cable. Larger wind machines may use aircraft cable, which has even greater strength. Towers use three to five guys at each level, and often require two or more sets. Four guys are used where the site or method of erection requires them. Tilt-up towers, by necessity, use four guy cables. Three guys are more commonly used because they do the job at the least cost.

Under special circumstances more guy cables and anchors may be necessary. Fayette Manufacturing used a novel guying layout on its mid-sized wind turbines. They used guyed towers of steel tube and anchored the guy cables by driving large screws into the ground. But soil conditions in the Altamont Pass were suspect. To lessen the risk that a screw anchor would pull out of the ground, they used two anchors and accompanying guy cables at each of three guy points. (The tower was guyed at only one level. There were three sets of two guy cables and anchors.) This lessened the loads on each anchor, reducing the chance that an anchor would fail. It also ensured that the tower would remain standing if an anchor or its cable failed.

Figure 7-3. *Guyed tubular tower. The use of four guy anchors and a hinge at the base permits a guyed tower to be tipped into place with a gin pole. Shown are two of eight Aerowatt turbines at a small wind farm in southern France. The 7-meter (21-foot) Aerowatts power a popular winery (Chateau Lastours) tucked into the foothills of the Pyrenees.*

The loads on a guyed tower that might snap a cable or pull an anchor out of the ground are determined by the thrust on the tower and by the guy radius—the distance from the tower to the anchors. The guy radius is a critical aspect of guyed towers and is dependent on the site, the loads imposed on the tower by the wind turbine, and the stiffness of the mast. Because guyed towers need a lot of space, there's a tendency to set the anchors too close to the tower. This can affect the dynamics of the tower and the loads it can endure. Consequently, some manufacturers specify the guy radius precisely.

Guy radius is limited by the compressive loads the mast can withstand before it buckles, and by anchor construction. For example, when the anchors are too close to the tower the mast may buckle in high winds or the anchors may fail. As a rule of thumb, the guy radius shouldn't be less than one-half the height of the tower. The guy radius can be as great as you want. The tension in the guy cables and the compression on the tower continue to decrease as distance from the tower increases. Usually there's no reason for going beyond three fourths of the tower's height, although Unarco Rohn, which builds lattice masts for guyed towers, recommends a guy radius of 80 percent of the tower's height.

Usually there are two or more guy levels on most towers using a lattice mast. One anchor is used to guy all levels. The topmost guy prevents the tower from overturning and the lower guys prevent the tower from buckling. When the tower is stiff like the tapered tubular towers used on some larger wind machines, only one guy level is needed. Lattice masts and long sections of steel pipe are "softer" in engineering terms than large-diameter tubular masts, and additional guy levels are used to keep them from bowing.

Unarco Rohn's 25-G lattice mast is suitable for self-furling wind turbines up to 3 meters (10 feet) in diameter. For masts 80-110 feet (24-33 meters) tall, Unarco Rohn suggests three guy levels spaced equally apart beginning one rotor radius below the top of the tower. This will allow sufficient clearance between the topmost guys and the rotor while giving the tower the necessary strength and stiffness.

Novel Towers

If you can think of it, it's been tried. Trees are a good example. Yes, trees! Wind machines have been mounted atop trees—but not successfully. First, for a tree to be of long-term use it must remain alive. After you finish hacking away at it, its days are numbered. Unless the tree has been cut and treated with a wood preservative (then it ceases to be a tree and becomes, instead, a wood pole), it's a poor idea for a tower.

Trees seldom occur right where you would like your tower. Nor is there usually one lone tree that reaches above all surrounding objects standing out in the middle of a large clearing. In addition, the turbine will be difficult to install and service, and the manufacturer will not issue a warranty for such an installation. They know what will happen.

Rooftop Mounting

Another equally troublesome idea is to mount wind machines on buildings. Forget it! Seems like a fine idea at first glance: the building gets the turbine above the ground and eliminates the need for a tall tower. The building, though, creates its own turbulence. The wind machine must be installed above this turbulent zone to perform well and to ensure long life. This often negates any potential savings on the tower.

Few who consider this approach ask whether the building can support the loads created by the tower and the wind turbine. Wooden roofs of single-family residences can not. A reinforced concrete roof on a commercial or industrial building might. Can the roof, then, handle the dynamic loads—the vibrations—that the tower will transmit from the wind machine? If the building is an unoccupied warehouse, the vibrations won't bother anyone, but if it's an office building they may prove annoying.

Much has been written about the small Jacobs windcharger that was once installed on a tenement in the Bronx. True, it was done once, and it can be done again. But what's the point? The Bronx project was intended as a challenge to Consolidated Edison Company, the local utility. It succeeded, but it's not been done since. Rooftop mounting has never been a practical option.

Farm silos also seem ideal for a low-cost tower. They're already in place, usually stand tens of feet above surrounding farm structures, and are relatively close to where the power will be used. Alcoa tried it and, like rooftop mounting, it hasn't been used since (see Figure 7-4).

Farm Windmill Towers

Though certainly not novel, farm windmill towers are ubiquitous. They can be found nearly everywhere in the United States and Canada. Because of their abundance, there's always a temptation to buy a used water-pumping windmill tower and adapt it for a small wind-electric system.

Figure 7-4. *Silo mounting. Alcoa experimented with mounting their small Darrieus turbine on farm silos. The concept was abandoned when Alcoa withdrew from the market.*

But their utility is limited unless you plan to use the tower as it was intended.

Towers for the American farm windmill are a special case of the freestanding category. Most farm windmills have been installed on light-duty truss towers. (Some water-pumping windmills have been installed in the Southwest on pipe towers made from readily available well casing as shown in Figure 7-5.) Farm windmill towers typically have less taper, are squatter, than the truss towers used for small wind turbines. (The height of a water-pumping windmill tower is proportionally about five times the width of the base. In contrast, the height of a small wind turbine tower is nine times the base width.) This design enables the tower to use less steel in the legs and braces and a less substantial foundation than required for a tower supporting a similarly sized modern wind turbine.

Farm windmill towers are also short. Most are no more than 40-50 feet

Figure 7-5. Free-standing tower of steel well casing supporting a farm windmill.

(15 meters) tall, particularly in the western United States. Ken O'Brock, who distributes water-pumping windmills along the East Coast, says farm windmill towers up to 80 feet (24 meters) tall can be found among Amish settlements in Pennsylvania and Ohio. These are the exceptions, and the Amish are not about to part with them.

The most commonly used farm windmill is only 8 feet (2.4 meters) in diameter. Towers used with these machines are not suited for larger wind turbines even if you found a rare 80-foot tower. Never use a lightweight tower designed for an 8-foot diameter farm windmill with a wind turbine 3 meters (10 feet) or more in diameter. A turbine 5 meters (16 feet) in diameter presents three times the frontal area of the 8-foot water pumping windmill and offers far too much thrust for the tower.

Light-duty towers, however, may be well suited for use with micro turbines up to 2.5 meters in diameter. Turbines of this size, because of their low cost, are often installed on short towers; the 40-foot farm windmill tower will work as well as any other short tower.

Wood Poles

In North America wood poles are as commonplace as farm windmill towers. And like farm windmill towers they're frequently considered a choice for a cheap tower. They're strong, rigid, and cheap when bought in quantity. Wood poles can be installed by crane or utility truck with a special boom (see Figure 7-6).

Wood poles are classified according to their circumference 6 feet (2 meters) from the butt end. Poles of a given class and length are rated to carry approximately the same load. They can handle even greater loads when guyed. Wood poles are rigid and are more like truss towers than they are like guyed towers or steel poles. Pole lengths suitable for wind systems are found only in Class 4 or better. A Class 4 pole is strong enough for small, self-furling wind turbines up to 5 meters (16 feet) in diameter.

Though abundant, wood poles are not commonly used as wind machine towers. They're unsightly, difficult to climb, and the heights available are limited to the length that can be conveniently shipped. Even if you don't mind their looks, others might—especially the local zoning officer. It's also difficult to route the furling cable for a small wind turbine down the side of a wood pole.

The inexpensive wood poles used by utilities are too short for most wind turbine applications. Utilities often use poles only 40-50 feet (15 meters) long. For the minimum 60-foot tower height needed by a small wind turbine, a 70-foot pole is necessary. (Ten feet of the pole is embedded in a concrete foundation.) Longer poles are available but the cost rises rapidly with lengths beyond the standard sizes used by utilities. Longer poles are also more difficult to transport. Despite these drawbacks, the more commonplace wood poles may be suitable for micro turbines.

Figure 7-6. Small windcharger used for cathodic protection of a pipeline atop a wooden utility pole. Note air brakes (buckets) on this 1930s-era design.

Steel Pipe

Well casing or water pipe is a frequent choice for guyed pipe towers supporting micro wind turbines. Steel pipe is readily available, inexpensive, and light enough that it can be moved by hand. Do-it-yourselfers often choose well casing for these reasons.

For their 1.5-meter turbine, Wind Baron recommends using steel pipe in a clever tilt-up tower that's inexpensive and easy to construct. They suggest using two 20-foot sections of 2-inch diameter, Schedule 40 grade steel pipe guyed at one level with ⅛-inch aircraft quality guy cables. This is probably the height limit for a tower of this construction. Taller towers require larger-diameter pipe for greater stiffness and more guy levels to prevent the tower from buckling. Towers of this type must be hinged for erection and service because they can't support the weight of someone climbing the tower.

The strength of water pipe or well casing comes from its thick walls. Thin-walled tubing can also be used, but it requires larger diameters for equivalent stiffness.

Other Considerations

Sometimes the site or local regulations may limit your tower options. For example, a small lot could preclude a guyed tower. Your choice would then be restricted to a free-standing truss or pole tower. The final choice will be affected by an evaluation of the relative cost, appearance, and ease of installation between the two.

Space

Guyed towers occupy more space than free-standing towers. Normally this isn't a drawback; where it is, guyed towers can be adapted to small lots by pulling the anchors in closer to the tower. If you must do so, check with the manufacturer or contact a structural engineer to run through the numbers for you. They may find that the tower has an ample safety margin with the shorter guy radius. On the other hand, if you're approaching the limits of the mast, you may be forced to use a free-standing tower after all.

Sites with a lot of traffic, either vehicular or pedestrian, will also limit your choice to free-standing towers. Many a guyed Wincharger tower was felled by a rancher on a tractor. Not only is there the danger of losing your wind system, but the guy cables themselves may present a hazard. To avoid tragedy, guy cables should be kept out of traveled ways. And they should be well marked.

Maintenance

Another factor to consider in tower selection is maintenance of the tower and of the wind machine. Will the tower have to be painted, for example? The steel used in most towers is galvanized. To provide corrosion protection, individual steel members or welded subassemblies are dipped into molten zinc. When shipped from the plant, galvanizing presents a shiny silver finish. It soon oxidizes outside, however, and takes on a dull gray luster. Galvanized surfaces do not require painting or other treatments.

Some truss towers and pole towers are assembled from Cor-Ten™ steel. Cor-Ten™ was developed for outdoor applications requiring low maintenance. Cor-Ten™ oxidizes to an attractive dark brown finish to protect the underlying steel from further corrosion. Essentially it develops a skin of rust and never needs to be painted or treated in any other way. Cor-Ten's only disadvantage is that rain running off the steel leaves a trail of rust for the first few years. This isn't a problem unless you plant the pole in the middle of a concrete lot. The concrete will soon look like the best Georgia red clay.

Access to the turbine shouldn't be overlooked either. The turbine will require at least occasional inspection if not maintenance. Some provision must be made for getting to it without the need to hire a crane. Truss towers can be ordered with climbing lugs or rungs. Guyed towers using lattice masts can be climbed using the lattice work (cross-girts). Tube towers, whether free-standing or guyed, require the addition of climbing lugs if they are not part of the package.

From a service perspective, hinged towers are advantageous for major repairs (see Figure 7-7). Instead of climbing up to the turbine and working with a crane, you bring the machine down to ground level. But hinged towers are not suitable for all wind machines. Those using a transmission that requires periodic oil changes are best left to fixed towers. Nor are hinged towers ideal for simple maintenance or inspections because of the difficulty lowering the towers. Hinged towers may be safer to work around once they're on the ground, but safely lowering them can be hair-raising. More than one installer has found his truck being dragged across the ground as tower and turbine came down with a thump. To work properly, hinged towers should be designed by professionals and used only where necessary.

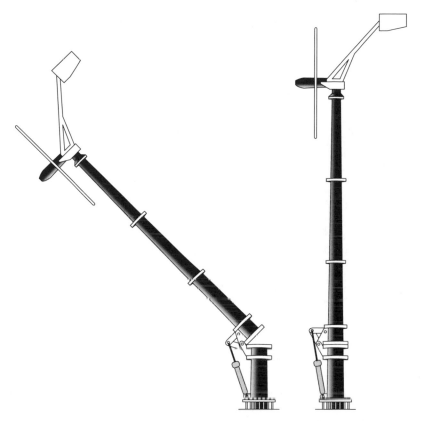

Figure 7-7. Hinged free-standing tubular tower. This hydraulically operated hinged tower was developed for use at inaccessible Norwegian telecom sites. (Roheico A/S)

Ease of Installation

Towers should also be easy to install. You may have found an inexpensive tower that will do the job you want, but if it takes a lot of effort to install, the savings may be offset by greater labor costs. If the tower sections are so large you need a crane or other heavy equipment just to get them off the truck, you'll need an experienced crew with equipment. You pay less for handling and shipping when the sections can be moved around by hand.

When considering how the tower is installed, also look at how the wind machine is mounted on top of the tower. A tower adapter, sometimes called a stub tower, is needed between the tower and the wind machine. One popular approach is to use a steel plate on the topmost section. The wind machine is mounted onto this plate with a series of bolts and leveling nuts.

(This plate is predrilled to accept the turbine's mounting hardware before it is shipped from the manufacturer.) The leveling nuts permit adjustment to ensure that the machine yaws correctly. If not level, downwind machines tend to stop in one position and yaw unevenly.

Another method is to use concentric tubes. For example, in one design, a tube extends from the bottom of the wind machine. This yaw tube is then slipped inside a split pipe at the top of the tower. The turbine is held in place by clamps compressing the sides of the split pipe against the turbine's yaw tube.

Tower Selection

Now let's put it all together. Let's assume you want to install a 3-meter turbine in your backyard. From Table 7-1, you realize that the Rohn 25-G tower, or its equivalent, will meet the turbine's structural needs. But you're not sure you want the guy cables cluttering up your yard.

You may have decided that a wood pole is too ugly to live with for the next 20 years. A steel pole tower is more attractive, but it's also far more costly (about twice the cost of the guyed tower) in the height desired. The truss tower costs only 50 percent more than the guyed tower but is more difficult to install. Though the truss tower is a good compromise between cost, aesthetics, and space requirements, you find the low cost and easy installation of the guyed lattice mast more appealing.

There's another plus to the guyed lattice tower. For small wind turbines, increasing the height of a guyed tower above 80 feet (24 meters) usually makes economic sense. The added costs are more than compensated through increased wind generation. That's not always true for truss towers. On a small turbine, the incremental cost of truss towers above 100 feet (30 meters) tall may exceed the expected gain in generation.

The cost differential between towers changes as wind machines or towers increase in size. The difference in cost between truss towers and guyed towers decreases with the increasing size of wind turbines. This trend becomes more marked with wind machines 7 meters in diameter and larger.

Towers also become a smaller part of the total cost of a wind system as the wind machine's size increases. For a small machine, the tower may contribute 30-50 percent to total installed costs. For larger wind machines, the tower is only 15-25 percent of total costs.

Because towers add so much to the total cost of small wind turbine installations, there's a tendency for do-it-yourselfers to build their own or

Table 7-1

Suitable Guyed Towers for Small Wind Turbines in North America

(For Self-Furling Wind Turbines Only)

Rotor Diameter

m	0.5	1	1.5	2	3	5	7
ft	2	3	5	7	10	16	23

Tower Height

m	10	10	10	30+	30+	30+	30+
ft	30	30	30	100+	100+	100+	100+

Type	pipe*	pipe*	pipe*	lattice	lattice	lattice	lattice

Size

nominal	1	2	2	Rohn 25G	Rohn 25G	Rohn 45G	Consult Mfg.
inch OD	1.5	2.4	2.4				
mm OD	38	61	61				

* Pipe = thick walled

use second-hand towers. Unless it's for a micro turbine, where the forces are limited, don't try to cut costs by skimping on the tower. It usually leads to trouble.

If you plan to use a tower not recommended by the manufacturer, here are a few questions to consider. Has this type of tower been used for wind machines by anyone else? If so, how has it performed? What, if any, were the problems encountered? What does the manufacturer think? Will they still honor their warranty? If not, why not?

As a final caveat, if at all possible, use a tower with a pleasing appearance. It isn't good enough to look around and point an accusing finger at other obtrusive objects (such as utility poles and transmission towers) on the horizon that have found acceptance. Wind systems shouldn't be an embarrassment to the community. You'll be happier in the long run and there will be fewer objections from your neighbors if the tower you select is aesthetically pleasing.

 **8**

Cutting Costs—

Not Corners

"You'll never get there from here," said the rancher in disbelief, "not without a four by four."

"We'll give it a try anyway," said the Easterners.

After an hour of fruitlessly searching the banks of the Powder River for a rumored windcharger, they were about to give up when they impulsively decided to follow a hunch. "If you were a homesteader, and you were settling this land in southeastern Montana, where would you plant yourself?" they asked themselves.

They pointed their small truck toward the horizon beyond the breaks and set off. Heading across the dry range land they bounced over pungent sagebrush and crashed down steep arroyos, wheels churning in the sand. Soon they could see an old farm windmill in the distance, and a sod house came into view. Finally the object of their search appeared, the pot of gold at the end of the rainbow—an old Jacobs windcharger. There it sat on a rusting tower, a home light plant with a shed full of glass batteries at its base, and a shoulder-high pile of antlers stacked nearby. It was absolutely still, and then a slight breeze caused the old mill to creak. A coyote howled in the distance.

That's how some in today's wind industry started their careers in the 1970s, leading expeditions to remote parts of Montana in search of once abundant windchargers. To bypass the high cost and poor reliability of the wind machines then being produced, they scoured the Great Plains, tracking down, buying, and rebuilding the windchargers of another era. It was a colorful and exciting period in America's reacquaintance with wind energy. Those days are gone forever. Nearly all of the pre-REA windchargers have already been claimed. But today there are other sources for used machines, as well as several other ways to cut the cost of buying and installing a wind power system.

The wind machine itself comprises 40-50 percent of the total cost of a small wind system, and about 65 percent of the cost for a wind machine 7 meters (23 feet) in diameter and larger. Towers become a less significant percentage of the total cost as wind machines increase in size. Installation averages 15-35 percent of the total cost. Because the wind machine accounts for so much of the cost, homeowners frequently try to save money by building their own.

Building a Small Wind Machine

If you're planning to build your own wind turbine, ask yourself some hard questions. First, why? Because you like to work with your hands, or solely to save money? Tinkering is a valid reason. You'll learn a lot about natural forces, mechanics, and Murphy's Law (if anything can go wrong it will). If you want to build your own wind machine as a financial shortcut to buying a commercially available wind system, forget it. You'll produce nothing but headaches.

Very few home-builts work reliably, and those few that do don't produce much usable power. They're more expensive than you might first imagine, and they can be dangerous. Manufacturers with teams of competent engineers have a hard enough time keeping their turbines operating. You must be exceptionally talented to do better with fewer facilities and no technical support. Building a wind machine that will work reliably and safely is beyond the skills of most homeowners. There are already too many home-built contraptions standing as derelict monuments to the mistaken belief that anyone can build a wind turbine.

This warning applies to towers as well. Numerous homeowners try to reduce costs by building their own towers or using whatever just happens to be lying around. Water-pumping windmill towers, for example, are unsuited for all but micro turbines. By building your own tower or using one unsuited to wind generators, you may not only be shortening the life of your wind machine, but you may also be endangering yourself and your neighbors. As Murphy himself would say, "there's never enough to do it right, but there's always enough to do it over."

The problem for home builders is dynamics. It's not too difficult to calculate the static loads operating on the wind turbine and tower. Anyone with a little background in math and physics can master the equations. Figuring out how to deal with the dynamic loads is altogether more involved.

There are several rotating or moving components that make up a wind generator and tower: rotor, transmission (where one is used), generator, yaw mechanism, and the tower. The interaction of all these moving components in varying winds is almost unpredictable. Dangerous harmonics can develop between these components, causing dynamic loads to exceed the static loads for which they were sized. Design teams try to predict when these harmonics will occur and how to prevent them from doing damage. You will be doing it solely by the seat of your pants.

If you must build your own wind turbine, choose a design where the forces involved are manageable, such as any design featuring sail wings. Keep it small (less than 1 meter in diameter) and keep it simple (use furling only, leave variable pitch to the pros). Blades using sails are inexpensive, can be made in any size, and are easy to work with. They are also unlikely to fly apart and cause damage or injury. A big plus for sail wings is that under severe loads (high winds), the sail cloth simply tears away, leaving the rotor intact.

Plans for a Cretan sail windmill can be obtained from the Centre for Alternative Technology in Wales (see Appendix H). The horizontal-axis rotor is patterned after those on the island of Crete and is used to drive an automobile alternator. The design calls for readily available materials and should cost less than $1000 to construct.

If you want to build your own, carefully review Jack Park's *The Wind Power Book*. He offers advice for do-it-yourselfers on materials and various designs. He also explains how to make the necessary calculations.

Used Wind Machines

Though building your own isn't a realistic way to cut costs, buying a used wind machine could be. Buying used wind turbines can be fraught with risk as well, but it gives you a better starting place, assuming you find a workable wind machine. Whether it's a used contemporary wind turbine or a used windcharger, you can find out how well it worked in the past and what problems you can expect, rather than starting from scratch with a home-built.

Today there are two principal sources of used machines: small residential turbines that were built during the late 1970s and early 1980s, and the much larger turbines used in California. Unfortunately, there was a lot of junk built for both markets. There are also a few pre-REA windchargers still in circulation.

Contemporary Models

Most observers expect a brisk used turbine market in the mid 1990s as thousands of first-generation turbines in California are replaced. Just as salvage windchargers from the 1930s fed the mushrooming market for wind systems in the 1970s, used turbines from California could propel a revival of low-cost wind power on the Great Plains. Already, used Danish wind turbines are finding their way to small wind projects in Minnesota and Iowa and to farms and ranches in the Canadian province of Alberta.

Unless you're in the scrap business, stay away from Storm Master, Century, ESI, Dynergy, and Windtech. These medium-sized wind turbines will be more trouble than they're worth. Professional remanufacturers may be able to salvage the Century and ESI turbines. But the others should be used only in museums. Look for Jacobs, Enertech, and the Carter 25-kilowatt model. There are some 1500 units in California that were built by these American manufacturers.

The Jacobs model was built during the early 1980s and shouldn't be confused with the 1930s version. The contemporary turbine uses a rotor 7-8 meters (23-26 feet) in diameter to drive an alternator mounted vertically in the tower. Unlike the other machines, which all use induction generators, the Jacobs model requires a synchronous inverter to produce utility-compatible power.

Nearly any Danish turbine would be a good find. There are 2800 Danish turbines in the 15-16 meter size class in California that analysts expect will eventually be replaced by larger turbines. There are also 300 German turbines (Aeroman) that would be suitable for relocation.

Wind turbines in California have seen heavy-duty use continuously for several years, some for a decade or more. They will need a complete overhaul before they're put back into service. They may also need design modifications to improve their reliability and safety. Even with these additional expenses, they could be a good buy. Rebuilt turbines from California wind plants should cost no more than $1000 per kilowatt installed or about $300 per square meter of rotor swept area. This is one-third of their initial cost during the early 1980s, but only 30-40 percent less than the price of new medium-sized turbines because of significant improvement in the cost-effectiveness of modern wind technology. Used wind turbines in California, "as is," should cost 20-40 percent of the installed cost for a completely rebuilt machine, or $200-400 per kilowatt.

In small turbines the names to look for are Bergey, Northwind, Jacobs, and Enertech in that order. All but Enertech built wind machines suited for both battery charging and utility power. Enertech turbines used only induction generators and are unsuited for remote power systems.

Many of Northwind's early versions relied on Gemini synchronous inverters for producing utility-compatible power. There are 800 of these inverters still in the field, and they can be upgraded to improve their reliability, according to Mick Sagrillo of Lake Michigan Wind & Sun.

Hummingbird and the Enertech 1500 are not worth salvaging. But Enertech's 1800, 4-kilowatt, and 5-kilowatt models can all be successfully rebuilt, says Sagrillo. Aeropower and Sencenbaugh are two small battery-charging machines that were built in the San Francisco Bay area. Few are available and most of those are found on the West Coast. The Sencenbaughs could be a good find, but the Aeropowers, which used a belt drive, will need constant babysitting. Avoid the Swiss Elektro and the French Aerowatt.

Rebuilt small turbines cost about $1000 per kilowatt without a tower, about 40 percent of their original price. Because of their sterling reputation, rebuilt 1930s-era Jacobs will command a higher price (about $1500 per kilowatt), but they're difficult to adapt for interconnection with the utility. These machines are best used as they were designed, for charging batteries.

Used Windchargers

Although the probability of finding an old windcharger hidden in a shed on the outskirts of some dusty cow town is now remote, hunting for one can still be a rewarding experience. It can also be a monotonous and tiresome task. The outcome depends on persistence, a thirst for exploration, and a good measure of luck.

Many areas of the United States were not served by utility power until President Roosevelt's REA brought federally subsidized power to the hinterlands. Some regions didn't receive power until well into the 1950s. As the rural electric cooperatives extended service to more and more remote locations, the home light plants previously used were no longer needed. Often they were taken down and sold for scrap. Some were sold to neighbors who had not yet been "electrified." Others fell to the ground as rust, disrepair, and violent storms took their toll.

In the regions where they were once used, windchargers can be found almost anywhere: packed away in the back room of an old store, buried beneath a farmer's junk pile, or hanging in the barn. Some have been found still in their original crates. All of the easily accessible generators have been bought, and if they haven't, their owners don't intend to sell, and no amount of persuasion will convince them otherwise.

Not all windchargers were created equal (see Figure 8-1). The most desirable of the old windchargers is the Jacobs home light plant. It was the most reliable and one of the largest generators built during the period. The most common brand, on the other hand, is the Wincharger. While the Jacobs may have been known as the "Cadillac of home light plants," the Wincharger was considered more akin to the Chevrolet. The Wincharger, though much maligned, can be profitably rebuilt and used in remote power systems according to Mick Sagrillo. More importantly, there may be some Winchargers still available.

Figure 8-1. Two of the most popular windchargers on the American Great Plains were Wincharger (left) and Jacobs (right).

During the heyday of salvage operations in the mid 1970s, this brand was often passed over in deference to the more valuable Jacobs generator. Wincharger produced a wide variety of models. The early ones used a two-blade wooden rotor counter-balanced by dual air brakes or buckets. These models ranged from 6 feet (1.8 meters) to 12 feet (3.7 meters) in diameter, with power ratings from 200 to 1250 watts. The smaller Wincharger models were direct drive: the blade was bolted directly to the shaft of the generator. Later Wincharger models became more sophisticated and sported four extruded aluminum blades 14 feet (about 4 meters) in diameter. Rather than air brakes, this model varied the pitch of two blades (not all four) to control rotor speed via two heavy weights in the governor. The rotor drove a 1-1.5 kilowatt DC generator through a single-stage transmission. Most models were painted yellow and used a triangular tail vane that distinguished them from water-pumping windmills and the Jacobs windcharger.

Jacobs took a different approach to building a home light plant, and the appearance of their turbines reflects this. Jacobs generators used a rotor with three wooden blades that spanned 13.5 feet (about 4 meters) in diameter. Rotor speed in high winds was controlled by varying the pitch of all three blades simultaneously. Early models used what Jacobs labeled the flyball governor. Later versions used a blade-actuated governor that took advantage of the centrifugal force acting on the blades themselves to change blade pitch. The tail vane is also distinctive.

Unlike Wincharger, Jacobs didn't use a transmission. Its direct drive generator is a massive affair of copper and iron. The turbine with blades, governor, generator, and tail vane weighs 500-600 pounds (225-275 kilograms), depending on the model. The generator came in several versions. Most were 32 volt, but a late model introduced to compete with REA generated 110 volts DC. The models were rated 1500-3000 watts. The Jacobs generator is the most sought after because it was built to survive the elements—and has. Some Jacobs generators have been running for over 30 years. Mick Sagrillo of Lake Michigan Wind & Sun found one on a farm in southern Minnesota that had been in operation for 60 years. Its reputation is well earned, and it is probably the best windcharger of the pre-REA era you can find.

Another windcharger worth seeking is the Parris-Dunn. It used a two-bladed, wooden rotor 6-12 feet (1.8-3.7 meters)in diameter. Very few of the larger models still exist, but you may stumble across one of the smaller ones. The Parris-Dunn was unique because of its approach to limiting rotor speed in high winds. Like the Jacobs, it too was direct drive. But the blades were bolted rigidly to the shaft of the generator. They didn't change pitch. In high winds the hinged generator and attached rotor gradually tilted vertically out of the wind. When the winds subsided, the rotor-generator combination fell back toward the horizontal running position. Though extremely simple, this system worked reliably. They're also easy to rebuild—a good reason to pick one up if you have the chance.

Ranchers, like everyone else, don't like to be rushed. Resist the urge to stuff a wad of bills into the owner's hand and haul away your windcharger. That machine you covet has probably been on the family homestead for 40 years or more, and the family might like to mull the deal over. The rancher's father may have installed that windcharger, and he might want to "leave it right there where Dad put it." Bargaining is half the fun of buying, and good bargaining requires consideration of more than your own interests.

Even after buying the generator, you are still a long way from erecting it in your backyard. First, you have to get it off the tower. To do that you will need pulleys, rope and cables, gin pole, safety harness, hard hat, utility belt, and assorted tools. Add to that list an adventurous spirit tempered with caution.

Lowering a wind generator from a tower is dangerous work. There's no chance to test the equipment under full load until the generator swings free. Everything must work then, or else. An excellent source of information on both how to safely remove and how to rebuild windchargers is Michael Hackleman's *The Home-Built, Wind-Generated Electricity Handbook* (see Appendix I). Hackleman devotes a whole chapter to the mechanics of raising and lowering windchargers.

Further Considerations

There's no escaping it, buying a used wind machine is risky, whether it's the risk of accident from working on bulky machinery in awkward places (atop a tower), or the risk of buying a wind machine that may not work reliably in your application. There's no simple way to tell how well a used turbine will perform or what's wrong with it until you've taken it apart. If you're averse to risk, buy a new turbine from a reputable manufacturer. When buying a used wind machine, whether or not it's from a dealer, the rule is buyer beware.

After years of shopping in supermarkets and department stores, few of us have well-developed bargaining skills. Our trust in products offered for sale has grown through extensive advertising and standardization. But not long ago, everyone haggled with the vendors at the local market. There, you examined the goods carefully, decided how much they were worth to you, and began the exchange. It's still that way with used cars today. Much remains hidden beneath the hood. So a great deal of faith is placed on the truthfulness of the seller with regard to its inner workings. The same is true with used wind machines. You must make not only a careful examination of the goods, but also a careful examination of the seller. Are the claims reasonable and verifiable?

Another important question to ask yourself is whether the wind turbine you're considering will meet your needs. Will the wind machine be used to heat water, produce line-quality power, or charge batteries? The answer will determine the degree of reliability and the type of generator required. If the wind machine will be used in a remote power system, then it must use an alternator or DC generator, and it had better work reliably. If it's line quality power you want, then a wind machine with an induction generator may be suitable.

Related to how you plan to use the wind system is why you want it in the first place. If you want a wind machine principally to tinker with (sure, you'd like it to be a paying proposition as well), then you're free to take more risks than if your primary need is a wind turbine that will consistently generate usable energy.

If you buy a used machine, use it as it was intended. Avoid mating a synchronous inverter with a battery-charging wind generator for instance. It's better to use it for charging batteries or heating domestic hot water through resistance heaters.

The first question to ask the seller of any used wind machine is, why is it for sale? Is it because the previous owner traded up, died, moved, or simply got fed up? If the latter, what's wrong with the turbine? What caused the problems? What were the headaches? Did the previous owner, for example, tire of climbing the tower every month to change the brushes or tighten some bolt? If so, how do you plan to deal with this problem?

Are parts readily available? Where can parts easily be found? How much do they cost? If parts are unavailable because the company went bankrupt, how difficult will it be to make them? Someone out there may be making parts for his own unit and could easily make a few more if need be. How will you find them?

The seller should be familiar with these questions and have answers for you or at least tell you where to look for the answers. The classifieds in *Home Power* magazine or other energy publications are a good place to check for spare parts (see Appendix I). If you can't find them there, you'll have a hard time finding parts, period.

Does the seller offer a warranty? If so, what kind: the manufacturer's, or the seller's own? How long does it last? A used wind machine may be bought directly from the previous owner or through a dealer. By buying the machine directly, you save by cutting out the middleman's markup. At the same time, buying direct forces you to install the turbine and tower yourself. By buying from a dealer you may pay more, but you gain some assurance that the turbine will work and will be installed properly. The dealer can also be more easily held accountable. In a direct sale, the previous owner may want to dump the wind machine and wash his or her hands of the whole affair as quickly as possible.

Take a hard look at the economics. Are you saving enough by buying a used machine to justify the risk that it will work, or work as well as it should? Check with local dealers. You may find that it doesn't cost much more to buy a new wind system instead.

Assembling a Kit

Kits are another way to cut costs. Kits for the serious wind enthusiast or do-it-yourselfer should contain—at a minimum—a professionally designed turbine. Because the design and construction of the wind turbine is the most demanding of all the components in a wind system, it's best left to those who know what they're doing. All you should be required to do is bolt on the blades and hang the tail (if one is used).

A complete kit should also contain a tower designed for the wind machine (not a string of lightweight TV antenna masts as some have advertised), all tower hardware, wiring, conduit, and electrical connectors needed, and a detailed assembly and installation manual. That's a tall order. No manufacturer or distributor presently provides such complete kits. (You're essentially asking for the same service and packaging a wind machine dealer receives from the manufacturer.)

Many of the smaller items, such as the wiring and the conduit, can be purchased locally. But unless you are familiar with the ins and outs of wiring and your local electrical code, you could run into problems. This is where a good installation manual becomes important. It not only tells you exactly what you need to do the job right but warns you where problems may develop and how to deal with them.

Buying a wind system kit is similar to buying a computer through the mail. You can save a significant amount of money, but you don't have ready access to someone who can help you correct a problem. When you buy a wind machine from a dealer, you always have someone to turn to for repairs.

With a professional kit you're not so much building a wind turbine as you are providing final assembly and installation. For example, Bergey Windpower offers an installation kit for their 3-meter turbine, including a guyed tower. The components can be handled easily and installed with a minimum of risk by following the instructions in their installation manual. The wind machine comes assembled, except for the addition of the blades and the tail vane. Because installation accounts for 25-35 percent of the total cost for this size wind machine, assembling and installing it yourself can produce considerable savings. Kits such as these are limited to wind machines less than 5 meters in diameter. The larger wind machines are more difficult and dangerous to install and manufacturers seldom offer them for owner installation.

Even on the larger wind machines, there may be aspects of the installation that you can do yourself. Depending on the dealer and the wind system you're planning to install, you may be able to do much of the site preparation yourself. You can clear the site and lay out the tower and anchor locations for a start. You may also be able to excavate the foundation, pour the concrete, and set anchors as needed.

Remember, though, that you assume liability for any work you perform. For example, say you prepared the guy anchor foundations, and one windy day an anchor pulls free and your nice new wind machine topples to the ground. The warranty provided by the dealer will not cover such an accident because you were responsible for properly installing the foundation and the anchors—even if you followed directions to the letter. The consequences are a pile of scrap metal and costly litigation.

Most of the cost in site preparation is in excavation, trenching, and pouring the concrete. You can cut costs by using your own equipment if you have it, by borrowing it if you can, or by finding a lower-cost contractor than that used by the dealer. Supplement the procedures outlined in Chapter 14 with detailed drawings provided by the dealer or manufacturer. Don't skimp. If the plans call for reinforcing bar, don't leave it out. Follow the plans religiously: your investment's in the balance.

Specifically, here is what you can do. Lay out and stake the position of the tower base pad, anchor locations, and cable runs. This may require accurately surveying the site where precise anchor location is required. Next, excavate for the tower pad and anchors. Build the forms for the concrete and lay in the reinforcing bar as specified by the manufacturer. Pour the amount and type of concrete called for. While the concrete is curing, you can dig a trench for a buried wire run from the tower to your house. (Aerial cable runs are less expensive and easier to install, but most dealers are moving toward underground runs because of the improved appearance. There's also less likelihood that the cable will be damaged by falling limbs and high winds.) The dealer may go so far as to permit you to lay the cables yourself. Most, however, will want to do this job themselves.

Depending on the size of the wind turbine, the type (variable speed or induction), and the location, you can save 5-10 percent of the cost by preparing the site yourself.

Don't be fooled by the easy sound of it. Preparing the site or installing the entire wind system yourself can be back-breaking work. Halfway through you may want to call in the cavalry, but this is one avenue for cutting costs that is realistic and attainable, and one where the savings are substantial.

Cooperatives—The Danish Approach

There's another way for individuals to reduce the cost of owning a wind system: cooperatives. It's an approach common in northern Europe, particularly Denmark. In Denmark cooperatives are a way for small investors to combine their financial clout. Joining together enables them to acquire the most cost-effective equipment possible, whether it's to process cheese, bake cookies, or generate electricity.

In the case of wind, Danish cooperatives buy a wind turbine, site it to greatest advantage, sell the electricity to the utility, then share the revenues. Cooperatives enable a group to buy the most cost-effective turbine available, even though it may generate more electricity than any individual may need. Because the cost-effectiveness of wind machines increases with size, it's often more economical to buy one 25-meter wind turbine than 100 3-meter turbines to produce the same amount of electricity.

Danish law encourages cooperative purchase of wind turbines by exempting them from revenue taxes on the portion of the wind generation that goes to offset an individual's domestic electricity consumption. Cooperatives and individuals are also paid well for their wind-generated electricity: 85 percent of the retail rate.

The cooperative approach has never been applied to wind energy in

North America, although agricultural cooperatives are found in midwestern dairy states. Cooperatives that sell their generation to the utility work only when the utility pays a fair price for the electricity. In much of the United States and Canada, Danish-style cooperatives may not work because of the low buy-back rate, often only 25 percent of the retail rate. This forces individuals to install wind turbines sufficient only for their own consumption.

Buying clubs are a related concept that may appeal more to Americans' individualism than do cooperatives. Buying clubs are also more adaptable to the small wind turbines that make the most sense when the utility's buy-back rate is low. A buying club pools money from individuals to buy a quantity of a particular product, in this case, small wind turbines. The bulk purchase gives the club more leverage when negotiating price with the manufacturer than an individual has. A club may be able to save 5-10 percent, possibly much more, by qualifying for a dealer's discount.

There are even greater discounts possible when installing multiple machines in one location. This is essentially the principle behind building wind power plants. Financiers pool money from multiple sources, negotiate a quantity discount from the manufacturer, save from economies of scale through mass installations, and negotiate with the utility for a higher buy-back rate from a position of greater strength than an individual can muster.

To summarize, if you want to cut the cost of installing a usable wind system, first and foremost find out how much of the installation you can do yourself. If possible, buy a kit from a reputable manufacturer and install the entire wind system. Or join with your neighbors and buy more than one if you can. If you are less interested in reliable performance, consider buying a used wind machine. But build the wind turbine yourself only if you're a risk taker and want to experiment.

9

Buying a Wind System

He that will not be counseled cannot be helped.

—John Clarke, *Paroemiologia*

Selecting a wind system entails gathering information (as you're doing now), sorting through it (weeding out the hype), and determining which combination of product, manufacturer, and dealer best meets your needs with the least risk and the best chance of success. Selecting the combination that's right for you is much like buying a car. You don't buy a car solely on what's under the hood or because of the transmission it uses. You look at the complete automobile. The same is true with a wind machine. You weigh the pluses and minuses of each component and consider the reputation of the manufacturer, not just the individual parts.

Choosing a Product

First determine the size range that meets your energy needs while remaining within your budget as explained previously. Next, find the product that offers the most for your money. Don't limit your evaluation only to energy output. Reliability, maintenance, and the soundness of the firm manufacturing it are equally important.

To gauge reliability, "ask the man who owns one," advises Jon Traudt of Windcatcher Company. Track down owners of the same wind machine or those owning models by the same manufacturer. Ask them how well it has performed. What kind of problems, if any, developed? Are they satisfied? If they had to do it over, what would they do differently?

Call the local utility for information, but consider the source when assessing their response. Most utilities in the United States, especially rural electric cooperatives, view wind turbines as competition. It's unlikely you'll hear a glowing endorsement of wind energy from your local utility company.

Contact the national trade association for help: the American Wind Energy Association in the United States, the Canadian Wind Energy Association in Canada, and so on. (For a list of trade associations in North America and Europe, see Institutions in Appendix H.) The association can alert you to any obvious cases of fraud. Remember, "If it sounds too good to be true, it probably is."

Locate any test reports on the product you can find. Some universities have tested specific wind machines, but most testing has been by government agencies. In the United States, the Rocky Flats Small Wind Systems Test Center conducted tests on small machines during the 1970s and early 1980s. Most of the wind machines they tested are no longer on the market, and Rocky Flats conducts little or no field work today. Rocky Flats and the U.S. Department of Energy's wind program have little up-to-date information on small wind turbines. West Texas State University is a good place to start in the United States for a frank appraisal of existing products. Fortunately, Canada continues to operate the Atlantic Wind Test Site on Prince Edward Island. This is the only place in the world that tests Canadian, U.S., and European products side by side at the same site. An excellent source of information on Danish wind machines is the Test Center for Wind Turbines near Roskilde, Denmark, as is ECN in the Netherlands for Dutch wind machines. The U.S. Department of Agriculture's Bushland, Texas, experiment station conducts tests of turbines in water-pumping applications. (For a list of test stations see Government-Sponsored Laboratories in Appendix H.)

Where test reports are not available, which is often the case, talk to the manufacturer or dealer. Ask what kind of tests have been run, for how long, and the highest winds experienced. Not all wind machines are built to the same standard. There was one case during the early 1980s where the designer sized his machine to withstand a maximum wind speed of no more than 90 mph in a parked condition. He asserted that no winds above that speed had ever been measured near his site in western Pennsylvania. This claim was false, and consequently the design of the wind machine was suspect, more so when considering that all other U.S. manufacturers design their products for a maximum wind speed of 120 mph (54 m/s).

With the revival of interest in small wind turbines during the early

1990s, some products have been introduced that are built only for light-duty service. They're simple but not rugged. Because they're inexpensive, they're proving popular. They may work fine—most of the time. But their longevity is site-dependent. One severe storm could wreck them beyond repair.

Knowing how long the tests were run or the length of time that a particular model has been in service is especially important. Unscrupulous manufacturers have frequently resorted to touting their products as "extensively tested" when they haven't been. In one case, the new product had only been in operation during a mild summer in Ohio, an area of moderate winds. The first time this "extensively tested" product was installed at windy West Texas State University it flew apart. In a notorious Utah case, the "extensive tests" were conducted on a bench-scale model. Not even a working prototype had actually been tested.

Reputable, well-tested products, in contrast, have been in unattended operation for years at numerous sites and have endured hurricane-force winds. Manufacturers with well-tested products, such as Bergey Windpower and Northern Power Systems, can provide documentation on the performance of their machines over time under harsh conditions. Other legitimate manufacturers in the United States and Europe can do the same.

When evaluating performance and reliability, don't be alarmed by occasional reports of defects. You're looking for trends. If every turbine built so far has thrown a blade and they are still throwing them, then there's a good chance the one you're looking at will too. But wind turbines shouldn't be held to any higher standards than we hold other machines. After more than 80 years of development, automobiles are still being recalled by the thousands for manufacturing and design defects. Yet we continue to buy and use them. We try to minimize the risk of buying a lemon by trying to select a model with the least potential for problems. Reputable manufacturers of wind machines make mistakes like everyone else. Your challenge is to find one that makes fewer mistakes than the rest.

Design defects usually show within the first year of operation. Like automobiles, new products must undergo a period of debugging. Unexpected problems will undoubtedly show up and must be corrected. These problems are greatest when the product is first introduced and decline thereafter. Ideally, you'd like a wind machine that has been on the market for several years, and one that operates successfully in a range of environments. This isn't always possible, so you have to rely on your own judgment.

Take a good close look at the wind turbine itself. How is it assembled? Does it look like it was welded together in a backyard shop? Do the parts fit snugly or have they been made to "press fit" with a hammer? Now step back and look at the wind machine as a whole. How do you feel about it? Is it something you would be proud of or will you have to put it "out back" so no one will see it? Wind machines are highly visible objects. Their appearance is important both for your satisfaction and for acceptance by the community.

Examine the promotional literature describing the wind machine. Are the estimates of energy output reasonable? Do they stress generator size while ignoring energy output altogether? Most manufacturers will present a list of parameters that succinctly describes how the wind machine performs, how it functions, and what it can be used for. This will include the AEO, power curve, and power form.

Estimates of annual energy output may ultimately replace the rated power at rated wind speed method prevalent in the United States to describe the size of a wind machine. More often than not the AEO is presented as a graph of estimated generation at various wind speeds at hub height. Occasionally the AEO may also be presented as a table of annual generation at two or three typical average wind speeds.

Power form indicates how the power will be used. For wind systems interconnected with the utility, this will be given as the nominal voltage and frequency of the electricity generated. For battery- charging wind systems, power form should indicate the DC voltage.

Most product specifications should also include: the average noise level, cut-in and cut-out wind speeds, maximum power, overspeed control, maximum wind speeds, and rotor speed.

Noise measurements should always specify the distance from the wind turbine, otherwise the information is meaningless. Most product literature in the United States doesn't include noise measurements. Specifications for most European turbines do. Cut-in, cut-out, rated wind speeds, and maximum power were discussed previously. Overspeed control is a concise description of the method used to protect the wind turbine in high winds or during a loss of load. The maximum wind speed is the speed the turbine was designed to endure unattended without suffering damage. For induction wind machines, the rotor speed is the average revolutions per minute from cut-in to maximum power. In a variable-speed wind machine, rotor speed is given as a range of rotor speeds from cut-in to maximum power.

The following is a summary of product specifications for a hypothetical wind turbine. The power curve and estimates of AEO are often presented separately.

> Power form: 3 phase, 220 volts AC, 60 hertz (50 hertz in Europe)
> Average noise level: 50 dB(A) at 8 m/s wind speed at 330 feet (100 meters)
> Cut-in speed: 10 mph (4.5 m/s)
> Cut-out speed: 50 mph (22 m/s)
> Maximum wind speed: 120 mph (54 m/s) design, 112 mph (50 m/s) tested
> Overspeed control: blade stall, brake, tip flaps
> Rotor speed: 200 rpm

Also ask to see a copy of the owner's manual. Is it well written and sufficiently detailed to tell you what's needed to maintain the wind system and to operate it safely? For example, are there instructions for starting and stopping the wind machine? Does it provide a parts list?

Engineers, when ordering expensive equipment, often ask for a copy of the operator's manual from each company competing for a contract. They then compare them. This gives the engineers a better understanding of the equipment they're buying, and it also gives them a feel for the manufacturer's approach to problems that may be encountered by the user. Don't expect the dealer to give you a manual gratis, but they should offer it to you at a nominal charge.

Now examine maintenance. What maintenance is required, for example? How often must it be performed? How difficult is it to perform? Must someone climb the tower or is a cursory examination from ground level adequate? If parts must be replaced, are they readily available or must they be specially ordered?

What kind of materials are used in the blades? Are they wood, wood composites, or fiberglass? What kind of corrosion protection is provided? Are exposed surfaces painted or galvanized? These are a few more questions to answer before evaluating the product.

Evaluating Vendors

After you have dissected the technology, you must evaluate less tangible factors such as the durability of the manufacturer and the reputation of the dealer.

Manufacturers

How long has the company been in the wind business? What's its track record? Does it have sufficient financial resources to honor its warranty commitments? There's no easy way to find answers to these questions. In most cases you'll be dependent on the dealer for information. Even when you do get the answers, it's hard to determine what's important and what isn't. For example, a well-established company that has been in business for several years is a better risk than one just starting out. Likewise, a company that's partially or wholly owned by a major national corporation usually indicates that it has ample financial reserves to survive a major warranty recall. Nevertheless, a corporate executive who has no personal stake in the company can much more quickly make a decision to cut her losses during hard times and cease production than the owner-entrepreneur who has put her own sweat and blood into the business. Only you can decide on which business you want to place your bets.

Dealers

There are few dealers in North America today. Most small wind turbines are sold direct, or through the mail.

The dealer you choose, if you use one, is determined primarily by the wind machine you want and where you live. Most dealers represent more than one company, to round out their product line. Even so, within a certain locale there will be only one dealer for each brand. (Manufacturers want to ensure a healthy dealer network, so they limit the number of dealers selling their product.) Proximity is important. If repairs or service are needed, particularly during an emergency, you don't want a dealer who lives on the other side of the continent.

Determine if dealers are reputable by checking with their previous clients. Have they been prompt in making repairs, or have they taken their time while hustling new sales? Dealers should have references available for such an inquiry. If not, are they willing to provide them? A dealer should also provide professional and character references if you ask for them.

Also call the manufacturer and check whether the "dealer" is authorized to sell its product. In one instance, a so-called dealer was selling a popular brand without authority to do so from the manufacturer. This dealer had declared bankruptcy previously, leaving a number of clients high and dry without spare parts or service for their ailing wind machines. In this case, he was selling a used wind machine—that is, until the authorized dealer blew the whistle. The whole sad affair could have been avoided by a single phone call to the manufacturer.

Don't be misled by membership in various organizations as a claim of legitimacy. Some dealers and some manufacturers use membership in a host of trade associations as a promotional tool. It's one of the oldest marketing tools in the book and most often used when no other credentials exist. Anyone can join an organization.

What to Expect

If you buy through a dealer or use a contractor to install the turbine, you have a right to expect that all work will be performed according to standard practices and to local building and electrical codes. The work should also be performed in a timely manner and all construction debris removed from the site before the job is considered finished.

Don't expect overnight miracles. If delivery of a component has been delayed due to circumstances beyond the dealer's control, you shouldn't hold the dealer accountable. The dealer should make a reasonable effort to expedite the installation of your wind system or its repair, but don't expect the dealer to jump at your every request. Keep in mind that dealers operate a business and that they may have other commitments. At the same time, the dealer should fulfill those obligations stated in the contract or implied during negotiations.

Contracts and Warranties

To ensure that you get what you pay for, put it in writing. Demand a written contract and warranty and consider having an attorney look them over. Installing a wind system is a major investment akin to buying a car—or a house. It's worth the added cost of getting good legal advice. You may need an accountant's advice as well.

You need to know specifically what is included in the price you have been quoted. If you plan to do any of the work yourself, the contract is necessary to spell out exactly where the dealer's responsibilities end and yours begin. The contract should also describe exactly what you must do to meet the terms of the warranty. Who has the final say, for example, as to how the work should be done? (Usually, the dealer does.) How will disputes be resolved should they arise? What is covered by the warranty? What isn't? How long does it last?

Most small wind machines come with a 1- to 3-year warranty. Extended warranties are sometimes available, for a price. Because of the difference in warranties between manufacturers, it's wise to read the fine print. Check whether the warranty is transferable or assumable by the manufacturer if the dealer goes bankrupt. Ask who pays for shipping or for the field work on warranty repairs and whether damaged or defective parts must be returned before replacement parts are shipped.

Another aspect is the terms of sale. The contract should state the amount, how, and when payments should be made. In general, you will pay for the wind machine and installation in advance. (You don't drive off the lot with a new car until you have handed over your check. Similarly, you shouldn't expect the dealer to install the wind system without your first paying for it.)

Terms vary from one dealer to the next. Usually a down payment is made to secure your order. Then full payment is required for the turbine and tower when they are ready to be shipped. Some dealers require payment for the turbine, tower, and installation in advance. In most cases 5-10 percent of the total contract is held by the buyer until the wind machine has been installed and operates properly.

In multiple machine purchases such as for a wind farm, the buyer has more leverage with the manufacturer and can obtain written assurances of performance that the wind machine will generate power as advertised and that it will be available to generate power a minimum percentage of the time. In Denmark, manufacturers have been held responsible for performance claims on turbines sold to individuals. Unfortunately, such assurances are often not offered to purchasers of small wind machines in North America.

Case Studies

In a classic example of how not to go about it, the rural cooperative in DuBois, Pennsylvania, bought a wind system from a manufacturer in nearby Clearfield, the forerunner of Fayette Manufacturing. The co-op didn't contact anyone about the company or its product. Nor did they investigate the company's claims. If they had, they would have found that the wind machine was a prototype (not a well-tested wind machine ready for commercial sale), that the manufacturer couldn't possibly build and install the wind machine for the contracted price (one of those "it's too good to be true" deals), and that the wind machine couldn't do what the manufacturer said it would. In short, neither the rural co-op nor the manufacturer knew what they were doing. The machine was installed and never worked. It stood for years along Interstate 80 as a testament to ignorance.

Some buyers take a more studied approach. Capitola Reece wanted a wind turbine that would pay for itself and work reliably. One more thing was certain: at 74 the retired teacher wasn't about to climb the tower and fix it herself. She also acknowledged she didn't know the first thing about wind machines or even where to buy them. In her research at the library she ran across an article about a fellow who was promoting them and wrote to him. After thoroughly reviewing his promotional literature, Cappy, as she's called, arranged for him to visit her site. She was ready. If it was going to cost her money to get him there, she was going to get her every penny's worth. She had the site picked out, copies of all her utility bills, and a notebook full of questions. "How much wind do you think I have? How much does it cost for an anemometer? For a wind turbine? What tax credits are available? Is it noisy? What maintenance is required? How many have been installed; how many have you installed? How well have they worked? How much energy could I produce here? What does the utility think about all this? How will it affect my taxes and insurance rates? When the utility lines go dead will it still work?"

After the inquisition, she took the dealer to her proposed site. Bad news. He would be glad to install an anemometer, but he would just be taking her money, because there were too many trees nearby. Though not tall, they were tall enough to block the flow at the anemometer. The results would be less than the wind speed at the nearest airport. He gave her his estimates based on the airport, and left. "Well," Cappy thought, "we'll just check this guy out." She called the state energy office, the manufacturer, AWEA, and a previous client with a similar wind turbine. He seemed all right. Still, she wanted her attorney to look over his contract and offer of warranty. She also wanted to talk to the township supervisor about the need for a zoning variance and to the utility about the interconnection. No variance was required for her rural site. The utility did not know a thing about the particular wind machine or the dealer, but did warn her that the few wind machines installed in their area had not worked well. That didn't deter Cappy. She thought they would be less than thrilled with the idea.

The attorney had some objections; so did her bank. The contract called for a sizable amount of money for a rather novel purchase and it called for most of it up front. "It won't do," said her attorney. He demanded changes in the contract and terms of payment. He wanted to pay after installation. The dealer balked—too great a risk for his small business. But a compromise was reached. A portion of the payment would be held in escrow by the attorney until the wind machine was installed and operating. The dealer agreed. His needs were met by knowing that the money was earmarked for him and was in safe keeping.

The three parties met and signed the contract. The dealer then ordered the equipment. The contract stipulated that the dealer had 90 days to install the wind machine and get it running. Within two months it was operating, but just before the final payment was made a problem developed. Because the escrow account hadn't been released, the dealer hustled to make the needed repairs. Cappy was satisfied and enjoyed years of nearly trouble-free service.

 **10**

Interconnection with the Utility

Imagine the following scene:

Night has fallen and the sky is clear. December's chill winds whip the Lake Erie shoreline. Drifting snow swirls about the fence posts and outbuildings of George McClain's small farm. The whistling wind rises in crescendo and then dies away in an unpredictable ebb and flow. A faint whirring, rhythmic and ever present, can be heard. Dark, saberlike shapes sweep the starry sky.

"Looks like it's going to be a cold one tonight," George predicts. His two kids, scampering around in their flannel pajamas, flee their mother as she readies them for bed. Darlene, both mother and partner in the McClains' dairy, responds, "George, don't you think we ought to turn the electric heat up? I feel like I'm coming down with something."

"Yeah, Daddy," the kids chime in, "just turn it up to seventy like we used to."

"Now you kids know better than that," he says. "Christmas will be here soon and we want to buy that new car we've been waiting for so long, don't we," he winks at Darlene. "We only get one more check from Pennelec before the new year and I want to sell them just as much power as we can. On a night like this everybody's going to be switching on their electric heaters. We need to save every kilowatt we can. The more we save, the more we can feed to Pennelec. I'll bet we can make fifty dollars by morning, more if this weather holds. Those turbines will really be turning out the juice in winds this high. Just listen to 'em hum."

Far-fetched? A family that awaits winter's winds and looks to the local electric utility as a source of income? Certainly not. The day has come when farmers such as the fictional McClains can sell a new cash crop: energy. But the McClains will not be alone. Anyone who has access to the wind, has the land, and can afford an investment in the future could find themselves selling power to the utility. Many already have.

Despite the important role of small wind machines in providing power to remote sites, more wind machines are interconnected with electric utilities than are used in any other application. In the United States alone there are more than 20,000 wind turbines generating utility-compatible electricity. Another 5000 interconnected wind turbines are working elsewhere in the world, mostly in northern Europe.

It's true. You can interconnect your wind system with the local utility. And yes, you can even sell power back to them. As you can imagine, though, it's not as simple as it first sounds. You may jump all other hurdles only to find the utility is slower than molasses in January when it comes to granting your request to interconnect with their lines. For most utilities, interconnection gets a low priority and your request may become buried under a stack of paperwork.

This was the case in the Pittsburgh suburb of Fox Chapel. The dealer, Bill Hopwood at Springhouse Energy Systems, had contacted the utility about the same time he began the zoning variance application. Yet even though the zoning board took months to make their decision, and it took another couple of months to install the wind machine, the utility still hadn't made up its mind what to do. Eventually they decided to let the interconnection proceed as a goodwill gesture toward the client, the Western Pennsylvania Conservancy. Since then the wind turbine has been in service for more than a decade—without mishap—as part of the Conservancy's renewable energy program.

PURPA

This strange state of affairs, where American homeowners can generate electricity in parallel with giant electric utilities, is the result of the Public Utility Regulatory Policies Act (PURPA), a part of the 1978 National Energy Act.

Many utilities in the United States had already seen the handwriting on the wall before 1978. It was inevitable, they felt, what with the public clamor over rising rates, that they would have to permit interconnections

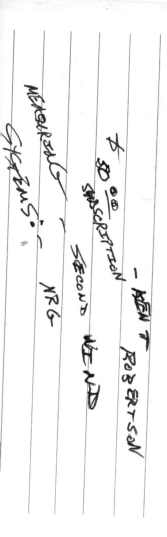

at least to improve community relations. Utili-
:apacity binds even welcomed these small power
. But a few fought every inch of the way and
ng them—willingly or unwillingly—into the oil

t revolutionary. (When pondering its potential,
er how a law such as this passed the congressional
ts. It's enough to revive one's flagging faith in
tifaceted, PURPA is most widely known for Sec-
: utilities must buy power from, and sell power to,
wer producers at reasonable rates.
trusted the Federal Energy Regulatory Commis-
sibility for drafting the regulations resulting from
were then implemented by state regulatory au-
ility or public service commissions have jurisdic-
ies and their compliance with PURPA.
y all utilities, both regulated (those that come un-
tility commissions) and unregulated (those utili-
jurisdiction). Investor-owned utilities (IOUs as
power corporations such as the Tennessee Valley
al cooperatives all have to comply with PURPA's
ere are important exceptions. Some state utility
direct authority over rural electric cooperatives and

removed two major barriers to more widespread
es and other alternative sources of electricity in the
xempts small power producers from restrictions of
Previously, a home wind system could have been
considered a utility and regulated as such by the state public utility commission. The paperwork burden alone would have buried many small
power producers. Second, PURPA assured wind system users of backup
power, and it stipulated how utilities were to charge for this standby supply.

Prior to PURPA some utilities charged discriminatory rates for backup
service. In effect, they said, "Sure we'll sell you power when you need it,
but, boy, are you going to pay for it." PURPA put a stop to such antics.
Utilities now can't penalize small power producers—overtly—by charging
unreasonable rates for standby service.

Through PURPA, Congress sought to encourage development of renewable energy by removing barriers to its use. PURPA also created a

powerful financial incentive that was absent before. Utilities must not only allow small power producers to generate electricity in parallel with their own system, they must also pay for that power.

Under FERC rules the buy-back rate must reflect the costs the utility avoids by not having to generate the power itself. This avoided cost could be more or less than the retail rate. During the early 1980s, the avoided cost in parts of the United States was near to or exceeded the retail price for electricity.

The regulations went even further. They allowed state utility commissions to take into account escalating costs over the life of the contract with a small power producer when ruling on a utility's buy-back rate. If the state utility commission chooses to encourage alternative energy, it may establish a levelized buy-back rate. In the case of a wind system designed to run for 20 years, the levelized rate would be much higher than today's avoided cost. For example, assume a buy-back rate today of 10 cents per kilowatt-hour and in the year 2000 of 20 cents per kilowatt-hour. The levelized rate could be set at 15 cents per kilowatt-hour over 20 years. Levelized rates offer much greater revenue in the early years than is available from rates based on escalating avoided costs. This accelerates payback and increases the return on investment.

PURPA fundamentally changed the way Americans look at power generation, conservation, and supplemental power sources. It encouraged decentralization and offered decentralized energy investment opportunities as well. Anyone who can afford a wind machine and has a good site can get into the utility business. As in this chapter's opening scenario, it alters our view of energy conservation from one of conserving to save money to conserving to make money.

PURPA and high buy-back rates also affected the size of wind systems homeowners or farmers could choose. As a supplemental power source, wind systems were looked on originally as a means of reducing utility bills, and they were sized accordingly. Sales to the utility at or near the retail rate encourage the user to seek the most economically sized wind system on the market. Because the cost-effectiveness of wind machines increases with increasing size, a homeowner is more likely to install a 7-meter wind turbine than a 3-meter machine, even though the larger turbine would produce more power than needed.

It is only a short conceptual step from buying one turbine larger than that needed to buying two, three, or even more turbines. Space, the level of risk one is willing to take, and the availability of capital are the only limits.

Like the McClains, farmers who began looking at the wind as a way of reducing their utility bills soon recognized another resource that could be tilled and a new crop harvested—a cash crop on contract at that. This concept gave birth to wind farms.

Direct sales to the utility also gave potential users more flexibility in siting. Under PURPA, users may purchase power simultaneously from the utility while also selling power to them. Let's say you own some rolling farmland. Your house and barn rest snugly at the base of a tall hill with trees all around. It's a beautiful setting but a lousy place for a wind machine. It just so happens that a utility line crosses the top of the hill, which has been cleared for a pasture. The top of the hill is ideal for a wind turbine, but a good distance from the house.

What to do? Install the wind machine on the hill and sell all your power directly to the utility. At the same time you will continue to buy power from them for your house and barn. The revenues from the wind machine will offset your bill much like it would have if you had installed it near the house.

Practical Application of PURPA

That's the good news. The bad news is that during the past decade utilities across much of the United States have convinced regulators to use their average cost of fuel to determine the avoided cost. The average cost often includes cheap natural gas, financially subsidized hydroelectricity, and environmentally subsidized coal. Originally FERC had intended that utilities pay small power producers their incremental cost of energy as well as capacity. By the early 1990s few utilities were paying for avoided generating capacity, because, they argued, there was no need for new power plants. That situation is bound to change during the remainder of the decade and avoided cost is expected to inch up.

Even in California, where 75 percent of the world's wind generation is produced, the calculation of avoided cost discourages the use of small wind turbines. During the early 1990s California utilities were paying only 30 percent of the retail rate for wind-generated electricity not under existing contracts. As a result there are few individual wind turbines interconnected with the utility system in a state world-renowned for its development of renewable energy.

The situation is different in Denmark, Germany, and the Netherlands, where utilities pay far more for wind-generated electricity from individual wind turbines (as opposed to those in wind power plants) than in the United States.

To summarize, PURPA gives the small power producer a little bargaining power where there was none before. If you want to install a wind system and it meets certain safety standards, then the utility must permit you to interconnect with its lines and it must offer you standby or supplemental power at reasonable rates. These rates cannot discriminate against you because you are using a wind generator. The utility must also buy any excess power you produce.

The specifics vary from state to state and from utility to utility. But there are provisions common to all. One of the more important is the regulation permitting the utility to charge for the interconnection costs they incur. This is normally in the form of a one-time bill for the installation of additional equipment. These charges vary, but typically range from $100 to $250.

Net Energy Billing

Output from a wind turbine fluctuates over time as the wind gusts and subsides. When you superimpose a wind machine's varying output onto the varying consumption of electricity in a typical home, there may be times when excess power is produced. This is more likely to happen at night during a winter storm when the wind is howling and there's little consumption of electricity. The excess power generated under such conditions will flow through the kilowatt-hour meter to the utility, running the meter backward.

Because most utilities in the United States pay less for the energy they buy back than for what they sell, they don't like their meters to run backward. They'll often send someone out to ratchet the meter, after which it will run in just one direction, measuring only the power you consume. If the output from your machine is substantially less than your consumption—say you're installing a 3-meter turbine at your all-electric home—this doesn't present a problem. You may never feed excess power back into their lines. But if you're using a larger machine and you want the utility to pay for the power you sell back to them, then at least one more meter will be needed: one to measure the energy you consume; and the other to measure the excess energy flowing back into the grid.

Ideally you would like to run your watt-meter backward, selling any excess energy at the retail rate. Regulatory commissions in eight states (Oklahoma, Texas, Minnesota, Wisconsin, Maine, Massachusetts, Rhode Island, and Connecticut) encourage just such an approach, called *net energy billing.* It essentially enables the user to bank excess wind energy with the utility until it's needed.

In principle, net energy billing allows you to run your kilowatt-hour meter backward. In practice, two ratcheted meters will probably be used so the utility can keep track of what's happening. The utility balances the account, usually every month, paying the retail rate for any excess energy you produce until your account's net is zero. That is, you can produce excess energy at the retail rate until the amount of energy sold back equals that consumed. In effect, you're storing excess generation with the utility instead of storing it in batteries.

Utilities in some states balance the account annually. This balances seasonal cycles of high and low winds, and is preferable to monthly accounting. When everything works properly (the wind system closely matches your needs), you won't owe the utility any money and they won't owe you. This allows the utility to settle the account without issuing checks for purchased electricity, saving them substantial administrative costs. It also gives you a better deal by minimizing sales to the utility at less than the retail rate. Under net billing, any deliveries to the utility in excess of net purchases will be purchased at a rate based on the utility's avoided cost, often only a fraction of the retail rate.

Germany has taken an approach that's a close cousin to net energy billing. In 1991 a federal law required utilities across Germany to pay 90 percent of their retail rate for purchases from clean energy sources such as wind machines. Denmark has a similar, though not identical, policy. Danish utilities buy excess generation at 85 percent of their retail rate.

Although the United States once pioneered the PURPA concept, its implementation has been retarded by state regulatory commissions responding to pressure from electric utilities. Denmark, with fewer wind turbines in residential, interconnected applications than in the United States (2000 individual medium-sized wind turbines in Denmark versus 4000 small machines in the United States), produces 20 times more wind-generated electricity, largely because of the higher buy-back rate. In Denmark, as in Germany, the higher buy-back rate enables homeowners and small businesses to buy larger, more cost-effective turbines that produce excess electricity for sale to the utility. Wind turbines destined for non-wind farm use in the United States must be sized to minimize the production of excess electricity. In most cases this limits the wind turbine to less than 10 meters in diameter.

Power Quality and Safety

Utilities were initially reluctant to interconnect their lines with small wind systems because of concerns about safety and power quality. To the chagrin of critics, there have been few problems with safety, voltage flicker, harmonics, or other technical issues.

Wind turbines in the United States have operated nearly one billion hours interconnected with local utilities and generated more than 15 billion kilowatt-hours of electricity without incident. Most of this experience has been with wind turbines driving induction generators, a technology utilities understand well. But more than 80 million hours have been accumulated with variable-speed wind turbines using "synchronous" inverters. Wind systems, whether using synchronous inverters or induction generators, now have produced line-quality power for years without endangering utility equipment or personnel. Wisconsin Power & Light, after a 7-year study of four small wind turbines operating on its system, found that "no safety problems were encountered at any of the sites." Despite this success, some utilities may still have questions that need to be addressed.

Utilities in the United States have a franchise from the state to supply electric power within a restricted territory. The company is required to provide reasonably reliable service to its customers by this franchise. They are responsible only to the point of delivery known as the service drop. This is where the utility's lines reach the building or premises. From this point on, the customer is responsible for the installation, operation, and maintenance of all other wiring—including that of a wind system.

The tariff under which the utility operates allows it to refuse service when the safety of their equipment or linemen is threatened or when they believe service to other customers may be interrupted or impaired. The utility's only means of ensuring a safe interconnection with a wind machine is to refuse service.

The safety of their linemen is the utility's principal concern, as it should be. They may fear that an interconnected wind system could "energize", deliver power to a downed line during a storm-related power outage and electrocute a lineman. This fear is not entirely unfounded. Linemen have been injured by improperly wired emergency generators.

Most synchronous inverters and all induction generators are line synchronized, that is, without the presence of the utility's line they can't generate power. Nevertheless, it's possible for generators, in rare circumstances, to "self-excite," that is, to provide their own excitation energy. Utilities require that all wind machines designed for interconnection with the utility must disconnect themselves from the utility's line during an outage and must not be able to self-excite. To preclude this, relays or electrical switches are placed on the utility side of the wind machine's inverter or control panel. When utility power is present, the AC relay, called a contactor, is energized, completing the electrical circuit. If utility power is lost for any reason, the spring-loaded relay is de-energized (turned off) and opens the circuit, disconnecting the wind system from the utility line.

Power Factor

Self-excitation is a problem only where capacitors are used on the wind machine side of the contactors. Wind turbine manufacturers and utilities frequently use capacitors to correct "power factor" (see Figure 10-1). Capacitors on the utility side of the AC contactor are disconnected from the wind machine whenever the contactor opens. Any capacitors used with a wind system for power factor correction should be placed on the utility side of the interconnection or should be designed to bleed down (discharge) after power is removed. This prevents self-excitation as well as any shock hazard to those servicing the wind machine.

Why does a utility or wind generator need capacitors to correct power factor? To understand the answer you need to know the difference between *true* and *apparent* power. (Yes, they both exist, and there is an important difference.) Power in watts, as you recall, is the product of voltage and amperage. This is generally true, but when we're dealing with alternating current we need to add another parameter to the equation: phase angle.

Don't let phase angle scare you. It merely describes the degree with which rising and falling voltage is in phase with the rising and falling current (see Figure 10-2). If current rises from zero at the same time voltage rises from zero, the two wave-forms are said to be in phase.

When this occurs, the cosine of the phase angle (0 degrees) equals unity (1), and true power is the product of voltage and amperage. In this case true power and apparent power are equal, and the power factor—the ratio of true power to apparent power—is 1. This is the ideal. Unfortunately, the real world seldom looks like this.

Loads on the utility's lines cause the current waveform to shift slightly. Current either leads (starts rising earlier) or lags (starts rising later) voltage. In rural areas, where power must be transmitted over long distances, the length of the line itself is sufficient to cause current to lag behind voltage.

Figure 10-1. *The utility uses capacitors to correct the power factor in its distribution system.*

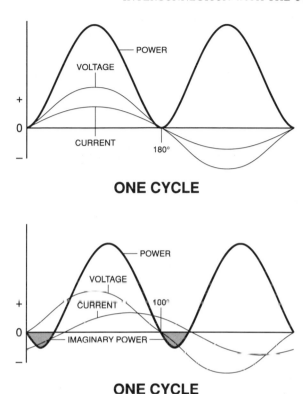

ONE CYCLE

ONE CYCLE

Figure 10-2. True and apparent power. Top: Power is simply the product of voltage and current when they re in phase. Bottom: When voltage and current are out of phase, true power is less than apparent power. The utility must deliver apparent power, but it gets paid only for true power. Thus, the utility is concerned about power factor the ratio of true to apparent power.

At this point you may cry out, "Who cares?" The utility cares. And they care a lot. When current and voltage are out of phase, true power decreases (the cosine of the phase angle becomes less than one), yet apparent power remains the same. The utility is called on to deliver apparent power. Their entire system from generators to transmission lines must be sized to meet the demand for apparent power. Generators, including wind generators, are rated by their ability to deliver apparent power in units of kilovolt-amperes and not, usually, true power in kilowatts.

The utility's trusty watt-hour meter measures only true power. Therein lies the dilemma. They must generate apparent power while getting paid only for true power. And true power is always less than apparent power (the power factor is either 1 or less than 1). They get shortchanged. The utility tries to correct this—to keep the power factor as close to unity as possible—by adding banks of capacitors to their lines.

The foregoing applies to wind systems as well as to electric utilities. It is of importance to you because the power factor of your wind generator may cause the utility to take corrective action by installing capacitors and then charging you for them. Power factor is also a favorite whipping boy for opponents of wind power within electric utilities.

If problems arise you need to remember that the utility pays only for true power when you sell to them. (You are now in the utility's shoes, that is, trying to maximize the production of true power by getting your power factor as close to 1 as you can. Values above 90 percent are desirable.) Don't think you're selling them horse meat and getting paid for prime beef, as has been implied by some utility spokesmen. Moreover, your wind system appears to the utility as just another load (albeit a negative load) calling for power factor correction. It's no different than if you added a couple of freezers or new power tools to your home as far as your impact on the utility is concerned. Utilities don't charge you for power factor correction when you install a new freezer, do they? No, of course not. And they shouldn't charge you for power factor correction when you install a wind generator either.

Voltage Flicker

Another problem arises with certain wind systems using induction generators. On some models the rotor is locked in place until the cut-in wind speed is reached. The rotor is then motored up to synchronous speed by drawing power from the utility. The effect on the utility is similar to the startup of a compressor motor in a refrigerator or freezer. There's a momentary surge of current until the rotor gets up to speed and the wind begins driving it. Because the power rating of wind generators is usually larger than most household appliances, the magnitude of the in-rush current can be large enough to cause a slight voltage drop in the line. The result may be voltage flicker. Your lights may dim briefly whenever the wind generator starts. This isn't a serious problem but can be annoying. The utility, though, may claim that an induction wind machine will detract from the level of service they offer other customers and may require a dedicated transformer to mitigate the problem. The transformer isolates the voltage drop to the customer using the wind machine and often eliminates the problem entirely.

In sparsely populated rural areas most homes or farms already have a dedicated transformer; that is, only one customer is served by the transformer. One transformer, on the other hand, may be used for several customers in the suburbs. You can tell if you have a dedicated transformer by taking a look out your window. The utility primaries, or high-voltage lines, are carried at the top of the utility pole. Leads from the primaries are attached to the top of the transformer. The low-voltage lines or secondaries are attached to terminals on the side of the transformer. The service drop to your home is always from the secondary (low-voltage) side of the transformer. If the secondaries are strung directly to your house and to no other, the transformer is dedicated to your service drop. But if the transformer is located several poles away and the low-voltage lines serve several other customers, the transformer is communal and your wind turbine could affect your neighbors.

Not all induction wind machines have this problem. Most don't motor the rotor up to speed. The rotor free-wheels instead. When there is sufficient wind to turn the rotor at synchronous speed, the AC line contactors are energized and the wind system is brought on line. There's little in-rush current, and there's often no voltage flicker. Consequently, the utility should not impose a blanket requirement for a dedicated transformer without first demonstrating the need for it.

Harmonics

Synchronous inverters are not without their power quality faults. They can produce current harmonics in various degrees. This does not affect most electric appliances but, theoretically, can cause electromagnetic interference with television, radio, and telephones. The degree of interference depends on the inverter, its size and location, and the level of electrical noise on the utility's lines with which it is connected. (Power from the utility is never free from harmonics itself.)

This interference may be noticeable but not necessarily objectionable. For example, one wind machine owner thought his machine was performing normally when his wife walked into the kitchen and wanted to know why he had turned the wind turbine off. Amazed, he asked her what she meant. His wife told him she had grown accustomed to the faint humming it made on her radio, but she couldn't hear it anymore. He proceeded to check the control panel and found the inverter was down with a blown fuse.

With nearly 2000 synchronous inverters operating successfully in the United States and several hundred more in northern Europe, current harmonics have not presented a significant problem. Again, a transformer can be used to isolate any disturbance to the customer's immediate vicinity. But as in the case of voltage flicker, the utility should not make a blanket requirement for a dedicated transformer without proving there's a need for one.

Dealing with the Utility

Utilities are not inherently evil, though they may often be depicted that way. They are a business and their employees are charged with making it profitable. Utilities are often large institutions. And like any big bureaucracy, the right hand frequently doesn't know what the left is doing.

Corporate policy, as expounded by company managers, may be adamantly opposed to the interconnection of small wind systems. Yet the word may not have drifted down to the lower levels where the work gets done. You may call up and be surprised to find the staff friendly and curious about your project. They may go out of their way to be helpful. This was the case with Metropolitan Edison, the once infamous Pennsylvania utility with nuclear reactors at Three Mile Island. In spite of management's public opposition to renewables, Met-Ed's staff were extremely cooperative with the interconnection of a small turbine near Harrisburg.

Of course, there's the other type: utilities that make supportive pronouncements about clean energy sources, brag about all they're doing to improve the environment but, when you contact them, tell a different story. Despite PURPA in the United States Mike Bergey reports that in some cases it has taken more effort to get utility cooperation than to build and install the wind system. In a particularly flagrant example, the Los Angeles Department of Water and Power simply told one wind machine dealer they didn't want small wind turbines connected to their lines. He chose to build a stand-alone system, instead of challenging the powerful publicly owned utility.

The fact that the utility is a bureaucracy also means that nothing happens quickly. You should notify them of your plans months in advance. They may have a clearly defined policy and may be able to give you the answers you seek promptly. More often than not the particular person handling your account won't have dealt with a small power producer before and will cover themselves every step of the way with time-consuming approvals. You need to appreciate the utility's point of view because you're dependent on their cooperation.

The difference today is that we now have a decade of actual operating experience. Many utilities have experience with small wind turbines on their systems. If not, they may know of neighboring utilities with interconnected wind turbines. They certainly have read about the success of wind energy in California and Europe. Utilities are far less likely to balk at interconnection today than 10 years ago. But as Bob Gates of Zond Systems, a successful developer of wind power plants, advises, "you don't argue with an 800-pound gorilla." Utility executives dread nothing more than a self-righteous customer strutting into the office and demanding that the utility kneel down and graciously grant a trouble-free interconnection. If you want a fight, there's no better way to find one.

Wind turbine manufacturers have experience dealing with utilities, and they can be helpful in obtaining a fair contract. They can also assist with any technical issues that the utility might raise. Often the manufacturer will talk directly with the utility engineer handling your application. This can facilitate a quick agreement because they speak the same technical language.

After you have passed the technical gauntlet there may be a few more obstacles to overcome: liability insurance is one. Some utilities, principally the rural electric cooperatives, may insist that the small power producer purchase liability insurance in case of an accident. Here's their rationale: if a lineman is injured by a wind machine, the utility's Workers' Compensation insurance pays the lineman's claim. The insurance company that covered the utility's compensation sues the small power producer for recovery of its money. (The lineman may also sue for damages.) The small power producer loses in court, can't pay, declares bankruptcy, and the insurer has to absorb the loss. The utility's insurance rates rise and you know who pays the bill in the end: ratepayers. This scenario hasn't taken place with a wind system, but electric utilities have been sued for a lot less when they were not at fault—and have lost. The rural co-ops, because they are small, can't afford mistakes. They're gun-shy when it comes to liability.

The cost for this insurance isn't prohibitive, but it can cut deeply into the total revenues of a small wind system. Before you acquiesce, determine how the utility treats customers with emergency or standby generators. Are they required to have the same insurance? If not, why not? If they are not required to have insurance, the utility may be discriminating against you. The line contactors on your wind system respond in much the same manner as automatic transfer switches on standby generators. (Technically, anyone with a standby generator would fall into this category whether using a manual or an automatic transfer switch.)

After a decade of experience and more than 1 billion hours of interconnected operation from wind turbines in California alone, there's simply no longer any reason for rural co-ops to impose liability insurance. Bergey Windpower, which has installed more than 750 turbines and tallied more than 30 million hours of interconnected operation, argues that small wind turbines have proven they don't pose a threat to the safety or reliability of electric utilities. In Oklahoma, where there are more dispersed wind turbines than in any other state (370 units), utilities have suffered no losses despite the state utility commission's ban on liability coverage.

Another contract provision that might raise your ire is the "hold harmless" clause. Read the contract's fine print. If you don't understand what it says, get an attorney to look it over. One contract for a Pennsylvania utility held the company harmless against all claims arising from damage or injury due to the interconnection, whether the fault of the small power producer or the fault of the utility. One dealer inserted the following to make the contract more equitable: ". . . so long as the act(s) giving rise to the claim were not due to the negligent conduct of (the utility), its employees and agents." In this way, the utility becomes responsible for any damage they cause.

As with transformers and capacitors, the utility may try to load the interconnection down with unneeded equipment and raise the cost of the interconnection. Another item they may add is a redundant disconnect switch. For example, they may require an "accessible and lockable" disconnect switch between their billing meter and your service panel. This enables linemen to "lock-out" the service to your home on your side of the meter when working nearby.

For a wind system with any kind of electrical storage, this is a reasonable measure. The utility's linemen need only throw the switch to ensure that no power can feed back into the line. However, it's an unnecessary burden on an interconnected wind system without storage, especially if a lockable disconnect switch has been installed on the wind generator's service to the building as suggested in Chapter 14. This requirement can also be discriminatory.

When the wind system is disconnected from the customer's service panel, no power can flow from the wind generator to the utility's lines. With the wind generator disconnected, the customer's service is like that of any other utility customer. If the small power producer is forced to install another disconnect switch, this one between the billing meter and the service panel, then it's only reasonable to assume that all other customers must do the same. Field experience has shown time and again that the utility's distribution system, particularly that of rural electric cooperatives, poses a greater hazard to the operation of small wind turbines than do wind turbines to utilities and their personnel.

Where installed, disconnect switches are seldom used to protect utility linemen. Madison (Wisconsin) Gas & Electric testified at a hearing in 1991 that they had never used the disconnect switches of wind turbines on their system. Disconnect switches have proven useful not to utilities but to dealers and installers for isolating a wind turbine's circuit when troubleshooting a problem.

Should a dispute develop, consult the responsible authorities. In the United States contact the state's public utility commission. They could mediate the dispute or bring action against the utility through the regulatory process. Above all, be patient. The regulatory bureaucracy is painfully slow. But sometimes all it takes is a phone call from the right person for the utility to suddenly see the light and mend its ways.

Cooperation is growing between utilities and the wind industry as more utilities gain operating experience with modern wind turbines. Some utilities in the United States, such as Pacific Gas & Electric, now openly promote wind energy. Bergey Windpower reports that most utilities they deal with require only one phone call from the manufacturer to grant permission for the interconnection. Some utilities have also allowed net energy billing even when they were not required to do so.

Sizing—What's Right for You

From the previous discussion of PURPA it's apparent that finding the right size turbine for an interconnected application in the United States isn't difficult. In most cases a wind machine that produces less than your total consumption is preferred. Where the utility buys wind generation at approximately the retail rate, install the biggest wind machine you can afford. But before you can estimate what size wind turbine is best, you must first determine how much energy you need.

How much electricity do you consume? Do you know? If you're like most people, you know how much you pay each month but you don't have the foggiest idea of how much electricity you use. To find out, you don't have to do anything more complicated than pull out your old electric bills. There are at least two items of importance on them. One is your consumption in kilowatt-hours. The other is the total cost. From this you can calculate how much electricity you use each year and what you're paying for each kilowatt-hour.

To encourage energy conservation, most utility bills clearly tell you how much electricity you consume each month and how it compares with past consumption. Some utility bills show your consumption over the past 12 months. If your bill doesn't have this feature (an electric use profile), you'll have to tally up monthly consumption yourself for the entire year.

Often the utility estimates your consumption from past usage and balances your account whenever they get around to reading your meter. If you have one of those so-called budget accounts, where you pay so much a month regardless of immediate consumption, the utility has estimated your monthly or annual consumption. There should be some indication on the bill of what the utility expects you to use.

The average U.S. household without electric heat uses approximately 25 kilowatt-hours per day or 750 kilowatt-hours per month. The typical household consumes about 9000 kilowatt-hours per year. Californians and people living in the Southwest consume less, about 6000 kilowatt-hours per year. European households use half the amount of the typical U.S. family. Dutch homes use less than 3200 kilowatt-hours per year. Canadian homes use even more than those in the United States.

Homeowners with electric heat use considerably more than those without. Depending on the climate, the size of the building, and how well it's insulated, an all-electric home will use from 15,000 to 80,000 kilowatt-hours per year.

If you plan to use the wind turbine to reduce your heating bill from other fuels, apply the same technique as above to bills for heating oil or natural gas. Say your primary concern is the cost of heating with oil. Find out how many gallons of oil you use each year and what it costs. You can then convert the oil consumed to kilowatt-hours of equivalent electricity as in the example given at the end of Chapter 5.

Use your utility bills to learn more about how and when you use energy. In the process, you will become more attuned to how you use electricity and you'll begin to find ways to use less. It's always cheaper to conserve energy than to generate it. Before installing any wind turbine you should have the utility perform an energy audit and install the conservation measures they recommend.

You can easily estimate how much an appliance is costing you by looking at its name-plate. Every appliance has a label giving its power rating in watts or the current it draws in amps. Use watts to calculate the amount of power the appliance consumes. If the label gives amperage instead, it will also list the voltage. The product of voltage and amperage will give you watts.

Consider an electric 110-volt space heater rated at 1000 watts. To consume this much power the heater must draw 9 amps. If you operated this 1-kilowatt space heater for 1 hour it would consume 1 kilowatt-hour of electrical energy.

If the space heater ran 1 hour per day, and electricity cost 10 cents per kilowatt-hour, in a month's time it would cost:

$$10 \text{ cents/kWh} \times 1 \text{ kWh/day} \times 30 \text{ days} = \$3.00$$

You can apply this same technique to any appliance or load in your home to gauge its effect on your total electric bill.

Consumption varies not only from month to month but also by time of day. For a typical home without electric heat, use peaks with peak activity around the house: early morning as everyone readies themselves for work or school, again at lunch, and then in the evening around dinner. Because wind speed also fluctuates throughout the day, output from the wind machine may not match consumption. The wind may howl at night when consumption is lowest and be calm during the day when use peaks. In an independent power system, this mismatch is tempered by storing excess energy in batteries. In an interconnected application, excess energy flows to the utility. During light winds, the utility makes up any shortfall.

Because most utilities in the United States pay substantially less for energy flowing into their lines than they charge for what they sell, it's often necessary to size the wind machine so that it rarely produces excess energy. If you expect that the wind machine will generate a surplus, you'll probably want to use as much as possible of the excess energy on-site where it's most valuable.

If you determine that the sale of excess energy is not in your interest, then a wind machine meeting 50-70 percent of your demand may be more appropriate than a larger one displacing all your current consumption. Wind turbines 3-5 meters (10-15 feet) in diameter will meet a sizable portion of the average residential utility bill. Larger wind machines, those greater than 7 meters (about 21 feet) in diameter, may be necessary to appreciably dent the consumption of all-electric homes. But stay flexible.

Instead of simply limiting yourself to the smaller turbines, another option is to use the larger and more cost-effective wind machine with a dump circuit to direct the excess energy to a standby load (see Figure 10-3). For

DUMP CIRCUIT

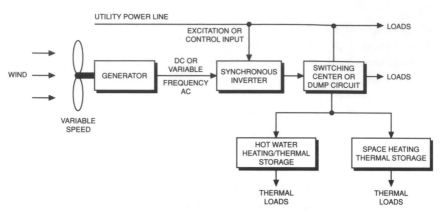

Figure 10-3. Dump circuit. Excess electricity is dumped or diverted to alternative loads, rather than being sold back to the utility at less than the retail rate.

example, you could install another tank for your domestic hot water. As in the wind furnace concept, wind-generated electricity is used to heat water in the tank, which is then fed to your hot-water heater. The preheated water reduces your consumption of conventional energy, whether you heat water with electricity or with natural gas. The extra tank allows you to dump excess electricity into thermal storage whenever you have it. Your water heater can then draw from this tank whenever it needs to. In this way, you have stored electricity as low-grade heat.

You may wonder why you need the dump circuit and an extra tank if you already have an electric hot-water heater. When the tank's thermostat calls for electricity to heat the water to a certain temperature, it draws automatically from the house's circuits, which the wind machine is supplying. That's the problem. It only draws power when needed. It's the boss. We want to use the wind-generated power when it's available, and that's not always at the same time when there's a demand for it.

Once you have mastered the concept of using excess electricity to preheat domestic water, you can turn your attention to other loads. If you heat your house with hot water, then you can preheat the water in your furnace. Just put in a bigger storage tank. *Voilà,* we are right back where we started with the wind furnace concept. You have the same advantages as before, plus the wind machine supplies all of your regular electrical loads.

If the utility pays about the same amount for the energy they buy as that they sell, you won't have to go through the trouble of adding the dump circuit. You can then shop for the wind machine that's most cost-effective. Economics improve with increasing size. So you'll probably be seeking a wind machine that meets as much of your consumption as your budget will allow.

11
Stand-Alone Power Systems

Prior to the development of interconnected wind turbines, wind generators were used for powering remote sites where utility power was nonexistent (see Figure 11-1). These home light plants used wind machines and banks of batteries sized to carry the household through winter winds and summer calms. Occasionally the dealer would throw a backup generator into the mix to charge the batteries during extended calms. The high cost, poor reliability, and maintenance requirements of these early systems discouraged all but the most hardy from living beyond the end of the utility's lines.

That's no longer true. Home power systems have become so "mainstream" says Wes Edwards, that homes using them now qualify for mortgages. Edwards, a licensed electrician who has lived "off the grid" in northern California since 1974, says that today's improved inverters, low-power appliances, and the widespread availability of photovoltaic (PV) panels have revolutionized stand-alone power systems.

Data compiled by Pacific Gas & Electric confirms Edwards' observation. In an early 1990 study of its service area, PG&E found the number of stand-alone power systems mushrooming at a rate of 29 percent per year. They expect the market to continue expanding as urbanites increasingly move to rural areas not currently served by the California utility. The business prospects looked so enticing that PG&E even toyed with the idea of providing the stand-alone home power systems itself instead of building additional power lines.

New technology has made all of this possible. Until PV's entry into the market, remote power systems were solely dependent on wind machines (in

STAND-ALONE WIND MACHINES

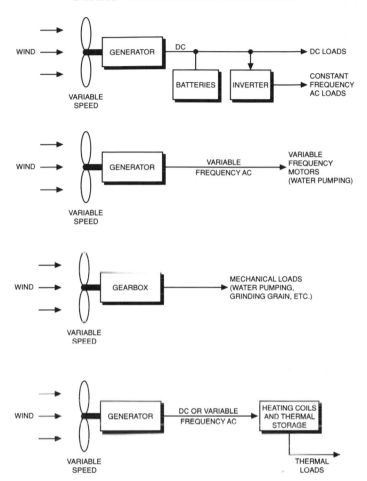

Figure 11-1. Means of using wind machines in stand-alone applications. Historically wind machines have been used to charge batteries (windchargers) in remote power systems, to mechanically pump water (farm windmills), or to grind grain (European windmills). Today, wind turbines can also be used to drive AC motors directly, without the need for inverters, in specialized pumping applications. Wind machines can also be used to generate heat.

some special cases, small hydro systems) and backup generators. The modularity of PVs has transformed the remote power system market by enabling homeowners to more closely tailor power systems to their needs—and their budgets.

Despite their other advantages, wind machines are less modular than PVs. When scaling up the output from remote power systems, it's easier to add more modules to a PV array than it is to add more wind machines. Although the output of some micro turbines is no more than that from most PV modules (50 watts), each additional turbine requires a separate tower and controls. Wind turbines quickly jump in size from 50 watts to 250 watts. Thus, PVs can be added in smaller increments than wind.

Wind is also far more site specific than solar. It's safe to assume that nearly everywhere on earth the sun will rise and set every day. Not so with wind. Wind follows daily and seasonal patterns that are less predictable. Throughout much of North America the winds are strongest in the winter and spring and weakest during the summer. Fortunately this pattern happens to coincide with the attributes of solar energy. Winds are generally strongest when the sun's rays are weakest, and winds are weakest when solar radiation reaches its peak. For this reason wind and solar are ideally suited for hybrid systems that capitalize on the advantages offered by each technology.

Hybrids

With advances in solar and wind technology, it just doesn't make sense today to design a stand-alone system to use only wind or solar. Hybrids offer greater reliability than either technology alone because the remote power system isn't dependent on any one source (see Figure 11-2). For example, on an overcast winter day when PV generation is low there's likely sufficient wind energy available to make up for the loss in solar electricity.

Wind and solar hybrids also permit use of smaller, less costly components than would otherwise be needed if the system depended on only one power source. This can substantially lower the cost of a remote power system. In a hybrid system the designer need not size the components for worst-case conditions by specifying a larger wind turbine and battery bank than necessary.

The hybrid concept is often carried one step further to include a fossil-fueled backup generator for the same reasons. In effect, stand-alone systems substitute the fuel and the maintenance of the backup generator for a larger solar and wind combination. Depending on the size of the backup generator and the power consumption at the time, the generator can top up discharged batteries and meet loads not being met by the combined wind and solar generation.

HYBRID STAND-ALONE POWER SYSTEM
WITH BACKUP GENERATOR

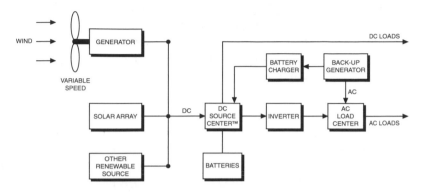

Figure 11-2. *Hybrid power system. With advances in wind and solar energy, remote power systems are no longer dependent on any single technology. The addition of a backup generator provides even greater flexibility in sizing the system's components.*

Despite advances by hybrid power systems in improving reliability and reducing the overall size of the power system, initial costs remain high. It behooves the potential user to reduce demand as much as possible to keep costs down. Advances in energy efficiency permit users to meet their energy needs from smaller, less expensive power systems than once possible. The development of compact fluorescent lights and energy-efficient appliances now makes this possible with little sacrifice in lifestyle.

Reducing Demand

Reducing consumption by conservation and efficiency measures improves the services that a renewable power system can provide by stretching each kilowatt-hour to do as much work as possible. This is particularly true at remote sites where you must spend your own time, effort, and money to generate electricity.

To reduce demand, find out what you're consuming now. As advised in the previous chapter, perform an energy audit of your lifestyle. Knowing how, where, and when you use energy is even more important for a stand-alone power system than for an interconnected wind turbine. Determine what appliances you plan to use at your remote site and estimate their consumption.

Use Table 11-1 as a guide. It presents the typical rate of electrical usage when each appliance is operating. You must still estimate how many hours each appliance will be in use to find the number of kilowatt-hours consumed. Conserve as much as possible. It's nearly always cheaper to save energy than to generate it with a hybrid power system. To maximize the value of your renewable power system and minimize its cost, carefully pare your electricity consumption to the minimum needed for the services you expect.

Decide if there are any electric appliances, such as electric hot-water heaters or electric stoves, that can be switched to gas or other fuels. It makes no sense to squander your hard-earned electricity on inefficient appliances or on uses where electricity isn't well suited. Heating is one of them. Heating with gas, oil, propane, or wood is far more economical at a remote site than heating with electricity.

Though cooking consumes little energy overall, electric stoves have high peak power demands that will affect the size of inverters and other hybrid components. Cook with gas or propane, or use a microwave oven instead.

Pacific Gas & Electric found in their study of stand-alone systems that most remote generation is used for lighting, refrigeration, and water pumping. (Remote sites are seldom served by municipal water sources.) Lighting is the easiest to tackle. Compact fluorescent lights can reduce lighting demand significantly. By lighting only those areas where light is needed, by using as much daylight as possible, and by simply turning off lights when they're not needed, you can cut lighting consumption by two-thirds.

Similar savings can be achieved with refrigeration, but at a price. Modern refrigerators use about 1000 kilowatt-hours per year, half that of 1970 models. Today's most efficient models use 700-800 kilowatt-hours per year in the sizes commonly found in American homes. Sunfrost refrigerators, the efficiency champions, use only 300 kilowatt-hours per year. Unfortunately Sunfrosts cost about $2500. In contrast, the most efficient conventional refrigerators cost only $500. If the 400-500 kilowatt-hour difference between a Sunfrost and a standard refrigerator pushes your energy budget over the edge, forcing you to add another wind turbine or several more solar panels, the Sunfrost may well be worth the added cost. This is just one of the many tradeoffs you must make when designing a stand-alone power system.

Depending on the size of the house and the climate, air conditioning can double the consumption of an otherwise energy-efficient home. If you

Table 11-1

Typical Residential Energy Consumption

	Therms/hour	kWh/hour		Therms/hour	kWh/hour
Heating			**Laundry**		
Small Gas Furnace	0.6		Electric Clothes Dryer		5/load
Large Gas Furnace	1		Gas Clothes Dryer	22/load	.5/load
Space Heater		1.5	Washing Machine		
Baseboard Heater		3	Cold		.25/load
Electric Furnace		10	Warm wash, cold rinse	.11/load	.25/load
Heat Pump		3-5	Hot wash, warm rinse	.33/load	.25/load
Air Conditioning			**Appliances**		
110-volt window unit		1.5	Stereo		0.03
220-volt window unit		2.6	Color TV		0.23
Central		4.5	B&W TV		0.07
Portable Fan		0.2	Vacuum Cleaner		0.75
			Microwave/5 minutes		0.1
Water Heating			Toaster/use		0.08
Electric		300-400	Toaster Oven		0.5
Gas	20 30		Electric Range		
Heat pump		175-225	Oven		1.33
			Surface		1.25
			Cleaning/use		6
Refrigeration			Gas Range		
16 cu. ft. frost-free refrigerator		100-150	Oven	0.09	
20 cu. ft. frost-free refrigerator		115-180	Surface	0.07	
10 cu. ft. manual defrost refrig.		35-60	Cleaning/use	0.5	
15 cu. ft. frost-free freezer		70-150			
Lighting					
General		50-200			

Source: Pacific Gas & Electric Co.

must have air conditioning, ask yourself whether an evaporative cooler will suffice. Swamp coolers use far less electricity and work well in arid climates such as the southwestern United States.

For energy efficiency ratings of appliances from washing machines to refrigerators, and for tips on reducing your energy consumption without reducing your comfort, read *Consumer Guide to Home Energy Savings* by the editors of *Home Energy* magazine. (See Appendix I.)

The key is to remain flexible. Sacrificing the lifestyle you desire isn't necessary, but some modification of behavior often proves beneficial for optimizing the performance of a hybrid power system. For example, cutting back on energy-intensive discretionary loads on days when the power supply is reduced extends battery life and leaves a little extra in storage available should you need it for those more important loads like pumping water or refrigeration. Not unlike our ancestors, learn to synchronize your lifestyle with the weather. Do the laundry when it's windy or on a bright sunny day. In this way you take full advantage of the fuel as it's available.

Turning off unneeded appliances isn't much of a burden for those who are energy conscious, it's already become second nature. But for those going "cold turkey" from a highly consumptive lifestyle where energy's undervalued, it can be a rude awakening. In such cases it might be wise to gradually reduce your consumption until you're ready to make the transition to producing your own power.

The average American household should be able to reduce its consumption to about 3600 kilowatt-hours per year or about 10 kilowatt-hours per day. This isn't spartan living. Most Europeans live comfortably on this amount or less. How much you're able to reduce your consumption will determine not only what size system you need, but also whether you should wire for DC or AC, and at what voltage you should operate your power system.

AC and DC Systems

All stand-alone power systems produce and store DC electricity. Photovoltaic arrays produce DC directly. Most wind machines produce AC, which must then be rectified to DC as in your car alternator. The DC is then stored in batteries until demanded by appliances in your house. The exception to this scenario is the backup generator, which can provide constant-frequency AC for AC loads while also recharging the batteries.

There are two sides to the stand-alone power system, generation (and storage) and loads. Since most of the loads will be supplied by the batteries most of the time, the choice becomes whether to feed the loads with DC directly or with AC through an inverter. Electronic inverters add complexity and slightly reduce the power system's overall efficiency. But conventional homes use AC, and if maintaining the resale value of your house is important, use AC in the conventional manner. Then if you ever want to add utility power or sell your house, you won't need to rewire from DC to AC. You may also find that a new home wired for AC is much easier to finance than one wired for DC. Though banks will finance—sometimes a bit reluctantly—a new home with a stand-alone power system, they prefer that the house conform to conventional wiring practices.

Don't discount DC entirely; there may be some major loads, like refrigeration or water pumping, where DC wiring is justified to limit inverter losses. Sunfrost refrigerators are built in both DC and AC versions for just such situations.

An equally important decision is the voltage of the generating system: the solar panels, the wind turbine, the batteries, and the backup generator. Low voltage requires heavier cable to conduct the same amount of power as that at higher voltages. For the same size cable, low voltage wastes more energy in resistance losses than higher voltage. Thus the voltage preferred for operating a remote power system safely and efficiently is directly proportional to the amount of power that has to pass through it.

Richard Perez, editor of *Home Power* magazine (not to be confused with *Home Energy* magazine), says that systems requiring less than 2 kilowatt-hours per day, such as vacation cabins, should stick with 12 volts; those using up to 6 kilowatt-hours per day should go with 24 volts; and those using more than 6 kilowatt-hours per day should opt for 48 volts.

Although some micro wind turbines are suitable for 12-volt power systems, most wind and solar hybrids use 24-48 volts. The initial cost of 120-volt systems, because of the large number of batteries needed, is generally prohibitive except for those systems designed to meet consumption approaching 20 kilowatt-hours per day (about 7000 kilowatt-hours per year). High-voltage systems do permit greater flexibility in siting your renewable power sources than do 24-volt or 48-volt systems. If you need to move your solar array into a more exposed position or move your wind turbine to a hilltop, the 120-volt system may be your only choice to keep line losses and cable costs to an acceptable minimum.

Sizing

At extremely small loads (a few lights, radio, etc.), a 12-volt PV system comprising one or two 50-watt panels that power DC appliances makes the most economic sense. For such applications as vacation cabins, PVs are easier to work with and less expensive than wind. But as loads increase above a few hundred watts, wind and solar hybrids become more attractive. At loads typically found in most households, wind is far more cost-effective than PV modules (see Table 11-2). Small PV arrays generally cost about $10 per kilowatt-hour. The generation from small wind systems at good

Table 11-2

Comparison of Photovoltaics and Wind

	PV			Wind		
Size	4 panels	8 panels	15 panels	1.5 m	3 m	7 m
Capacity (kW)	0.21	0.42	0.79	0.25	1.5	10
Output (kWh/yr)	387	767	1451	800	3,000	20,000
Storage Cost	$1,400	$1,400	$2,800	$1,400	$2,800	$14,000
Total Cost	$4,200	$6,700	$11,600	$3000	$10,000	$36,000
$/kWh	10.9	8.7	8.0	3.8	3.3	1.8

Assumes 6 m/s average wind speed at hub height and 5 sun hours per day.
Installation not included.
PV prices; Real Goods.
All systems use same quality battery; $1/amp-hour
Storage = 80% of rated capacity usable.

sites costs one-fourth to one-half that of an equivalent PV array with the same inverter and batteries.

Like wind, the output from a solar array isn't constant over time, season, or from one region of the country to another. Real Goods, a mail-order dealer of PV arrays and energy-efficient appliances, estimates PV systems in the southwestern United States and Plains states produce at their rated output about 5 hours per day. A 200-watt system will generate 1 kilowatt-hour per day, and 365 kilowatt-hours during the year. To estimate the potential generation from a PV system in the United States, multiply the array's rated capacity by the effective sun hours per day for the region closest to your site in Figure 11-3.

Because they're more modular and far easier to install, it's always simpler to add additional PV modules than to add another wind turbine (see Figure 11-4). If, after installing your stand-alone power system, you find that the backup generator is running too often and you want to reduce its fuel consumption, consider adding more solar panels or consider using a tracker for the existing array. These racks support the solar panels and "track" the sun as it moves across the sky.

Most PV modules are mounted on the roof and, for much of the United States, are simply tilted at an angle equal to the latitude plus 15-20 degrees. Like installing wind turbines on taller towers, PV generation can be improved by mounting the array on a tracker. In winter the tracker boosts performance only a modest 10-15 percent. But during the summer, when

PEAK SUN HOURS PER DAY
ANNUAL AVERAGE

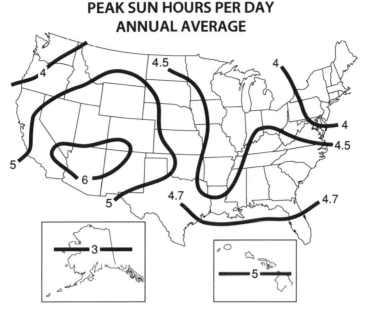

Figure 11-3. Annual average peak sun hours per day. To estimate the potential generation from a photovoltaic array, multiply the array's rated capacity times the peak sun hours per day for the region nearest you. For example, a 200-watt array will generate about 1000 watts per day in a region with 5 peak sun hours per day, or 365 kilowatt-hours per year.

the winds are most likely to be light, trackers really shine, giving 40-50 percent more energy than a fixed array. For more information on sizing solar arrays see Joel Davidson's book *The New Solar Electric Home.*

Inverters

Whether you're using a PV array, a wind machine, or a hybrid stand-alone system, you will require an inverter to operate conventional AC appliances. Most inverters produce a modified AC sine wave that can serve a wide variety of AC loads from sensitive electronics (computers and stereos) to washing machines.

To determine the size of the inverter needed, add up the demand, in watts, from all appliances that are likely to operate at the same time. The inverter should be sized to handle both the surge requirements of the induction motors in refrigerators and washing machines, and their continuous demand when operating for extended periods. Small appliances often

Figure 11-4. Photovoltaic or solar cells are built in modules. Multiple modules are ganged together in an array. If more power is needed, it's a simple task to add more modules. Some arrays used in telecommunications, such as this on a mountaintop in California, have been operating for nearly 20 years.

demand 1.5-2 times their rated current when they first start. Large appliances such as washing machines and refrigerators can draw three to four times their rated current when first switched on. For example, an electric motor using 500 watts will require 1500-2000 watts when starting.

Inverter ratings vary, but all manufacturers list both their continuous and surge capacity. Refer to the *continuous output rating.* This is what the inverter can actually supply over a long period without failure, keeping in mind that few loads operate continuously, and those that do draw little current. Give yourself some room. Sandia National Laboratories recommends sizing the inverter to 125 percent of the expected load.

A 2-kilowatt inverter should run most minor loads and some major loads, like a washing machine or a microwave, when operated singly. A single 3- to 5-kilowatt inverter or two 2-kilowatt inverters may be necessary if there's any chance the refrigerator, washing machine, well pump, and microwave will operate simultaneously. Microwaves, hair dryers, and similar loads draw a lot of current (1000-1500 watts) but are operated only for short periods. They may influence the size of the inverter needed, but they contribute little to total energy consumption. Jim Davis at Energy Transfer Systems specifies 3- to 9-kilowatt inverters for families accustomed to all the creature comforts. Davis, a California installer of remote power systems, likes to build a power supply "big enough to give them a comfortable place to live."

The inverter should also provide fused protection from the various sources of generation and should offer power factor correction for inductive loads. Ideally the inverter should also have disconnect switches on both the AC and the DC sides, and load management switches to limit certain loads from exceeding the inverter's capacity.

Never try to operate an electric dryer, electric hot-water heater, or electric stove on a stand-alone power system unless you're using them as a dump load for excess generation. Use bottled gas (propane) instead. These electric appliances consume inordinate amounts of electricity and place an unreasonably high current demand on the inverter.

Some inverters also offer optional battery-charging functions. If you don't use one packaged with the inverter, you'll need to add a battery charger to your system that's capable of handling multiple inputs: those from the renewable sources, as well as the backup generator. Without a battery charger, there's no way to ensure that the batteries stay properly charged.

Batteries

No electrical generator works 100 percent of the time, even those of the utility. Batteries permit a renewable power system to coast from one spurt of power to the next, from windless night to sunny day. They're integral to a successful home power system. They also add significantly to the cost. Conventional lead-acid batteries, the type most commonly used, cost about $200 per stored kilowatt-hour. The principal alternative, nickel-cadmium, costs from $350 per kilowatt-hour for rebuilt batteries, to as much as $1500 per kilowatt-hour for new, heavy-duty cells.

Because batteries are so widespread there's a bewildering array of sizes, types, and quality. For a battery with a long life, avoid selecting batteries purely on price. Batteries for stand-alone power systems must be capable of numerous deep discharges. Cheap automotive or truck batteries ($50 per kilowatt-hour) might be suitable for a tinkerer, but not for a remote power system. Though popular with do-it-yourselfers, golf-cart batteries are still less than ideal. Forklift or nickel-iron batteries are a good choice when available, advises Mick Sagrillo of Lake Michigan Wind & Sun. Above all else, use batteries that are designed for the frequent charge-discharge cycles found in remote power systems.

When considering the cost of batteries, don't overlook shipping. Lead and cadmium are among the densest materials known, and shipping batteries made of them any distance at all incurs a hefty freight charge.

Batteries are finicky and require special care. First, they don't like to be over- or undercharged. To prevent permanent damage, lead-acid batteries shouldn't be discharged to more than 80 percent of their capacity. Nicads are more tolerant: they can be fully discharged without damage. Second, lead-acid batteries don't tolerate temperature extremes well. The capacity of lead-acid batteries decreases markedly with temperature. They're least effective in the dead of winter, when lighting loads are greatest. They can even freeze, particularly when discharged. One way to avoid damage in cold climates is to store batteries in a cellar below the frost line. But don't bring them indoors without proper ventilation.

Batteries should always be isolated from living areas and from electronics (see Figure 11-5). The gases given off by lead-acid batteries are not only highly corrosive, they're highly explosive. Jim Davis prefers sealed, maintenance-free batteries for this reason. They may not be the optimum for someone who's willing to nurture their batteries, but sealed batteries are preferred by those who want a "hands-off" system.

It's a good policy to separate batteries from inverters, switches, and service panels. This prolongs the life of the electrical components while guarding against a spark igniting the hydrogen given off during high charge rates.

Though more costly than lead-acid batteries, nicads offer several valuable advantages. Nicads are better suited to cold climates because they're less susceptible to freeze damage. Nicads also require less maintenance and last longer than lead-acid batteries. Nicads can also be fully discharged without harm. Unlike the nicad batteries used in computers, those designed for power applications don't suffer from "memory effects."

Size batteries and other fixed hardware to the size of system you eventually want. Avoid undersizing batteries, inverters, and wiring. With a lead-acid battery bank, you're locked in if you find that storage is insufficient after a few years of operation. More batteries can't be simply added to the system in small increments. It's a fixed entity, so additions must include a complete new bank of batteries of the proper voltage wired in parallel.

Undersize the PVs, if anything. Make up the difference in lost production with the backup generator. You can incrementally add PV modules, or a larger wind turbine as you can afford them. It's much more difficult to go

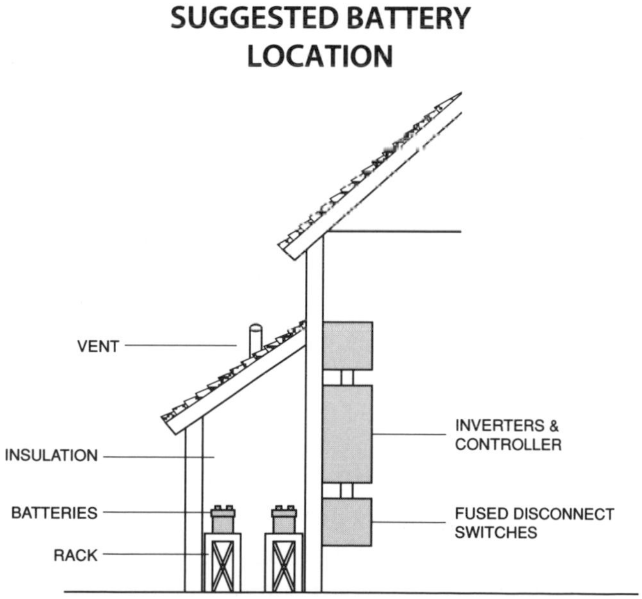

SUGGESTED BATTERY LOCATION

Figure 11-5. Suggested battery and inverter placement. Where possible, batteries should be housed in a well-ventilated room separate from the living quarters.

back and rewire your house, replace your battery bank, and switch out your inverter than it is to add one more solar panel.

Backup Generators

In a properly sized power system with ample battery storage, the backup generator may not be used at all, particularly if you're willing to adapt your usage to the resource. When the wind is blowing or the sun shining, operate those discretionary loads. (Do your laundry, for example, only when there's a surplus of power.)

Because most remote sites need propane for heating, cooking, and domestic hot water, it's relatively simple to use the same propane to power a backup generator. Propane is modular, portable, and offers good utility.

A backup generator, or gen-set, allows you to design a system with less battery storage. You can substitute fuel costs for extra batteries. PG&E found in their survey of California stand-alone systems that 80 percent of the respondents used a backup generator. But the gen-sets provided only 2 percent of the total energy, suggesting that most remote systems had enough storage for almost all conditions. The backup generators, even if seldom used, provided a valuable service: peace of mind.

Because generators operate most efficiently near full load, it's best to run them only after lead-acid batteries have fallen to about 20 percent of their full charge. The generator can then run at full output until the batteries are brought back up to 80 percent of full charge. The free fuel sources, wind and solar, can then top up the batteries with long duration charging cycles. This limits the overall running time of the generator, extending its life. It also keeps fuel consumption down. Automatic controls are available for monitoring the batteries' state of charge that will start and stop the generator as needed.

For long life, look for generators that operate at 1800 rpm, not 3600, and that have self-starting capabilities. Avoid the inexpensive portable generators used at construction sites. They're not well suited for remote power systems.

Utility Backup

An alternative is to use utility power, where available, to make up for discharged batteries (see Figure 11-6). An automatic transfer switch senses the load and switches between inverter-supplied power and the utility. This system permits users to gradually wean themselves from utility power by adding components incrementally.

STAND-ALONE WIND SYSTEM
WITH UTILITY BACKUP

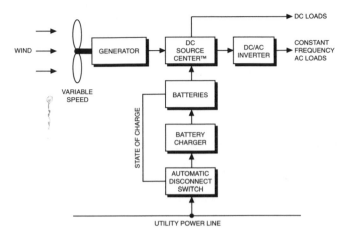

Figure 11-6. *Utility backup. This configuration for a semi-autonomous home is useful when the utility is uncooperative with a direct interconnection, or when homeowners want to wean themselves gradually from utility power. Electronic controls monitor the battery bank. If the batteries become depleted the automatic disconnect switch closes, directing utility power to the battery charger.*

If the renewable power sources consistently fail to provide enough power and utility service is nearby, you can bring in utility service to charge the batteries. You substitute utility power for the backup generator. On the surface this may seem to defeat the purpose of a stand-alone system. But keep in mind that the independent power system still serves all the loads, not the utility. The stand-alone system will still provide service if utility power is interrupted. This approach makes a lot of sense in areas of the world where utility power is unreliable.

Utility backup is not unlike the uninterruptible power systems widely used to protect computers from power outages. Many critical computer systems operate continuously from batteries. The utility is used only to charge the batteries. The loads draw current from an inverter that's constantly fed by the batteries.

Typical Systems

Remote power systems using a micro turbine and a few solar panels have proven popular for those who want to use renewable energy in a limited application. These entry-level systems are suitable for vacation cabins and recreational vehicles. The low cost of their components also makes them perfect for hobbyists and backyard experimenters. They're also well adapted for rural second homes that could someday be upgraded into permanent residences. These low-power hybrids can be packaged with small DC-AC inverters to power consumer electronics, but many are used with DC appliances obtainable through specialty houses serving the recreational vehicle market.

Micro turbines are the wind energy equivalent of solar walk lights: they're inexpensive and easy to install. They're light enough that you can pick one up and carry it home in your arms. Because most micro wind turbines are used for vacation cabins where there's less stringent demand on performance, they're often installed on shorter towers than their big brothers. The Aerogen, Marlec, SOMA, and Wind Baron Neo models are often installed on guyed masts using readily available galvanized steel pipe.

Table 11-3 examines the cost of a micro hybrid package assembled from the PV and wind systems listed in Table 11-2. This hybrid uses Wind Baron's Neo micro turbine, four 50-watt solar panels, a small inverter, and sufficient batteries to provide 6 kilowatt-hours of storage on a 24-volt system. Though easily installed with rudimentary tools, this hybrid system will generate about 2.5 kilowatt-hours per day or slightly less than 1000 kilowatt-hours per year.

Table 11-3

Cost of Micro-Power Systems

	kW	kWh/dy	Storage kWh	Cost
Solar system	0.2	1.0	6	$4,200
Wind system	0.2	1.6		$3,000
Less duplicate batteries				(1,400)
Total		2.6	6	$6,000

For bigger systems, Mick Sagrillo has found that for much of the United States 80 percent of the hybrid's generation should come from wind, and the remainder from solar. The exceptions are in the Southwest and Southeast, he says, where intense sunlight and light winds make the optimum proportions closer to 50-50. In Sagrillo's experience, wind will produce more power than PV in the northern tier of states and Canada for an equivalent capacity.

Let's assume that you want to assemble a hybrid system that will meet your 3500-kilowatt-hours-per-year-load. From Table 11-2, select the 800-watt PV array and the 3-meter wind turbine. Though hybrids enable downsizing storage requirements, both the PV and the wind system in Table 11-2 have been priced with the bare minimum of storage. For our case we'll assume that we want at least 2 days of storage. By combining the two systems we can eliminate one inverter. But we'll want a backup generator sufficient to handle the loads for a system this size. Table 11-4 puts all this together in a system where the wind turbine provides two-thirds of total generation and there's sufficient storage for 2.5 days.

Table 11-4

Cost of Stand-Alone Hybrid Power Systems

	kW	kWh/dy	Storage kWh	Cost
Solar system	0.8	4	13	$11,600
Wind system	1.5	8.2	13	$10,000
Less duplicate inverter				($1,400)
New refrigerator				$500
Backup generator & charger	6			$4,500
Total		12.2	25.6	$25,000

Stand-Alone Economics

Now let's look at the economics of this hybrid system. In general, users of remote power systems don't expect their wind and solar generation to compete with the cost of utility-generated electricity. They typically install a remote power system because the cost of extending utility power to their site is even more costly. California utilities charge new customers about $10 per foot for overhead line extensions (50 percent more for buried lines). PG&E charges as much as $60,000 per mile to bring in utility power. The situation's no different in Europe. Electricité de France charges rural residents in France FFr 100 per meter ($32,000 per mile) to extend utility service.

Under these conditions a stand-alone system can pay for itself in the first year if it's more than ½ mile or 1 kilometer from the utility's lines. If you're considering a stand-alone power system in the United States on purely economic grounds, in general it's cheaper to bring in utility power if your home is less than 1000 feet from existing utility service. Most U.S.

utilities provide free line extensions up to 1000 feet for houses that will use an electric range, clothes dryer, and water heater.

Extending the line may not be the only cost you incur. Many utilities require a minimum purchase of electricity to justify extension of the line. In Pennsylvania, West Penn Power Company requires a minimum monthly payment of $100-200 for a period of 5 years from customers requesting line extensions.

Table 11-5 examines the costs of installing and operating the hybrid system described in Table 11-4. The batteries are replaced in years 8 and 16. In current dollars this system will cost about $36,000 over its lifetime.

Now compare the costs of the hybrid system to the cost of extending the utility's line 2500 feet at $10 per foot. Assume that the utility will also require a minimum purchase of $100 per month for 5 years. Table 11-6 summarizes the cash flow based on these assumptions. Over 20 years utility service will cost about $39,000 in current dollars. As a rule of thumb, you'll find that if the cost of bringing in utility service approximates the cost of a hybrid system, the hybrid is cheaper over the long term.

Table 11-5

Economics of Small Hybrid Power System

Total Cost	$25,000		Rate Escalation	0.1
AEO (kWh/yr)	4,500		Mortgage Cost	0.1
Rate ($/kWh)	0.10		Inflation Rate	0.05
O&M, Ins.	0.02		Tax Bracket	0.3
Batteries	$5,600			

Year	Gross Savings	O&M	Battery Replacement	After-Tax Mortgage Cost	Net Revenues
1	$450	($500)		($2,186)	(2,236)
2	$495	($525)		($2,200)	(2,230)
8	$877	($704)	($7,880)	($2,311)	(10,017)
16	$1,880	($1,039)	($11,642)	($2,603)	(13,404)
19	$2,502	($1,203)		($2,784)	(1,485)
20	$2,752	($1,263)		($2,856)	(1,368)
					($59,000)
Net present value of expenditures					($36,000)

Table 11-6

*Economics of Bringing in Utility Power
2,500 Feet from Existing Utility Service*

Cost of Service	$25,000		Rate Escalation	0.1
Min. Purchase	$1,200			
AEO (kWh/yr)	4,500		Inflation Rate	0.05
Utility Rate	$0.10			

Year	Service Cost	Purchased Power	Net Payment
1	$25,000	$1,200	$26,200
2	$0	$1,200	$1,200
19	$0	$2,503	$2,503
20	$0	$2,753	$2,753
			$54,035
Net present value of payments			$38,500

Other Stand-Alone Power Systems

The use of wind machines in hybrid power systems for telecommunications and village electrification are essentially variations of remote power systems for residential use, but each has unique requirements that distinguish them from home light plants.

Telecommunications

Telecommunications demand reliability. Wind machines used in telecommunications encounter more extreme weather, operate more often (sometimes in excess of 7500 hours per year), and must function unattended for much longer periods of time than typically found in home power systems or even in commercial wind power plants. Only robust wind machines using fully integrated, direct-drive designs perform satisfactorily in the rugged environments characterized by telecom sites (see Figure 11-7).

A site in McMurdo Sound, Antarctica, illustrates the severe conditions that wind turbines must endure to serve telecom applications. Shortly after installation, a Northern Power Systems' model HR3 operated for 12 hours in the furled position during a fierce Antarctic storm. The radio station eventually went off the air when the exhaust stack for the backup generator blew away. After the worst of the storm had passed the HR3 dropped back

Figure 11-7. Hybrid power system for telecommunications. This 1-kilowatt wind turbine and adjoining photovoltaic array have been powering a remote telecom site in Wyoming for more than a decade. (Bergey Windpower Co.)

into its running position, recharged the system's batteries, and brought the station back to life. Twice during the first 2 years of operation anemometers at the site blew away, once after recording a wind speed of 126 mph (56 m/s). Since then the site has endured even stronger winds. The project has been so successful that it has been expanded to include three HR3 turbines.

Two U.S. manufacturers, Northern Power Systems and Bergey Windpower, meet the telecom industry's requirements for reliability and low maintenance. These manufacturers suggest using the turbines listed in Table 11-7 to meet the continuous loads found in telecommunication use at sites with average annual wind speeds of about 6 m/s (14-15 mph). The balance of the system would include an array of PV modules and batteries. Telecommunication companies often use thermoelectric generators to provide backup power. But a conventional propane, diesel, or gasoline generator could be used as well.

Hybrid systems using the wind machines in Table 11-7 have been able to substantially reduce fuel consumption at telecom sites in Canada and the United States. At a 14-mph (6.3-m/s) site on Calvert Island off the coast of British Columbia, two Northern Power Systems' HR3 turbines, in conjunction with a 1.2-kilowatt solar array and 84-kilowatt-hour battery bank, were able to cut the diesel generator's operating time substantially. Overall the hybrid system reduced fuel use nearly 90 percent, at half the maintenance cost of a conventional diesel system. At Norway's Hamnjefell telecom station above the Arctic Circle an HR3 turbine has met 70 percent of the site's loads since 1985.

Village Electrification

Between 1.5 and 2 billion people live without utility power. Extending utility service from the cities to remote villages in developing countries

Table 11-7

High Reliability Wind Turbines for Telecommunications at Sites with 6 m/s Average Wind Speed

Turbine	Continuous Load (kW)
NPS HR1	0.4
Bergey 1500	0.4
NPS HR3	1.0
Bergey Excel	3.5

where most people live, is costly, difficult to finance, and takes years of struggle. To surmount these problems some villages use small diesel generators. But they are expensive to operate and often unreliable. More and more developing countries are turning to renewable energy as a less expensive, more reliable, and quicker way to meet the electrical needs of rural areas.

Village power systems must meet standards for ruggedness and reliability similar to those in telecommunications. Though the weather may not be as demanding as that found on a wind-swept mountaintop, Third World villages are distant in both time and space from the technical support and spare parts found in the developed world.

The benefits of providing even small amounts of power to remote villages are magnified because so little electricity is needed to raise the quality of life. Two 7-meter turbines, which would supply only two homes with electric heat in the United States, can pump safe drinking water for a village of 4000 in Morocco.

The typical village system might use two or more wind machines, batteries, inverter, and backup generator (see Figure 11-8). And like hybrid home light plants, village power systems could also include a solar array. The key is to use as much power as possible directly, instead of storing it in batteries and running it through an inverter. This reduces both initial cost

VILLAGE ELECTRIFICATION

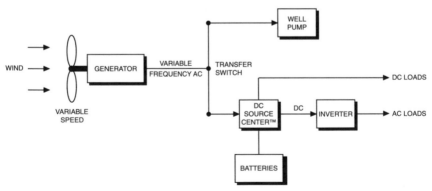

Figure 11-8. One scheme for using small wind turbines to serve a variety of loads in a village electrification program.

and complexity, while delivering more of the wind system's energy to do useful work.

Consider the Mexican village of Xcalac on the Yucatan Peninsula. There Bergey Windpower installed a hybrid power system using both solar and wind energy. Bergey erected Mexico's first wind farm, a 60-kilowatt array of six wind turbines, each 7 meters in diameter. They tied the wind turbines and a 12-kilowatt PV array into a large battery bank, and fed output from the batteries to a 40-kilowatt inverter. The entire $500,000 system offset the construction of a proposed $3.2 million power line to the remote village.

If power is used directly to pump water, grind grain, or run other loads not dependent on utility-grade electricity, the need for batteries is diminished. The batteries and inverter then need to be sized only for those loads that must use constant-frequency AC. In a concept conceived by Bergey Windpower the output from the wind turbine is manually switched from the direct loads, such as water pumping, to the batteries and inverter as needed. For example, the operator monitors the water level in a storage tank and the batteries' state of charge to determine where the power should be directed. The operator is also responsible for starting the backup generator when the power system can't meet demand. Eliminating automatic switches decreases the likelihood that a minor component could fail and imperil the entire system. It also ensures that one person is always responsible for operation of the power system.

In the next chapter we'll look at another way to use wind turbines in a stand-alone application: for pumping water.

 12

Pumping Water

For those in industrialized countries served by community water systems, where water (like electricity) is available on demand, it's hard to imagine the importance of a water pump to the world's rural population. Lifting or carrying water accounts for much of the energy expended in Third World villages. It's also a major load for North American homes beyond the reach of utility lines. In fact, settlement on the Great Plains wasn't feasible until experimenters developed a reliable means of pumping water from deep wells. The technology that made settlement possible was the American water-pumping windmill, and it forever changed the face of the landscape.

The European or Dutch windmill, which was used to some extent on the East Coast, wouldn't work effectively on the Great Plains. Unlike the shallow surface sources for which the Dutch windmill was well suited, water on the Great Plains was found deep below the surface. The mechanical farm windmill, for the first time, enabled settlers to pump the relatively low volumes of water needed for domestic uses from deep wells.

The farm windmill was such a perfect match between the needs of settlers and the abundant winds found on the Great Plains that it spawned a huge domestic industry. During the past 100 years U.S. manufacturers have built more than 8 million water-pumping windmills, most for watering livestock. The design proved so successful that it has been widely copied around the world. Today nearly 1 million remain in use, mostly in Argentina, the United States, Australia, and South Africa. Some have been in quasi-continuous operation for more than 80 years. Because it works, the

American farm windmill remains popular even today. After a century of use, it's certainly a proven technology.

Mechanical Wind Pumps

Farm windmills, like modern wind turbines, have their own arcane vocabulary and obscure methods for estimating performance. The following section first examines the factors affecting the pumping capacity of farm windmills, then compares this with that available from modern wind turbines. The description of the American or *classic* multiblade windmill technology uses English units. However, tables for estimating the pumping capacity of modern wind turbines use both the English and the metric system because the technique is applicable worldwide.

Pumping Head

The energy needed to pump water is a function of the volume, and the height, or *head*, it must be lifted (see Figure 12-1). It takes as much energy

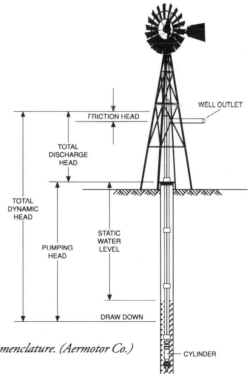

Figure 12-1. Pumping head nomenclature. (Aermotor Co.)

to lift 10 gallons of water 10 feet as it does to lift 1 gallon 100 feet. Sizing a wind machine to meet the need for electrical energy is simply a matter of estimating electrical consumption. But sizing a wind pumping system requires not only an estimate of how much water is used, but how far the water must be pumped from the well to its point of use.

The *total dynamic head* includes not only the distance the water must be lifted from the well, but also the height the water must be lifted to fill any above-ground storage tank (the *discharge head*). The dynamic head also includes any energy lost due to friction in the pipes (*friction head*). Bear in mind when estimating head that the static water level in any well will be lowered or drawn down once pumping begins. The depth the water level falls in the well depends on the pumping rate. Thus, the pumping head is always greater than the static head.

Friction in pipes is directly proportional to the rate of flow. You can transfer a given amount of water with large-diameter pipes and a low flow rate, or small-diameter pipes and a high flow rate. For example, 1-inch pipe may be less costly than 2-inch pipe, but a pump must overcome more than 20 times more friction in the smaller pipe at a flow rate of 10 gallons per minute. Use Table 12-1 to estimate the approximate friction head in the pipes serving a rural water system in North America.

For example, let's assume that you plan to install a farm windmill on a well 100 feet deep. You expect to pump about 10 gallons per minute. At

Table 12-1

Friction Head for Iron Pipe in the United States

Loss of Head in Feet per 100 Feet of Pipe

Water Flow Rate (gallons/ minute)	Pipe Diameter (inches)				
	1	1.5	2	2.5	3
5	3.25	0.4	-	-	-
10	11.7	1.43	0.5	0.17	0.07
15	25	3	1.08	0.36	0.15
20	42	5.2	1.82	0.61	0.25

Loss of head in feet for 90-degree elbows					
	6	7	8	11	15

(Adapted from Aermotor Co.)

this rate the pump will draw down the static water level 5 feet. Let's say you plan to store water in a tank with an inlet 5 feet above the ground up a small hill with a 5-foot gain in elevation. Now add the final details: a 100-foot run of 1-inch pipe that will need three 90-degree elbows to route the water into the storage tank. Table 12-2 sums the total dynamic head for this wind-pumping system.

Table 12-2

Sample Calculation of Total Dynamic Head

	Head (feet)
Static head	100
Drawdown	5
Discharge head	10
Friction head	
pipe	12
elbows	18
	—
Total dynamic head	145

Unless the tank is below the well outlet, you'll need to install a *packer head* to pump water into a storage tank. The packer head enables the farm windmill to pump water above the height of the well. In the past, farm windmills used for providing domestic water sometimes had storage tanks built right into the towers; others delivered water to a nearby tank on a raised platform. In either case, a packer head is needed when the outlet is above the well opening.

Estimating Farm Windmill Pumping Capacity

Early farm windmills directly coupled the wind wheel (rotor) via a crank to the sucker rod. The sucker rod lifts the column of water in the well. Thus, the windmill would lift the sucker rod every revolution of the rotor. In light winds the weight of the water in the well would often stall the rotor, bringing it to a halt.

Today, most if not all mechanical wind pumps are back-geared (see Figure 12-2). These windmills use a transmission to increase the rotor's mechanical advantage in light winds. Most back-geared windmills lift the sucker rod once every three revolutions. This increases the farm windmill's complexity but enables it to pump water more reliably in light winds.

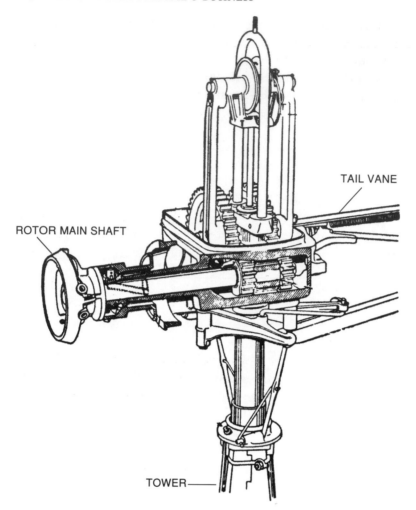

TAIL VANE

ROTOR MAIN SHAFT

TOWER——

Figure 12-2. Most farm windmills are back-geared, that is, the rotor turns several revolutions per pump stroke. (Aermotor Co.)

American farm windmills are available in sizes ranging from 6 feet to 16 feet (about 2-5 meters) in diameter. Australia's Southern Cross can be ordered in sizes up 25 feet (about 8 meters) in diameter. The most common size in the United States is the 8-foot mill, which is capable of pumping less than 10 gallons per minute (2270 liters per second) from depths of 100 feet (30 meters). How much they can actually pump has, until recently, remained a mystery.

Estimating the amount of water that a farm windmill might deliver is even a darker art than estimating the electrical output from a wind generator. Standard windmill pumping tables are based on instantaneous wind speed, pump cylinder diameter, and the depth of the well. These tables, which were probably derived during the late nineteenth century (no one knows for sure), give you little idea how much water can be delivered within a given wind regime. They do illustrate, though, that the performance of the farm windmill is strongly influenced by the relationship between the windmill, the pump, and the pumping head. Too big a pump and the windmill will stall; too small a pump and the windmill will operate less efficiently than it might otherwise.

Table 12-3 is adapted from Aermotor's table of standard pumping capacities and pumping heads for winds 15-20 mph when the windmill is set for the maximum stroke. The common 8-foot windmill, when matched with a well cylinder 2 inches in diameter, will pump about 3 gallons per minute from a well about 140 feet deep in winds 15-20 mph. According to Aermotor, the pumping capacity remains the same for the same size well cylinders among rotors 8-16 ft in diameter. By varying the pump's *stroke*—its vertical travel—Aermotor uses the increased power from the larger diameter rotors to increase the height through which the water can be lifted. A 12-foot farm windmill will pump water at the same flow rate as the 8-foot rotor, but will lift the water through more than twice the total dynamic head.

Table 12-3

American Farm Windmill Pumping Capacity

(15-20 mph wind speed)

Pump Cylinder Diameter (inches)	Flow (gallons/ minute)	Flow (liters/ minute)	Maximum Total Pumping Head in Feet for Wind Wheel Diameter (feet)				
			8	10	12	14	16
2	3	12	140	215	320	460	750
2.5	5	20	95	140	210	300	490
3	8	30	70	100	155	220	360
3.5	11	40	50	75	115	160	265
4	14	52	40	60	85	125	200
4.5	18	66	30	45	70	100	160
5	22	82	25	40	55	80	130

Adapted from Aermotor Co.

For any given rotor, the stroke can be adjusted to vary the proportions between the volume pumped and the head. Adjusting the windmill to the short stroke decreases the volume that can be pumped by one-fourth, but enables the windmill to work against a one-third greater head. For the same head, the shorter stroke will permit the rotor to start pumping in lighter winds.

Nolan Clark at the U.S. Department of Agriculture's (USDA's) experiment station in Bushland, Texas, undertook a series of tests to determine just how much water the farm windmill could pump under standard conditions. He and his associates found that on average the typical 8-foot diameter windmill can pump about 1-2 gallons per minute (4-8 liters per minute) under conditions found on the Great Plains.

Table 12-3 is the farm windmill equivalent of a wind generator's power curve. Unfortunately, it's for only one wind-speed range, not for the full range of wind speeds needed to estimate production anywhere the wind isn't blowing constantly at 15-20 mph. However, Alan Wyatt at the Research Triangle Institute has devised a simple formula for calculating the potential output from mechanical wind pumps operating at various average wind speeds and pumping heads when the windmill is properly matched to the pump.

The traditional farm windmill, because the blades are not true airfoils, is much less efficient than modern wind turbines. The overall operating efficiency of farm windmills is only 4-8 percent. Assuming an overall operating efficiency of 5 percent and a Rayleigh speed distribution, Wyatt's formula is,

$$\text{daily or monthly volume in m}^3 = \frac{0.4 \times D^2 \times S^3}{H}$$

where D is the rotor diameter in meters, S is the average daily or monthly wind speed in m/s, and H is the total pumping head in meters. Table 12-4 summarizes the calculations for an 8-foot diameter farm windmill. For a site with an average wind speed of 11 mph (5 m/s) at a hub height of 40-50 feet (12-15 meters), an 8-foot mill will pump about 2600 gallons (10 cubic meters) per day from a well approximately 100 feet (30 meters) deep. Similar tables for rotors 10-16 feet in diameter are given in Appendix F. Like the estimates of AEO using swept area, these tables of expected pumping capacities of farm windmills are only crude approximations of what may actually occur.

Table 12-4

*Approximate Daily Output for
American Farm Windmill*

8-foot Diameter Rotor in cubic meters/day (gallons/day)

		Average Annual Wind Speed m/s (approx. mph)								
		3 (7)		4 (9)		5 (11)		6 (13)		7 (16)
Pumping Head										
(m)	(ft)	(m³) (gal)		(m³) (gal)		(m³) (gal)		(m³) (gal)		(m³) (gal)
10	30	6	1,700	15	4,000	30	7,900	51	13,600	82	21,600
20	70	3	800	8	2,000	15	3,900	26	6,800	41	10,800
30	100	2	600	5	1,300	10	2,600	17	4,500	27	7,200
40	130	2	400	4	1,000	7	2,000	13	3,400	20	5,400

Source: Center for International Development, Research Triangle Institute
Assumes overall efficiency of 5 percent.

New Technology for Mechanical Wind Pumps

The American farm windmill no longer has a monopoly on wind pumping. Today there are more options available, from modern mechanical wind pumps to modern wind-electric pumping systems.

Experimenters have developed two devices for improving the operation of the traditional farm windmill. The simplest is a counterbalance to the weight of the sucker rod. Farm windmills tend to speed up when the sucker begins its downward journey. On the upstroke the rotor slows down as it lifts both the weight of the rod and the weight of the water in the well. (This is easiest to see in light winds.) The change in speed changes the tip-speed ratio of the rotor and its efficiency. To steady the speed of the rotor and maintain a more optimum relationship between the rotor and wind speed, some designers have added weights or springs to counterbalance the weight of the sucker rod.

Another approach tackles a more fundamental problem with wind pumps. As noted before, the power in the wind increases with the cube of wind speed. But the pumping rate of mechanical windmills varies linearly. For a given pump size and depth, if the stroke of the windmill is adjusted for optimum production in high winds, the windmill will perform poorly, if at all, in low winds. Ideally the stroke should vary with wind speed to more closely match pumping capacity with the power available. In tests a variable-stroke design by Don Avery of Hawaii has doubled the output from the traditional farm windmill.

Though these innovations sound appealing, neither mechanism is widely used. Manufacturers haven't adopted the technology and they continue building farm windmills the same way they have for the past 100 years. Dutch researchers, on the other hand, have successfully designed modern versions of the farm windmill. The modern wind pumps developed by CWD in the Netherlands use only 6-8 blades of true airfoils in contrast to the 15-18 curved steel plates found on the American farm windmill. As a result they use fewer materials to do the same job, and are simpler and less costly to build than the American farm windmill. Both aspects are important for Third World countries such as India. Though nearly twice as efficient as the traditional design, modern wind pumps haven't proved as rugged. They may be best suited for regions with light winds.

More promising than advancements in mechanical wind pumps has been the development of new wind-electric pumping systems. For the same rotor diameter, today's wind-electric pumping systems can deliver about twice as much water as the traditional farm windmill depending on how well each pump is matched to its load.

Electrical Wind Pumps

Previously the only means for using the wind to pump water without utility power present was installation of a multiblade farm windmill, or a complete battery-charging wind system and electric well pump. But Bergey Windpower, in conjunction with the USDA, has spearheaded the commercial development of an innovative variant to wind-electric pumping systems (see Figure 12-3).

Their approach frees the wind turbine from the need for batteries and inverter by coupling a modern wind turbine directly to a well pump. When wind is available, the wind turbine drives the pump at varying speeds, pumping more in high winds than in low winds. This system stores any energy surplus during high winds as water in a storage tank rather than as electricity in batteries.

Direct wind-electric water pumping simplifies matching the aerodynamic performance of the wind turbine to water pumping by varying the load electrically instead of mechanically changing the stroke of the farm windmill. There's also greater siting flexibility. The farm windmill must be

Figure 12-3. *Wind-electric pumping. Modern wind generators can be used to pump water like their mechanical forebears. The 10-kilowatt wind turbine shown here at Ain Tolba, Morocco, replaced a diesel engine. (Bergey Windpower Co.)*

located directly over the well. This usually isn't the best location for a wind turbine. In hilly terrain water is found at the bottom of swales, whereas the wind is often found on hill crests nearby. In a wind-electric system the wind turbine can be sited where the wind is strongest and the turbine will perform best. The gain in performance more than offsets the cost of the electrical cable to the well.

Like battery-charging home light plants, wind-electric pumping systems are also capable of providing power to multiple loads. These loads may use either utility-compatible power from conventional battery-inverter systems, or the untreated, low-grade electricity produced by the wind turbine. Specialists in village electrification are finding increasing uses for low-grade electricity to run conventional motors at variable speed not only for pumping water but also for grinding, cooling (vaccine refrigeration), and freezing ice (fish storage).

Wind-electric pumping systems are also competitive. Even at low-wind sites, those with 9-mph (4-m/s) average speeds, small wind systems can deliver water at less cost than either PVs, diesel generators, or farm windmills. Table 12-5 compares the installed cost of two wind-electric pumping sys-

Table 12-5

Water Pumping Installed Cost Comparison

	Wind Electric		Wind Mechanical		PV
Size	3 m	7 m	3 m	7 m	20 panels
Flow (gal/day)	6,100	54,000	4,200	25,000	5,600
Cost	$6,200	$22,000	$7,000	$16,000	$8,000
Cost/gallon ($/gal)	1.00	0.40	1.70	0.60	1.40

Assumes average wind speed = 5 m/s, pumping head 100 ft (30 m), 3-meter turbine, 60-foot tower; 7-meter turbine, 80-foot tower.
Assumes 5 sun-hours/day, 50 watts/panel.
Installation and maintenance not included.

tems with two farm windmills and one solar system. The PV system was sized to pump the same amount of water as the small wind-electric system. Based on USDA's tests, Nolan Clark says the 3-meter Bergey, which costs about as much as an 8-foot diameter, water-pumping windmill, will pump nearly twice as much.

Bergey's 7-meter turbine can pump upward of 800 gallons per minute (53 liters/second) or pump against heads of 750 feet (225 meters). It's well suited for the high-volume applications that might be found in Third World villages. Two 7-meter Bergey turbines installed for an international development project in northeastern Morocco now provide four remote villages with three times more water than they ever received in the past from a diesel-powered system. As in the Moroccan project, where one turbine is insufficient to meet the need, several machines can be used at different points in the water distribution system or ganged together in a mini-wind farm.

Table 12-6 lists the pumping capacity of the Bergey Excel for a well 115 feet (35 meters) deep within various wind regimes. Table 12-6 also illustrates the importance of the well pump in determining the volume of water the wind turbine can deliver.

High-speed wind turbines such as the Bergey, Northern Power, and others may be twice as efficient as the farm windmill in pumping water, but they have higher startup wind speeds than mechanical wind pumps due to their low-solidity rotors. In extremely light winds they may deliver less water than mechanical wind pumps. There are also only a few models of wind-electric pumping systems available. Mechanical water-pumping windmills are manufactured in a range of sizes. American farm windmills are available

Table 12-6

Wind-Electric Pumping System

7-meter, 10-kW wind turbine
Total pumping head: 35 m (115 ft)

Annual Wind Speed		Daily Water Delivery			
		Pump A		Pump B	
(m/s)	(mph)	(m³)	(gal)	(m³)	(gal)
4	9	76	20,100	96	25,300
5	11	162	42,800	152	40,100
6	13	260	68,600	204	53,900
7	16	349	92,100	245	64,700
8	18	416	110,000	272	71,800

Source: Bergey Windpower Co.

in 2 foot increments from 6 to 16 feet in diameter. There's even a broader spectrum of PV arrays suitable for water pumping. PVs can be tailored much more closely to the exact water demand than either mechanical or wind-electric wind pumps.

Yet simplicity and versatility favor the wind-electric systems. Modern integrated wind turbines use far fewer moving parts than mechanical wind pumps. And they never need their oil changed. One Bergey Excel has been pumping water at USDA's Bushland station for 3 years without service. Whether they will last as long as their mechanical counterparts, only time will tell.

Estimating Water Use

Finding the right size wind machine for a water pumping application is much like sizing a stand-alone power system to meet your need for electricity. Sizing a wind-pumping system requires an estimate of both average and peak demand. Average demand corresponds to the total volume of water required and is similar to the overall electrical energy generated in a wind-driven remote power system. Peak demand corresponds to the maximum flow rate needed to meet simultaneous water uses, and it resembles the peak power demands that determine the size of inverters in a home light plant.

Water consumption is as much a reflection of lifestyle as electricity, and each person's use varies with habit, culture, climate, and the availability of water. Third World villages consume as little as 10 gallons per person per day because water must be pumped by hand and often carried some distance (see Table 12-7). Urban dwellers, because water is often more plentiful, may use three times as much as Third World villagers. Americans use considerably more.

Table 12-7

Total Average Daily Water Use

	liters/person	gallons/person
Third World village	40	10
Third World urban	100	30
U.S. average	570	150
U.S. arid regions	950	250

As late as the mid-1970s, design manuals suggested that a rural homestead in the United States could manage with 50-100 gallons per person per day. By the late 1980s average consumption had risen to 150 gallons per person per day. In arid regions of the United States, like southern California, planners assume each household will use no less than 250 gallons per person per day, not including the water needed for lawns, gardens, and ornamental plants. Watering a lawn or garden is little different than irrigated agriculture. Both require a lot of water. Watering with just one garden hose may use 300 gallons per hour. Livestock watering and other farm uses may also add dramatically to water requirements (see Table 12-8).

To estimate your average water requirements, sum the typical uses in Table 12-8 and multiply by the number in your household, the number of livestock you expect to raise, and the size of any lawn or garden you plan to water frequently. The USDA estimates that a typical farm may require up to 6000 gallons per day, including domestic uses.

Most homesteads without livestock or intensive gardening won't require nearly as much. Yet a family of three in an arid region of the Great Plains could quickly exhaust the daily supply from a standard farm windmill during the light winds common in late summer. If each person uses 250 gallons per day, and the family tills a large garden (100 feet by 100 feet) during the summer months, the household will need about 1400 gallons per day. An 8-foot farm windmill (see Table 12-3) at a 9 mph (4 m/s) average wind speed could pump only about 1300 gallons daily from a well 100 feet deep. There's little margin for error in such a system, and storage is necessary to ensure an adequate supply.

Water use by remote homesteads becomes self-limiting. Families learn to limit their water consumption in part because they don't have unlimited power for pumping, or unlimited storage. Designers of rural water supplies have observed that people use far less water if they have to pump it by hand or carry it from a well. Those using a stand-alone power system become more sensitive to water conservation if they have to start a noisy backup generator to water their lawn. Like electricity, it pays to conserve. The less water used, the smaller the wind-pumping system necessary to meet your needs, or the more surplus that can be stored for days of light winds. For wind-electric pumping and for conventional remote power systems where batteries power the well pump, if less energy is devoted to water pumping, more can be devoted to electrical appliances. Many nonfarm families today live comfortably beyond utility lines on less than 100 gallons per person per day. These modern-day pioneers have demonstrated that you can enjoy life in the country with modest water and power requirements.

Table 12-8

Daily Water Use

Farm Animals	liters/animal	gallons/animal
Horses	50	13
Dairy cattle	70	19
Steers	60	16
Pigs	20	5.3
Sheep	10	2.6
Goats	10	2.6
Chickens	0.3	0.08

Household Uses	liters/ person	gallons/person
Bath tub/filling	130	35
Shower/use	90–230	25–60
Flush toilet/use	10	2–7
Dish washing, hand/day	80	20
Dish washing, auto./day	40–80	10–20
Water softener/use	570	150
3/4-inch hose/hour	1,130	300
Laundry/use	110–190	30–50
Misc. uses/day	90	25
Lawn/Garden (1000 ft²)	2,270	600

Adapted from Sandia National Laboratories, Heller-Aller Co., and *Planning for an Individual Water System*, AAVIM

In the example of the family given earlier there's clearly a need for storage. Storing water in a large tank will provide a backup supply when winds are light and, just as importantly, will provide an emergency supply for fire protection.

Storage

Storing water in a tank is cheaper and more reliable than storing energy for pumping in batteries (see Figure 12-4). A storage tank also provides the fire protection that batteries can't. For rural areas in the western United States, fire protection is a critical requirement for any water system. Fire protection will often determine the storage needed and the maximum instantaneous demand placed on the water supply.

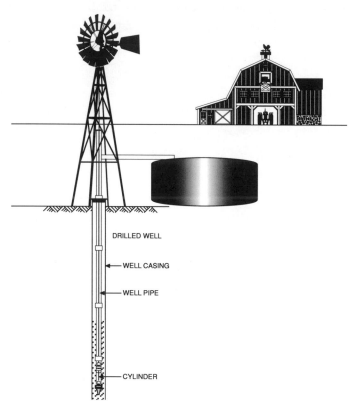

DRILLED WELL

WELL CASING

WELL PIPE

CYLINDER

Figure 12-4. Farm windmill and storage tank. Storage is a necessary part of a wind pumping system whether it's for watering livestock or serving domestic needs. (Aermotor Co.)

Proper water system design requires both the average water usage and the water demand, or the flow rate at any one instant. The number of faucets that you expect to use at any one time plus other uses, such as a shower or washing machine, governs the maximum flow rate the water system must be capable of meeting. For stand-alone power systems in the past, pumps were sized accordingly. The pump would then operate whenever water was needed, drawing power from the batteries.

In mechanical systems, the farm windmill alone can't reliably provide the pressure to meet the flow requirement because wind speed and the pumping rate vary. In the past, more or less constant pressure was provided by storing water in a tank on an elevated platform. The tank provided a modest amount of pressure by gravity for the limited water uses common in the pre-RFA era.

Today nearly all rural water supplies in the United States use accumulators or pressure tanks to provide the constant pressure demanded by contemporary consumers. Where fire protection isn't necessary, the storage provided by pressure tanks reduces cycling of the well pump, allowing the pumps to operate more efficiently than if they operated continuously whenever water was required. Pressure tanks in the United States are available in a range of sizes, storing as little as 2 gallons to more than 1000 gallons.

Because pressure tanks must be charged by an electric well pump, they're not well suited for a true gravity-fed water supply typically used with mechanical windmills. But a farm windmill can be used to pump water as available into a storage tank. The rural water system can then draw water from storage and pump it into a pressure tank with an electric pump driven by a stand-alone power system (see Figure 12-5). Some have adapted electrically powered pump jacks to the reciprocating well pump used by the farm windmill. Thus, if water needs to be drawn from the well and the winds are light, the remote power system switches on and drives the pump jack.

Ample storage shouldn't be avoided. Some building codes and insurance carriers may require it for fire protection. Richard Perez of *Home Power* reports that in northern California fire codes may even require the addition of engine-driven pump jacks to ensure an adequate flow during emergencies.

Galvanized steel tanks are a common site at rural homes in arid regions of the United States. Homesteads on the Mojave Desert of southern California store a minimum of 2500 gallons in unpressurized above-ground tanks, many prefer 5000-gallon tanks (about 20 cubic meters). In colder climates, cement cisterns are buried below the frost line to prevent freezing.

HYBRID WIND PUMPING SYSTEM

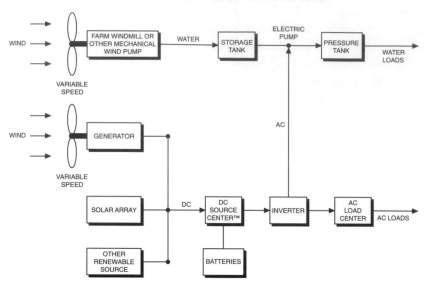

Figure 12-5. *Hybrid wind pumping system. Most homes today require greater pressure than that provided by gravity-fed water systems.*

Ken O'Brock of O'Brock Windmill Distributors says this is a common practice among the Amish of Pennsylvania and Ohio where operating farm windmills are still common.

For a comprehensive guide to designing a rural water system, consult *Planning for an Individual Water System* by the American Association of Vocational Instructional Materials (see Appendix I for details).

Irrigation

Many applications for pumping water, notably irrigation, require considerably more capacity than the farm windmill alone can supply. Irrigated crops, intensive gardening, and large lawns use substantial volumes of water as seen in Table 12-9. High-discharge irrigation wells on the Texas Panhandle, for example, may pump upward of 1000 gallons per minute (3.8 cubic meters per minute) from depths approaching 400 feet (120 meters). The American farm windmill isn't big enough to dent such a load.

Modern medium-sized wind machines can be used for irrigation by either pumping water mechanically or by producing electricity to run a

Table 12-9

Daily Irrigation Use

Irrigated Crops	m³/ha	gal/acre
Village farms	60	6,400
Rice	100	10,700
Cereals	45	4,800
Sugar cane	65	6,900
Cotton	55	5,900
Lawn/Garden	240	26,100

Adapted from Sandia National Laboratories, and *Planning for an Individual Water System*, AAVIM

large well motor. Because the wind is intermittent, and irrigation requires such large volumes of water that storage isn't often practical, the wind machine typically drives the well pump in conjunction with a conventional energy source. Coupling wind turbines with conventional sources is a fairly simple task with an electric well pump, but it isn't quite as easy with the engine-driven pumps commonly used on the southern Great Plains.

West Texas State University, in cooperation with USDA, has developed an ingenious device for mechanically coupling the varying output of a wind machine with an irrigation pump: the overrunning clutch (see Figure 12-6). When the wind is strong and the wind machine is producing at full output, it mechanically drives the well pump entirely on its own, via the overrunning clutch. When the winds are weaker, the wind machine assists in driving the pump. The conventional power source makes up the difference. During a calm spell the conventional power source operates the pump alone. The overrunning clutch ensures that a constant volume of water is pumped regardless of wind speed. It also harnesses the wind to reduce the consumption of conventional sources. This concept has been successfully tested at USDA's experiment station near Amarillo, Texas, for several years.

Using the wind to assist irrigation pumping is even simpler when the pump is driven by an electric motor. Wind machines producing utility-grade electricity can be connected directly to the well motor in a configuration identical to that for an interconnected wind turbine at a home or business. When wind is available, the wind machine offsets consumption of electricity from the utility. If the wind turbine produces a surplus of energy during periods when pumping loads are light, the excess is sold back to the utility.

At USDA's Bushland experiment station a 13-meter Enertech E-44

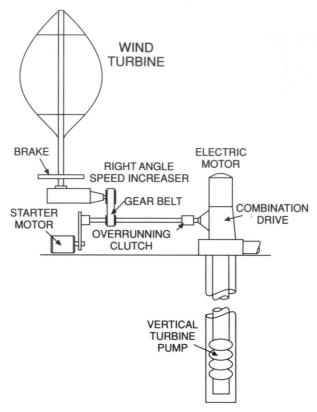

WIND TURBINE

BRAKE

RIGHT ANGLE SPEED INCREASER

ELECTRIC MOTOR

GEAR BELT

STARTER MOTOR

COMBINATION DRIVE

OVERRUNNING CLUTCH

VERTICAL TURBINE PUMP

Figure 12-6. Wind-assisted irrigation. The Darrieus turbine mechanically assists a well pump via an overrunning clutch. (West Texas State University)

generates about 90,000 kilowatt-hours per year with wind speeds at hub height averaging 7.1 m/s (16 mph). This turbine has been in use at Bushland since the early 1980s. Nolan Clark estimates that wind machines, like the E-44, in a wind-electric pumping system will operate 65 percent of the time in the Texas wind regime. With the addition of solar-electric pumping, the hybrid could provide water about 85 percent of the time.

Where any single turbine isn't capable of meeting irrigation demand, turbines can be ganged together for collectively driving one or several well pumps. In an interesting application of wind-electric pumping, Bergey Windpower installed four of its 7-meter turbines at a community college near Laredo, Texas, where they power a 40-kilowatt, drip irrigation system.

 **13**

Siting

Farmers, ranchers, and other rural residents should encounter few barriers to erecting small wind turbines on their property. Often siting in these circumstances is simply finding the most exposed place for the wind turbine. Problems can arise in suburban and more densely populated areas where some neighbors may not share your enthusiasm for wind energy. With care, consideration, and a good measure of patience, you should be able to meet any neighbor's concerns. Nevertheless, you should always make an honest appraisal of your site. You may find it unsuitable for wind energy because of physical constraints (too many trees and tall buildings, for example), or because of legal restrictions on how you can use your land.

Physical Restrictions

The foremost question is, do you have enough room? Not only must there be sufficient space for the tower but you must also have the room to install it. Manufacturers of small wind turbines in the United States generally recommend at least a 1-acre (0.4-hectare) lot for 10-kilowatt machines and larger. Smaller turbines need far less space. For example, guyed towers for small turbines have been installed on city lots so small that the anchors were placed in each corner of the backyard, and free-standing towers have been installed in equally cramped quarters. In one case the crane lifted the tower over the house to set it on the foundation. All of which is to say you can install a wind machine almost anywhere. But there are limits, and it's wise to know what they are.

Exposure and Turbulence

Wind turbines should always be located as far away from trees, buildings, and other obstructions as practical to minimize the effect of turbulence, and maximize exposure to the wind. *Turbulence* (rapid changes in wind speed and direction) is caused by the wake from buildings and trees in the wind's path, and resembles the eddies swirling around a rock in a stream. The buffeting by turbulence can be damaging to modern wind turbines because they use long slender blades operating at high speeds. Turbulence can wreak havoc on a wind machine, rapidly shortening its life.

Buildings and trees also drastically reduce the energy that is available to a wind machine. One overriding lesson that has been gleaned from nearly two decades of working with modern wind turbines is that you can't overlook the effect of obstructions whether buildings or vegetation. Though seemingly more transparent to the wind than a building, trees, shrubs, and even low hedgerows can rob energy from the wind.

When you're uncertain about the amount of turbulence over your site—go fly a kite. It's a practical and inexpensive means for detecting turbulence. Tie streamers to the kite string and note how they flutter in the wind (see Figure 13-1). Check turbulence this way for winds from several different directions until you're satisfied you've covered all possibilities. Note the site where turbulence is minimal. That's where you'll want the tower.

Locate the tower either far enough upwind or far enough downwind to avoid the turbulent zone around nearby obstructions (see Figure 13-2). When this is impractical, use as tall a tower as possible to elevate the wind

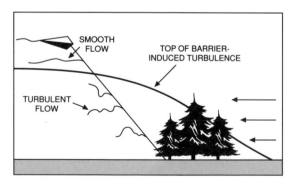

Figure 13-1. Trailing streamers from a kite is a simple yet effective way of detecting turbulence. (Battelle PNL)

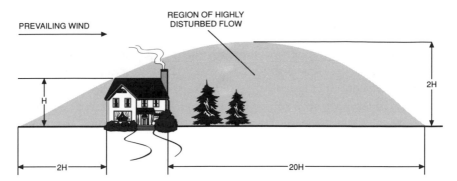

Figure 13-2. *Zone of disturbed flow. Wind speeds decrease and turbulence increases in the vicinity of obstructions. The effects are most pronounced downwind but also occur upwind as the air piles up in front of the obstruction. The flow over a hedgerow or group of trees in a shelter belt is disturbed in a similar manner.*

machine above the turbulence. If neither approach alone is sufficient, use some combination of siting and a taller tower.

From years of experience, installers have evolved a general rule of thumb: the entire rotor disk of the wind turbine should be least 30 feet (10 meters) above any obstruction within 300 feet (100 meters). If you've determined, for example, that a group of trees along a fence row are 60 feet (20 meters) tall you'll need at least a 90-foot (27-meter) tower. To ensure best performance you should use an even taller tower.

By all means avoid sites at the bottom of creeks, draws, or ravines and at the foot of hills. If there's a hill on your property with a well-exposed summit, site the wind machine there instead of lower on the slope, even if the summit is some distance from your house.

For more information on siting, consult Battelle Pacific Northwest Laboratory's *A Siting Handbook for Small Wind Energy Conversion Systems* (see Appendix I).

Power Cable Routing

Once you've selected the area where the tower will be erected, note how the power will be delivered to your house. Will you bury the power cables or run them through the air on poles? At this stage, you need to anticipate any problems that may develop later. They're easier to avoid than to solve. For example, a buried telephone line crossing your path may preclude digging a trench for an underground cable or at least make it more complicated.

Similarly, you may have trouble stringing an aerial cable if the electric utility's lines cross your path.

Ideally, the electric service from the wind machine should enter the building near where the utility's lines enter. If the service entrance and meter are on the other side of the house from where you are planning to erect the tower, how will you route the power cables? It's easier to route them around an obstacle in a trench than it is when stringing them from poles.

Are there any sidewalks, driveways, or roads in your path? How will you cross them? These are important questions because the answers affect the cost of installing the wind system. They also determine how difficult it will be to meet certain institutional restrictions such as the National Electrical Code™ in the United States. If you're planning, for example, to string the power cables across the roof of a building, they must have a specified clearance on all sides to meet the code's requirements. (Wiring and code requirements are treated more thoroughly in Chapter 14.)

Institutional Restrictions

Equally as important as finding the optimal site for the wind system is determining what legal requirements your locale places on structures such as wind turbines. In the United States land use zoning, building codes, and protective covenants may all apply.

Zoning

Most who have installed small wind machines in the United States have had little, if any, problems with zoning restrictions. Either their property was not covered by zoning ordinances, or permission was quickly and easily obtained where it was. Many rural areas are not zoned at all, and where they are, there are practically no restrictions on land that is zoned for agriculture. The situation changes as you near cities, small towns, and residential neighborhoods. There the right to swing your fist ends where your neighbor's nose begins.

Zoning (or more broadly, restrictions on land use) is a responsibility entrusted to local governments by the public to protect the general health and welfare. Officials will want you to show how your use conforms to the public's general agreement on what can and can't be done on land within your zone. Zoning officials have an obligation to treat you fairly. Above all they shouldn't discriminate against you because they're unfamiliar with wind turbines. Treat them cordially. One thing is certain, if you need a building permit or a zoning variance, you want them on your side.

Building Permits

Where zoning ordinances apply, you must conform to the law—period. Find out what the requirements are in your area by calling the local building inspector, board of supervisors, or planning office. You want to know what is necessary to obtain a building permit (where required) and who is responsible for issuing it (usually the building inspector). Get details. Whoever is responsible should provide a list of what you must do: the forms to fill out, the fees to pay, where and when to file, and any other information you must supply. Then methodically go through it.

The intent of this process is to determine conformance with the regulations governing your zone and to alert the public to your project. Take the initiative and contact anyone who might be affected, especially your neighbors. You have a responsibility to tell your neighbors what you're planning and why. Speak to them early in the project so that they feel consulted, rather than pressured into backing you. It's much better to talk with them informally over the back fence than in court or in a shouting match at a public hearing. If you get along well, there should be few problems, but if you've driven over your neighbor's prize rose bush for years, you'd better make amends. Objecting to your building permit is a great opportunity to even the score. You can head off conflict by respecting the needs of your neighbors. Treat them in the same way you would like them to treat you.

The building inspector will probably require a plot plan (map) showing the dimensions of your lot and where the tower will be located. You can prepare this yourself. Drawings of the wind machine, tower, and foundation with their specifications may also be required. The dealer or manufacturer can supply these.

Zoning ordinances follow either of two approaches. One allows you to do whatever you want, unless specifically prohibited. The other approach prohibits you from erecting any structure unless it is specifically permitted. Where the latter approach is used, your building permit could be denied simply because no one has ever installed a wind machine before.

In communities where this is the situation, you can sometimes get permission for a wind machine by bringing it under a permitted category such as lightweight radio towers, TV antennas, or chimneys. The building inspectors may be empowered to make such a determination. If not, formal action of the zoning appeal board is necessary. They must determine if your use conforms to the intent of the ordinance. Where it doesn't or where the ordinance specifically excludes wind machines or similar structures, you must obtain a variance from the regulation.

The zoning appeals board or board of adjustment is the arbiter of zoning disputes. They're a political body, and if there's a public outcry, they'll respond accordingly within the limits of the law. Variances—variations from the law—give the zoning appeals board flexibility in meeting the goal of zoning: the protection of the common good without undue restrictions.

They'll want to know whether your wind machine detracts from your neighbors' use of their land and whether it poses a threat to your neighbors or the community. You must convince them that it won't. The burden of proof is on you, the petitioner.

Often the granting of a variance is little more than a formality. You may not even need to be present. But if the board has questions that you have not answered previously, or if the variance is contested, you'll need to be present and you'll need to be well prepared. On occasion the unfamiliarity of wind turbines will fuel wild speculation about what they will do to the neighborhood. Often these fears can be quickly dispelled with the facts. Sometimes they can't. When contested, the public hearing can take on the appearance of an expensive courtroom battle with opponents bringing in their own "expert witnesses" to counter your assertions. It can be rough, even humiliating, if you're unprepared.

In suburban housing developments or planned communities there may also be restrictions (covenants) in the deed on how the land can be used. These restrictions are intended to preserve the identity of the neighborhood. Take a look at your deed. Or call your attorney, realtor, or mortgage company for information. If there are any restrictions, they'll know how to best deal with them. For example, the restrictions may be unenforceable.

Also note the location of any easements on your property for utility rights-of-way. In the United States, easements transfer use of the land without transferring outright ownership. Easements are commonly used for a host of public purposes: power lines, underground telephone cables, pipelines, future roads or sidewalks, and so on. These could all limit further use of the land. You may be unable to encroach on these easements with your wind turbine even though you own the land, there are no restrictive covenants, and you obtained all the proper zoning approvals.

Height Restrictions

The most frequent zoning problems in the United State involves height restrictions, setback requirements, and neighborhood concern about safety and noise. In most residential zones, there's a limit to the height of structures, usually 35 feet (11 meters), a relic of the days when fire brigades had to pump water by hand. Variances to such ordinances can be obtained by pointing out other structures taller than the limit that have been allowed under the zoning ordinance: radio towers, chimneys, or utility poles. (Local officials seldom have control over utilities.)

When you're within a mile from an airport or heliport in the United States, the Federal Aviation Administration (FAA) may impose lighting or special painting requirements. Normally, your wind turbine's potential interference with aircraft is never questioned—wind machines operate well below the height of aircraft—and no lights or painting are needed. You are required to report the existence of your tower to the FAA only when it exceeds 200 feet in height or when it's within the flight path to a nearby airport. If there is any doubt, the building inspector may forward a notification to the FAA or advise you to do so.

Safety

In some communities, towers must be set back from the property line a distance equal to their height for reasons of safety. If your lot is too small to permit this, you may want to reconsider wind power. Unfortunately, this restriction discriminates against wind machines in comparison to other common structures.

We think nothing of other man-made and natural hazards that pose a risk similar to if not greater than a wind turbine. We've all seen homes sheltered beneath the branches of an old oak tree where occasionally a storm-weakened limb crashes down onto the roof. We accept this hazard as the price we pay for the benefits the tree provides (shade and appearance).

The same is true for lightweight radio and TV towers. In many ways they are similar to towers for wind machines. They are made of metal and extend visually above the roof line. The public has grown to accept them, and because their failure rate is so low, users often mount them adjacent to occupied buildings.

You have the right to install a wind system on such a lot just as you also have the right to grow a tree of equal height without setback restrictions. If this is your desire, you must convince the zoning board that the wind machine does not pose any unusual hazard to you and your neighbors.

The zoning board will be concerned that your tower could collapse. You must show them that the tower meets certain design standards (as indicated by the manufacturer), and applicable building codes, and that similar towers operate throughout North America in a host of severe environments without incident. Though towers have failed, the occurrence is rare and far less frequent than that of falling trees or utility poles.

The board may also be concerned that the wind machine could throw a blade. Once again, you must convince them that the wind machine has been designed and built to accepted standards and that there's little likelihood that it will throw a blade into the midst of a neighbor's lawn party. You can best reassure them that your wind turbine won't become airborne by citing the number of similar turbines operating elsewhere, and the number of years these turbines have operated without incident. Thus, it behooves you to select a wind system with a proven track record: one where a host of units have operated reliably in a variety of applications for several years.

The fear that the tower could become an "attractive nuisance" is a related concern. Generally a property owner is not liable for accidents to trespassers, but a different test is applied to the acts of children. Swimming pools are thought to entice or attract children to trespass. Because children cannot discern the hazard presented by the pool, the community views it as a public nuisance, and if an accident occurs, a court can hold the owner liable. Zoning ordinances permit attractive nuisances when they have met requirements designed to prevent accidents. Swimming pools must be fenced, for example. The same ordinance may require that climbable towers, such as wind turbine towers, be fenced as well.

Fencing isn't the only way to prevent someone from climbing a wind turbine tower. Electric utilities seldom use fencing. On their transmission towers they simply remove the climbing rungs to a level 10 feet (3 meters) above the ground. You can do the same. Or you can wrap the base of the tower in sheet metal or wire mesh. These alternatives should be acceptable to a zoning board, and accomplish the same goal as fencing less obtrusively and at less cost. Utilities seldom erect fences around their utility poles or transmission towers. Imagine the outcry if every utility pole required a fence.

Aesthetics

For some the appearance of a wind machine on the skyline is symbolic of responsible stewardship—a step toward a sustainable future. To others it's a call to arms. Concern about the visual effect wind machines may have on a neighborhood should not be dismissed lightly.

Try to incorporate the community's wishes when you're considering the type of tower to use. There's much less objection to the clean lines of a tapered tubular tower than to a wooden utility pole. Likewise, a tubular tower is more pleasing in foreground views than a truss tower. However, truss and lattice towers should be considered. In distant views, guyed lattice towers and truss towers become nearly invisible. From an aesthetic perspective, the type of tower that's most acceptable (whether truss, guyed lattice, or tubular) depends on the viewpoint and the distance between the observer and the tower. There are no definitive aesthetic guidelines for small wind turbines. (There are for wind power plants, but the two applications are quite different.)

If someone objects on aesthetic grounds, point out similar structures on the horizon that we've learned to tolerate. You have as much right to erect a wind machine as the local radio station to install a tower or the utility to string a transmission line across town. While it's true that the utility's power lines and the radio's broadcast tower provide community-wide benefits, each person benefits individually. Your installation of a wind system differs little from the utility building a power line to your house. The appearance may differ, but the purpose remains the same.

Noise

Noise is another common concern. This concern is fueled in part by old reports of noisy wind turbines that were installed in California's San Gorgonio Pass during the early 1980s. Happily, the machines that were the source of the problem are no longer built. Still, there's no escaping the fact that an operating wind machine is audible to those nearby. Whether it's noisy or not is far more difficult to determine. Wind turbine noise, unfortunately, is a field where the technical and the subjective meet head on.

First, some background. Noise is measured in decibels (dB). The decibel scale spans the range from the threshold of hearing to the threshold of pain. (See Table 13-1 for the sound level of common noises on this scale.) Further, the scale is logarithmic, not linear. Doubling the power of the noise source—for example, by installing two wind turbines instead of one—increases the noise level only 3 dB. This fact alone causes more confusion about noise than any other aspect, because a change of 3 dB is the smallest change most people can detect. Thus doubling the noise content from a group of wind turbines may appear horrific on paper, yet it will be

Table 13-1

Typical Sound Pressure Levels (SPL) in dB(A)

	Distance from the Source		
Source	ft	m	SPL
Threshold of pain			140
Ship siren	100	30	130
Jet engine	200	61	120
Jack hammer			100
Inside sports car			80
Freight train	100	30	70
Vacuum cleaner	10	3	70
Freeway	100	30	70
Small (10 kW) wind turbine	120	37	57
Large transformer	200	61	55
Small (10 kW) wind turbine	323	100	55
Wind in trees	40	12	55
Light traffic	100	30	50
Average home			50
300 kW wind turbine	650	200	45
30-300 kW wind turbines	1640	500	45
Soft whisper	5		30
Sound studio			20
Threshold of hearing			0

Sources: *Handbook of Noise Measurement,* General Radio,
European Wind Energy Assoc., and Bergey Windpower Co.

barely perceptible to nearby residents. To double the perceived loudness, the noise level must increase nearly 10 dB.

For most discrete sources, such as wind machines, the distance to the listener is just as important as the noise level of the source. The location must always be specified because sound levels decrease with distance. For example, say the manufacturer provided a reference noise level at 330 feet (100 meters) from the wind turbine, and you planned to install the machine 650 feet (200 meters) from your nearest neighbor. Because the sound level decreases 6 dB for every doubling of distance from a point source, you can expect that your neighbor will hear a noise level about 6 dB less than that measured by the turbine's manufacturer.

The perceived loudness varies not only with the sound level but also with the frequency, or pitch. Our hearing detects high-pitched sounds much better than those low in pitch. The sound of a complex machine like a wind turbine is composed of sounds from many sources including the wind over the blades and the whirring of the generator. Each sound has a characteristic pitch making it distinctive from the others.

When measuring noise we try to take into account the way the human ear perceives sound by using a scale weighted for those frequencies we hear best. The A scale in decibels, dB(A), is most commonly used. This scale ignores those frequencies we can't hear, and emphasizes those that are most noticeable.

Impulsive sounds also elicit a greater response than sounds at a constant level. Kick a garbage can or bang a pair of cymbals and you'll get someone's attention a lot quicker than you will by humming a tune. Wind machines using two blades spinning downwind of the tower make a characteristic "whop-whop" as the blades pass behind the tower. This impulsive sound and its effect on those nearby may be missed by standard A-weighted measurements. Many of the complaints about wind turbine noise in California have been directed at two-bladed, downwind turbines. If this if the type of wind turbine you're planning to use, you may need to consider the effect from low-frequency impulsive sounds.

Another element is time. Simpler noise ordinances specify a maximum noise level that must not be exceeded. Others weigh the amount of time the noise occurs at various levels. This complicates the task of estimating a wind turbine's noise impact. Unlike the noise from trains or airplanes, which emit high levels infrequently throughout the day, a wind turbine may emit far less noise but do so continuously for days on end. Some may find this aspect of wind energy more annoying than any other. In windy regions the sound may seem incessant. The literature of life on the Great Plains is full of references to the ever-present sound of the wind.

Our perception of what is and is not noise is also affected by other subjective factors. If your neighbors object to your wind machine because you never invite them to dinner, they're more likely to find the sound produced by it objectionable than you are. On the other hand, if your community has fought rate increases with the local utility, the sound of your wind machine whirring overhead may warm their hearts.

The noise generated by a wind turbine must always be placed within the context of other noises around it. If you live near an airport or a busy highway, your wind machine will hardly create a noise problem no matter how loud it is. Another example is the wind itself. We site wind machines where it's windy. There's always going to be noise from the wind whenever the wind machine is operating because the wind rustles the leaves in nearby trees or sets power lines humming.

Studies of the ambient or background noise from trees have found noise levels of 51-53 dB(A) at 40 feet (12 meters) in winds of 15 mph (7 m/s). At this level the noise from nearby trees can mask the noise from a wind turbine operating in the same winds. For example, consider the problem faced by Bergey Windpower when asked to measure the noise from one of their 10-kilowatt wind turbines. They measured an ambient noise level of 53 dB(A) at wind speeds of 25 mph (11 m/s) near Norman, Oklahoma. At 323 feet (100 meters) from the turbine they measured 54-55 dB(A). Within 600 feet (200 meters) the wind turbine noise of 53-54 dB(A) had approached that of the background.

Despite the masking effect of high winds, a wind turbine will still be audible to those nearby, particularly when they're sheltered from the wind. The sounds emitted by a wind turbine are easily distinguishable from those of the wind. These sounds may not be objectionable, but they are noticeable. Some have compared this situation to that of a leaky faucet. Once you know it's there, you always hear it. The generator or transmission may produce a noticeable whine, for example, or the passage of the blades may generate more discrete sounds. The "swish-swish-swish" of three-bladed rotors is a common sound. Being audible, however, isn't the same as being objectionable. The whir of the compressor in a refrigerator is audible, but few find the sound objectionable.

Manufacturers provide noise measurements in *sound pressure levels* at some specified distance from the wind turbine. In the United States the norm is to report the reference noise level in dB(A) 100 meters (328 feet) from the wind turbine.

Local ordinances typically state the acceptable sound pressure levels in dB(A) at the property line or nearest residence. The European Wind Energy Association estimates that the noise from a 300-kilowatt wind turbine in an 8 m/s (18 mph) wind will drop to 45 dB(A) within 200 meters (600 feet). The aggregate noise from a small wind power plant of 30 such turbines will drop to 45 dB(A) within 500 meters (1600 feet).

But no wind turbine, no matter how quiet, can do better than the ambient noise. It is the difference between ambient noise and wind turbine noise that determines most peoples' response. It should be the objective of the planning authority to limit the increase in total noise. They shouldn't ask for the impossible by demanding that your wind turbine meet a standard that's lower than the ambient noise level.

To summarize, first find the noise level either emitted by the wind turbine or estimated at some distance from the machine. Next determine who will hear the noise, and their location. Then estimate the sound pressure level they will hear. Compare this with any applicable regulation, and the ambient noise under conditions when the turbine will most likely operate. Consider the disposition of your neighbors toward your proposal and anticipate the response you can expect.

Bergey Windpower suggests that if there's any doubt whether or not your wind turbine might disturb someone within ¼-½ mile (500-1000 meters), be a good neighbor and contact them in advance. Advise them of your plans and ask for their comments. Answer their questions as forthrightly as you can and try to incorporate their concerns when designing your installation. Bergey has found that the community's reaction to the noise from a small wind turbine declines after people have had a chance to acclimate to its new sounds.

Wind turbine manufacturers have made great strides in quieting their machines since the early 1980s. They've cut aerodynamic noise on wind turbines of all sizes by sharpening the trailing edges of the blades and employing new tip shapes. They've lessened the mechanical noise of larger wind turbines by isolating the gearboxes from the nacelles and installing sound-deadening insulation. These newer, quieter wind turbines can be good neighbors when sited with care.

TV and Radio Interference

Neighbors sometimes worry that a new wind turbine will disrupt their radio and television reception. This may have been a legitimate concern in the days when some wind turbines used metal blades. That's no longer true. Wind turbines today use blades made only from wood, plastic, or fiberglass.

Of the 25,000 wind turbines in the United States and Europe only a few have had any effect on communications whatsoever. Interference has proven a problem only in remote areas where TV and radio signals are extremely weak. And even in these rare cases the effects have been localized.

In fact, small wind turbines are used extensively worldwide to power remote telecommunication stations for both commercial and military uses. The turbines would never have been selected if there was any hint of interference. There are also more than 5000 wind turbines lining the ridges of Tehachapi Pass, a major corridor for microwave telephone links between northern and southern California. Even in the Tehachapi Pass there hasn't been any interference with telecommunications.

Birds

News reports originating from California may alarm critics that wind turbines produce more dead birds than electricity. Fortunately, that's not the case. It is true that wind power plants in California's Altamont Pass, which has the most severe problem of this kind anywhere in the world, are known to have killed birds, including raptors. While not belittling the problem or the concern of those trying to protect birds, it's helpful to put the issue in perspective.

Based on the sketchy data available, one wind turbine in the Altamont Pass will kill a bird every 20-40 years. According to Don Aitken of the Union of Concerned Scientists, other human activities kill birds in far greater numbers than wind turbines. At the present rate, says Aitken, it would take the 7000 wind turbines in the Altamont Pass 1000 years to kill as many birds as killed by the Exxon Valdez in Prince William Sound.

Despite the problem in the Altamont Pass, there's little evidence that single turbines or small clusters of machines kill birds in any significant number, though the question does occasionally arise. Consider the case of the Western Pennsylvania Conservancy and the Audubon Society of Western Pennsylvania. They manage a nature center in a Pittsburgh suburb and operate a small wind turbine as part of a display on solar energy. During the mid 1980s they found a dead duck at the base of the tower. Greatly disturbed, they called the dealer. He was speechless. The next day he inspected the wind turbine, which he found unscathed, for any tell-tale signs. (A bird the size of a duck would have severely damaged the small turbine.) A few days later, a neighbor called the nature center searching for his pet peacock. Meanwhile visitors had begun sighting a fox on the grounds. These reports prompted the center's naturalist to reexamine the dead bird, and the mystery was soon solved. He concluded it was the missing peacock, not a duck. And after finding signs of the fox near the tower, the center cleared the wind turbine of any crime.

Perspective

Where it exists, criticism of wind energy results largely from the fear of the change this new technology may bring to the community. Wind machines are as uncommon today as utility lines were 100 years ago. Just as we grew to accept (and now demand) the utility's intrusion on the landscape, it's likely that by the end of the decade we will have grown to accept wind machines in much the same way for much the same reasons.

Though it may fear this technology, the community should not apply more stringent standards to wind machines than it applies to any other similar structure or device now standing. Proponents of wind turbines need not ask for special treatment of wind energy, but they are at least entitled to equal treatment.

Whatever you do, don't bypass the building inspector or zoning official. You have a responsibility to comply with the rules even if you don't agree with them.

In New Cumberland, Pennsylvania, an unthinking homeowner bought a wind machine to install in his backyard. Then, to his chagrin, his application for a building permit was rejected. Not only was wind energy a nonconforming use in his residential neighborhood, his lot was too small. He hired an attorney and then engaged in a lengthy and expensive effort to get a variance. His neighbors objected vociferously. Then amidst the glare of television lights and a packed hearing room, his permit was denied. His troubles didn't end there. The dealer then refused to buy back the wind machine and the homeowner had to sell it at a loss. He didn't do his homework and it cost him dearly.

This unfortunate homeowner can be excused because of his enthusiasm for wind energy and his ignorance of the planning process. The same can't be said for some so-called wind farm developers who have committed similar blunders. The difference is the sums of money involved: not thousands as in the homeowner's case, but hundreds of thousands.

One group of self-styled professionals was planning to erect several unreliable wind machines in a New Jersey residential neighborhood—without a building permit. They were about to begin construction when the local news media broke the story. (There was an exciting mix of New Jersey–style, back-room politics involved.) The scheme was killed in a boisterous public hearing before the local zoning board.

These cases are the exception. They illustrate how not to install a wind machine. There are literally thousands of examples where the appropriate approvals have been obtained in an orderly and businesslike manner and the wind machine successfully installed. Consider the example of an upper-income suburb of Pittsburgh.

Fox Chapel Township has a reputation for strict interpretation of its zoning ordinances. "They'll never let you put one here," some said. Yet the dealer, Springhouse Energy Systems, and the client, the Western Pennsylvania Conservancy, were both respected and thoroughly prepared. (They had to get a variance just to erect the anemometer. So they were familiar with the process.) They answered all questions forthrightly, allayed the zoning board's fears, and, to the surprise of cynics, were granted the variance. The wind machine, a Bergey 1000, was installed without incident and has now operated successfully for a decade.

 14

Installation

He loosened the last bolt. The generator was now ready to swing free.

"All ready?" He yelled.

"Yeah, let 'er rip," replied the ground crew.

"You sure that pulley's secure?" he asked, his voice less certain now.

"Yeah, it's not going anywhere. Let's get this one down and go for a beer."

The old generator rocked on its saddle. Slowly it rolled off toward the gin pole. Suddenly there was a loud twang and the squeal of steel cable over pulleys as the 400-pound mass of copper and iron whizzed by to crash through the platform next to him.

He looked about in dazed silence.

"Are you all right?" they asked from below.

He glanced at his feet. Yep, all still there. Then to his hands. They were too, as were all his fingers. "Lucky this time," he thought.

"I'm okay, what the hell happened anyway?"

"That pulley broke loose from the tower."

This incident actually took place. It happened to an experienced crew working professionally. Though it occurred removing rather than installing a wind machine, it illustrates what can happen without thorough planning, preparation, and—equally as important in this case—execution.

If you're handy with tools, don't mind heights or hard physical labor, and have plenty of time, this chapter will offer you guidance on installing a wind system yourself. You gain by replacing the skill, time, and expense of the dealer-installer with that of your own. Working with a wind generator, tower, or (it seems) any of its subsystems requires muscling around some hefty components in awkward places. But it can be good exercise (a few trips up a tower will tone muscles you never knew you had) and a rewarding experience. You will develop a sense of accomplishment in doing it yourself, and you will learn more about your wind system than in any other way. You will know its strong points and also what can go wrong, where, and how much effort it will take to fix it. The process will also give you an appreciation for the effort and the skills required to reliably produce your own electricity.

Small wind machines, especially those less than 5 meters in diameter, may be installed by the homeowner or hobbyist who has basic construction skills. The work can be dangerous, as the previous account warns, but no more so than other projects around the home or farm. With proper respect for the hazards involved and close attention to detail, you can safely install a small wind turbine.

Wind machines greater than 7 meters in diameter entail proportionally more risk. The components are heavier and may require special equipment and techniques unfamiliar to most do-it-yourselfers. It's usually best to leave the installation of these machines to experienced dealers.

The following sections provide general information required by any installer of small wind turbines in the United States. Although the materials suggested may differ in other countries, the principles and techniques remain the same.

Always consult the manufacturer's installation manual for more detailed descriptions of anchoring and installation techniques. If you plan to install the turbine yourself, the information in this chapter will help you to select the tower, anchors, and erection methods that best suit your talents and the conditions at the site. After reading this chapter you may choose to hire a contractor instead of doing it yourself. The information gained, however, will enable you to track the progress and evaluate the performance of the contracted installer.

Whatever route you choose, thorough planning is essential. You must anticipate what will be needed at each step along the way, the problems you may encounter, and how to respond when you do. You must coordinate the schedules of your subcontractors, suppliers, and erection crew to keep the project moving smoothly. You must also choose an erection procedure that best suits you, your site, and the crew you'll be working with. If you lack any of the required skills you must find someone who has them.

Pace yourself. Assume it will take you twice as long as you expect. A skilled two-man crew can install a small wind machine in one day. It may take a novice a week, if you're cautious. Climbing a wind turbine tower is tiring, particularly if you have never done it before. Do it several times in one day and you may have had enough. As you tire you begin to make mistakes. Don't take chances on making mistakes because you tried to do too much in one day. Give yourself plenty of time.

Prepare for the installation by collecting the parts, fittings, and tools for the job. Learn how components are assembled, in what way, and with what tools. Make sure you have met all legal requirements and that you're insured for any accidents that may occur. If you're installing the wind system yourself, it's a good idea to check whether your insurance will cover hospitalization and liability for your friends who lend you a hand.

Without proper execution all your planning and preparation may be for naught. You may know the right way to do a task and have the right tools to do so, but if you don't follow through under the press of time and conditions (you may be tired and a trip down the tower to check a pulley as in the story told earlier may seem like more trouble than it's worth) the results can be disastrous.

Parts Control

Installation can be hindered and operation of the wind turbine prevented by components damaged in shipment. Before accepting delivery from the freight carrier, examine the invoice or billing form to determine the number of crates shipped. Make sure all are present and then carefully examine the crates for external damage. If damage is found, open the crates and look at the contents. Hold the shipper until you have determined the extent of the damage, if any, to the contents. The crates are designed to take some abuse while still protecting the product inside. Note any damage as precisely as possible and immediately contact the dispatcher at the freight company. Instant photographs can be helpful in verifying claims.

Tower sections and sensitive electronic components are the most easily damaged during shipment. The welded lattice tower sections commonly used in the United States for guyed towers can be crushed or bent from other goods on the truck. Check for proper alignment and any bent cross-girts. (It's hard to install a straight and true tower when a section is twisted or bent.) Damage to control boxes and synchronous inverters is much harder to determine. The best you can do is look for loose parts rolling around inside.

Catalog the serial numbers on the generator, blades, control panel, synchronous inverter, and tower. If you have to make a warranty claim, the numbers are much easier to find in your files than at the top of the tower. Serial numbers will aid troubleshooting if problems develop.

Make an inventory of all parts received as soon as possible. Many manufacturers provide a parts checklist for this purpose. Use it. The time to realize that an important bolt is missing is prior to installation, not while you're hanging from the top of the tower.

For those wind systems where the manufacturer doesn't also build the tower, the wind turbine and tower will be shipped separately. Often they will be delivered by separate carriers. Unless you have a special reason for removing the contents from the crates (for an inventory possibly), leave them as delivered until you're ready for the installation.

Foundations and Anchors

The tower and guy cables (where used) must be kept clear of vines, trees, and shrubs. It may be necessary to clear the site of any plants that could eventually interfere with the tower or guy cables. The site doesn't have to be level, so there's no need for grading.

Anchors are used to prevent guyed towers from overturning. Anchors resist uplift. Piers, on the other hand, resist loads in compression. On a guyed tower, for example, the anchors hold the tower upright and resist the forces trying to knock the tower over. The pier beneath the central mast supports the weight of the tower and wind turbine and resists the reactive forces trying to drive the mast into the ground. On free-standing towers, the legs act alternately as piers and as anchors, depending on the direction of the wind.

The type of anchor or pier used is contingent on the tower and the site. If you plan to install a guyed tower there are several anchoring options to choose from: concrete, screw, expanding, and rock anchors. The best choice for your site is determined by the engineering properties of the soil, the depth to bedrock, and the power equipment available in your area. For a free-standing tower, the choice is limited to concrete.

Anchors

Anchors must withstand the static and dynamic loads acting on the wind system, under all weather conditions, for the life of the system. They must do so without appreciable creep toward the surface or settling. The holding power of anchors depends on the area of the anchor, its depth, the soil in which it is embedded, and the soil's moisture content. Weight is a factor as well, but it's not as important as most think.

Soils vary tremendously in their ability to resist creep. Resistance to creep and to some extent settling is controlled by the soil's shear strength: the resistance of soil particles to sliding over one another. Shear strength is a function of soil type and whether the soils are wet or dry. Shear strength ranges from a maximum in solid rock to a minimum in muck or swampy soils.

One anchor manufacturer divides the shear strength of soils into two broad categories: cohesive and noncohesive. Cohesive soils, such as those with a high clay content, stick together; the particles cling to each other. These soils have a high shear strength. Noncohesive soils are generally those with a high sand content. In such soils the soil particles slide right by each other. Wind system manufacturers specify that their standard anchor designs are intended only for normally cohesive soil, those soils with a high clay content.

Engineers have created several classification systems to describe the properties of soils. A. B. Chance, an anchor manufacturer, has devised a soil-grading system for the application of their prefabricated anchors by the utility industry in the construction of power lines. Chance's soil groups range from a Class 0 for solid rock to a Class 8 for peat and muck soils that form in swamps. Solid rock provides the greatest holding capacity, peat and muck the least. In the Chance system, Classes 3-5 are considered normally cohesive soils.

Anchor holding capacity also decreases as the moisture content increases. Creep can be troublesome in saturated soils because the soil particles become fluid and tend to flow around the anchor. Water also increases the buoyancy of the anchor, reducing its weight. The holding capacity of anchors can be reduced 50 percent in wet soils. Wherever possible, anchors should be placed below the level of periodic saturation from heavy rains but above the water table.

Frost heave causes similar problems. When soil freezes it expands slightly (ice occupies a greater volume than water). If the anchor is not below the frost line, the cycle of freezing and thawing will heave or jack the anchor toward the surface. This is more of a problem for anchors than piers because the existing load acts to pull the anchor out of the ground. The forces on piers act counter to frost heave. The frost line varies from year to year and depends on the severity of the winter and the soil cover. Bare soil freezes more quickly and to a greater depth than a soil with a grass cover. (The grass and the organic soil it grows in act as an insulator, slowing the soil's winter heat loss.)

To determine the soil-holding capacity at your site you can test the soil with a probe or examine nearby road cuts. Better yet, talk to people who work with soil. In the United States the Soil Conservation Service (SCS), the county extension agent, or the office of the conservation district should be able to help. Explain your plans to them. Describe what it is you want to know and why it's important (you don't want the anchor pulling out of the ground). They will be able to tell you not only what kind of soil you will be working with but also the depth to the water table and the average frost penetration.

The SCS has prepared soil maps of much of the United States. It will be helpful if you can find your site on these maps. Local excavation companies are another good source. They have a feel for subsurface conditions since they work with them daily. They are in business to make money, though, not to give out free information. If you want their help, you should hire them.

The requirements for piers are less stringent than those for anchors. Most soils are strong in compression. With an adequate bearing surface, concrete piers of standard dimensions are used throughout North America.

Working with Concrete

The most common method for anchoring a tower or constructing a pier is to excavate a hole and partially (sometimes completely) fill it in with reinforced concrete.

Concrete is literally man-made rock, conglomerate, to be specific. It's strong in compression (as when squeezing an accordion), weak in tension (as when expanding an accordion). Thus, it works well as a pier or foundation but poorly as a beam. Tensile strength is improved by reinforcing the concrete with steel bars commonly called rebar (reinforcing bars).

Concretes are rated by their compressive strengths. Most construction in the United States uses concrete that obtains a minimum, ultimate strength of 3000 pounds per square inch after curing for 28 days. Strength is a function of the water-cement ratio and the degree to which curing has taken place. The lower the water-cement ratio (the more cement in the mixture), the stronger the concrete. Strength also increases with curing time.

Curing is rapid in the first few days. (Concrete *sets* or becomes rigid within an hour of adding water.) Hydration doesn't go forward if too much water evaporates in hot weather, or if the concrete becomes too cool in cold weather. Curing should take place for a minimum of 7 days before any load is placed on the concrete. The concrete will continue to gain strength if moisture and temperature conditions remain favorable for complete hydration. Heeding the above precautions, you can place concrete year round.

Concrete can be bought in bags and mixed by hand or bought directly from a ready-mix plant. The ready-mix plant will deliver, or in some communities you can pick up the concrete with a small trailer. In the United States, concrete from a ready-mix plant is sold by the cubic yard or yard. The cost includes delivery within a limited distance of the plant. If they have to haul it further, it costs more. For any installation requiring one yard or more, ready-mix is the best choice. It's far easier than mixing it by hand, it's quicker, and it offers a better way to control the quality of the concrete if you are inexperienced. Don't skimp on concrete.

Installation drawings invariably show nice neat anchors and piers that look like they were made with a cookie cutter (see Figure 14-1). Except where an anchor or pier is exposed at the surface, this precision isn't necessary. Where the soils are stiff and will not collapse into the hole, the concrete can be poured in place. For anchor blocks below the surface this is superior because the concrete acts directly on undisturbed soil. Forms are necessary where the hole is larger than the anchor or pier desired, where the concrete will extend above the surface, or when the anchor is in sandy soil. Mick Sagrillo has found that 55-gallon drums with their ends removed make suitable forms.

Installation instructions usually call for the concrete to be placed over a grid or cage of rebar to give the necessary tensile strength. In the United States, rebar is designated by its diameter in eighths of an inch. Thus, specifications calling for a rectangular cage of number 5 vertical and number 4 lateral will use ⅝- and ½-inch rebar, respectively. The rebar is tied together with wire so it won't move when the concrete is placed over it.

Rebar must be covered by 3-4 inches of concrete. When closer to the surface, acid-laden water enters the concrete and corrodes the steel. The rebar expands slightly, causing the concrete to spall or chip. For long life, the concrete must sufficiently seal the rebar from corrosion.

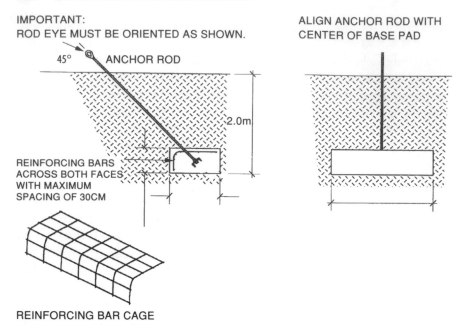

IMPORTANT:
ROD EYE MUST BE ORIENTED AS SHOWN.

45°   ANCHOR ROD

REINFORCING BARS
ACROSS BOTH FACES
WITH MAXIMUM
SPACING OF 30CM

2.0m

ALIGN ANCHOR ROD WITH
CENTER OF BASE PAD

REINFORCING BAR CAGE

Figure 14-1. Concrete anchor detail. (Bergey Windpower Co.)

To ensure that the rebar stays where you want it when the concrete is poured, it should be staked down in the excavation or tied to the forms so it won't "swim" around. Pieces of rock or brick can be used to keep the rebar cage off the bottom of the excavation. This keeps the rebar from being too close to the concrete's bottom surface.

Forms can be built from heavy plywood and a wooden frame. The frame, when staked to the side of the excavation, will hold the plywood form in place while the concrete is hardening. Where the soil is stiff and the excavation is no larger than the pier desired, a short form can be made from wooden planks set in the excavation to a depth that gives the finished pier its correct height above the surface. Cylindrical forms can be purchased for pouring concrete columns.

Before placing concrete, moisten the forms or the excavation to prevent them from absorbing water from the surface of the concrete mix and reducing its strength. Avoid pouring concrete from a height greater than 4 feet or the aggregates will begin to separate. Once in the form, work the concrete to eliminate air pockets by poking a board or shovel into the concrete. Work it around the rebar and along the sides of the forms. Don't overwork, or the aggregates begin to settle. Special gasoline-powered vibrators can be rented for working the concrete after it has been placed.

You can simplify the whole process by hiring a contractor to excavate the hole and pour the concrete. Get firm quotes before you do, and make sure they understand what the concrete will be used for. The fact that a lot is riding on their work may discourage them from cutting any corners.

Guyed Towers

Because they're frequently used with small wind turbines, let's first examine the requirements of guyed towers. To support a guyed tower you'll need a pier for the mast and at least three anchors to keep the mast erect.

The pier is simple to construct (see Figure 14-2) by excavating a hole about 4 feet deep. See Table 14-1 for the dimensions of a square pier suitable for most small wind machines up to 5 meters in diameter. In weak soils or with an extremely tall tower a larger pier may be necessary. (As with all other aspects of installing a wind machine, follow the manufacturers' suggestions. They know what works best for their particular product.) It may also be necessary in some areas to extend the pier deeper to get below the frost line. Strengthen the pier with a rebar cage before placing the concrete.

Guyed towers also need a means to keep the mast in place so it won't scoot off the pier. Before the concrete sets, insert a pin or threaded rod into the center of the pier. Most masts have a base plate that slips over this pin.

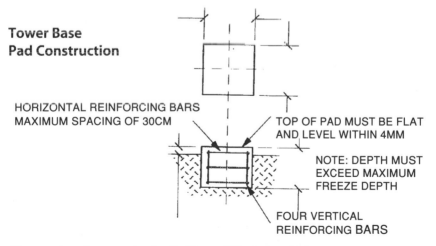

**Tower Base
Pad Construction**

HORIZONTAL REINFORCING BARS
MAXIMUM SPACING OF 30CM

TOP OF PAD MUST BE FLAT
AND LEVEL WITHIN 4MM

NOTE: DEPTH MUST
EXCEED MAXIMUM
FREEZE DEPTH

FOUR VERTICAL
REINFORCING BARS

Figure 14-2. Concrete pier detail. (Bergey Windpower Co.)

Anchors are a little more complex. The type of anchor used depends on several factors: soil strength, the depth to bedrock, and your access to power equipment. In normally cohesive soils you can choose from concrete, expanding, and screw anchors. Concrete anchors are the most popular, followed by screw anchors.

Though screw anchors are widely used by electric utilities, building inspectors and contractors are more familiar with concrete. The backhoe needed to excavate the hole for a concrete anchor is also more readily available than the line truck used to drive a screw anchor. Concrete anchors can also be adapted to weak or soft soils by simply expanding the anchor's bearing surface, that is, lengthening and widening the anchor. If you have any doubts about the holding capacity of screw or expanding anchors, opt for concrete.

Backhoes work best at digging trenches, so they're well suited for making the anchor excavation. The concrete can usually be cast in place. Backhoes can also excavate the pit for the pier. Table 14-1 gives the dimensions of concrete anchors for wind turbines mounted atop a 100-foot (30-meter) tower in normally cohesive soil.

An easier and quicker method than using a backhoe is to auger the holes for the pier and anchors. Where soils are not rocky and bedrock is well below the surface, a truck-mounted power auger, like that used by linemen to set wood utility poles, can drill holes for the pier and anchors in a matter of minutes.

For the pier, a 36-inch (1-meter) auger is best. It excavates a hole large enough to form the pier. Set a short form into the top of the hole so that it extends above the surface, place the rebar, and cast the concrete in place.

After the pier, the auger can then be used to drill holes for expanding anchors (see Figure 14-3). Expanding anchors work much like toggle bolts and similar fasteners used in plaster walls. Once the arrow-shaped anchor is inserted into the hole and expanded, the barbs resist being pulled back out.

The strength of expanding anchors is controlled by the soil type, the size of the anchor, and the firmness of the backfill. Expanding anchors hold more in heavy, stiff soils, less in sandy or swampy soils. The larger the diameter of the anchor, the greater its holding power. To reach its rated strength, the anchor must be fully expanded into undisturbed soil, and the hole backfilled properly.

See Table 14-2 for the dimensions of expanding anchors suitable for small wind turbines. These anchors are capable of meeting the anchoring loads on a 100-foot (30-meter) tower for soils ranging from Class 7 (sandy

Table 14-1

*Concrete Piers and Anchors for 100-foot Guyed Tower
in the United States*

(Self-Furling Wind Turbines Only)*

Turbine Size (m)		Dimensions (ft)	Depth (ft)
Piers			
	3	2 x 2 x 4	—
	5	2.5 x 2.5 x 4	—
	7	3 x 3 x 4	—
Anchors			
	3	3 x 3 x 1	4
	5	3 x 3 x 1.5	4
	7	3 x 6 x 2	6

*For most cohesive soils. For taller towers and less cohesive soils, refer to the manufacturer's specifications.

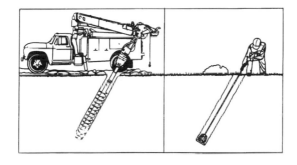

Figure 14-3. Expanding anchor installation. Augering hole for anchor, inserting anchor into hole, expanding the blades of the anchor with tamping bar, backfilling hole, and tamping. (A.B. Chance Co.)

or swampy soils) on the low end to Class 3 (representing hardpan, dense clay, and hard silts) on the upper end. The anchors are sized to give a margin of safety in the weaker soils. They're more than strong enough in the heavier soils.

After drilling a hole at the correct angle, set the anchor at the bottom. Strike the anchor with a heavy bar to force the leaves into undisturbed soil. (You can check whether the blades are expanded by shining a flashlight down the hole.) Once the anchor is fully expanded, attach the anchor rod. Gradually backfill the hole, and compact the soil with a tamping bar. That's all there is to it.

The development of screw anchors has further simplified the installation of anchors for guyed towers (see Figure 14-4). It's not unusual to install the three anchors needed for a small wind turbine in less than 30 minutes.

Table 14-2

Expanding Anchors for 100-foot Guyed Tower in the United States

(Self-Furling Wind Turbines Only)*

Turbine Size (m)	Anchor Size (in)
3	8
5	12

*For most cohesive soils. For taller towers and less cohesive soils, refer to the manufacturer's specifications.

Many truck-mounted augers have been adapted to drive screw anchors by replacing the auger bit with a special tubular wrench. The hydraulic boom controls both the angle and the rate at which the anchor enters the soil. Unfortunately, screw anchors can't be used everywhere. Rocky soils, in particular, can thwart the anchor from advancing.

Screw anchors are sized by the diameter of the screw and the number of helixes on the anchor shaft. Their holding strength is once again based on the cohesiveness of the soil. See Table 14-3 for screw anchors suitable in the most cohesive soils (those from Class 7 to Class 3).

Soil strength greatly affects the holding capacity of any anchor. For a 3-meter turbine in a soil of Class 3, the 8-inch screw anchor has a holding capacity over three times greater than the expected maximum load on the anchor. The 8-inch expanding anchor has a safety factor of five. But in the

Figure 14-4. *Screw anchor installation. Truck-mounted power augers can be adapted for driving screw anchors. Utilities frequently use screw anchors because they can be installed in the least time and with the least effort. (A.B. Chance Co.)*

Table 14-3

Screw Anchors for 100-foot Guyed Tower in the United States

(Self-Furling Wind Turbines Only)*

Turbine Size (m)	Anchor Size (in)
3	8
5	12

*For most cohesive soils. For taller towers and less cohesive soils, refer to the manufacturer's specifications.

weaker Class 7 soils, the safety factor drops to less than two. In heavy soils the 8-inch anchors are more than sufficient. In weaker soils, or if there's any doubt about the holding capacity of the soil at your site, check with the wind turbine manufacturer. Where necessary you can double the anchors at each guy. As mentioned previously, this was the approach taken by Fayette Manufacturing on their 1500 turbines in the Altamont Pass. Fayette installed their 10-meter turbine on guyed pipe towers using screw anchors. Because of soil conditions in the Altamont Pass, Fayette used two 12-inch screw anchors at each guy point. After a decade of operation there have been few problems with the anchors.

If you're unfortunate enough to encounter solid rock at or near the surface, none of the preceding anchoring methods can be used. You'll have to drill a hole at the proper angle with an air drill and compressor (see Figure 14-5). A rock anchor and rod are then inserted down the hole and wedged in place. These anchors have a high holding capacity, but installation is time-consuming and expensive.

For small wind turbines one anchor rod is sufficient for all guy levels in most installations. To minimize bending, the anchor rod must depart the ground at an angle that coincides with the resulting angle of tension in the guy cables. Bending weakens the anchor rod over time and can cause it to fail. The angle of departure depends on the height of the tower and the guy radius, and it's usually 45-70 degrees. In the field it's easiest to use a 45-degree angle of departure by measuring a rise of one over a run of one, but it's always better to follow the manufacturer's recommendations on the appropriate guy angle. The anchor rods most likely to be used with small wind turbines in the United States are given in Table 14-4.

Figure 14-5. Rock anchor installation. Drilling hole, inserting rock anchor, and then expanding anchor by turning the anchor rod. (A.B. Chance Co.)

Table 14-4

Anchor Rods for Guyed Towers in the United States

(Self-Furling Wind Turbines Only)

Turbine Size (m)	Rod Diameter (in)	Length (ft)
3	5/8	5
5	3/4	7
7	1	10

Free-Standing Towers

Free-standing towers, whether a truss or tapered tube tower, require an excavation and the placement of concrete. The easiest method, where the depth to bedrock permits, is to use an auger. For tubular towers a hole is drilled for a central pier to support the entire tower. On truss towers holes are drilled for a pier at each leg of the tower. In sandy or swampy soils, piers are insufficient for truss towers. Instead, they must rest on a massive concrete pad. All three techniques have been used in California on medium-sized turbines. (The holes augered for the tubular towers on some of the turbines near Tehachapi were big enough to swallow two pickup trucks end to end.)

For small machines 3-4 meters in diameter, the foundation can be built quickly by augering a hole 3 feet in diameter by about 10 feet (1 by 3 meters) deep. The rebar is placed in the hole along with a short tower section and the concrete poured. This technique works well with both truss towers and tapered tubular towers for wind turbines of this size. On larger machines atop truss towers, the piers for each leg may be up to 3 feet in diameter and up to 10 feet deep. Often the holes are left unfinished, that is, in their circular form. The rebar and anchor bolts for anchoring the tower leg are then added, and the concrete poured.

A backhoe can be used to excavate a rectangular hole for footers when augering isn't practical. When using a backhoe, formwork may be necessary, and a greater amount of concrete will be needed than that for a cylindrical pier.

You must ensure that the base tower sections or anchor bolts used in the foundation don't "swim" around when the concrete is placed. They must also accurately fit the foundation template provided with the tower. Otherwise, you could have a rude awakening when you go to set the tower on the base—it may not fit.

Novel Foundations

If power-installed screws work well as anchors, why couldn't they also be used for the pier supporting the mast of a guyed tower? In theory at least, they can. The foundations for light standards and transformers at substations are now being installed in this way. No one, however, has adapted this technology to wind systems. When they do so, an installer will be able to drive the anchors and pier in just minutes instead of waiting days for the concrete in the pier to cure. The whole wind system could then be erected in a day or less.

Using power-installed screw anchors to secure the legs of a free-standing truss tower is another possibility. There may be engineering limitations particularly in weaker soils, but the advantage of quick and easy installation justifies a look into whether it's possible. Fast-setting plastics, such as high-density foam, also offer promise if their curing time is less than that of concrete for an equivalent strength. Again, this technology has not yet been adapted to wind systems.

Assembly and Erection of Guyed Towers

The guyed lattice masts commonly used for small wind machines can be assembled a section at a time with a tower-mounted gin pole, or the entire tower can be assembled on the ground and hoisted into place with a crane. The method you use depends on whether a crane is available, can get to your site, and is affordable. In either case the guy cables must first be cut to length and attached to the guy brackets that fit around the mast.

Guy Cables

Only extra-high-strength (EHS) cable is used for guying the tower. It's three times stronger than common grade steel cable and 40 percent stronger than high-strength cable. If you're building your own tower, don't economize on the size and quality of the guy cable. Refer to the manufacturer's installation manual for the size of guy cable needed for the height of the tower you plan to install.

Steel cable is shipped in coils or on reels. Kinks damage the cable beyond repair. Avoid kinks by rolling the coil along the ground to lay out the cable. If the cable is on a spool, use a spool stand to unreel the cable.

If the manufacturer doesn't specify the length of the guy cables in the installation manual, calculate cable length by using the Pythagorean theorem:

$$\text{Guy Length} = \sqrt{(GR^2 + GH^2)}$$

where *GR* is the guy radius and *GH* is the guy level or height above ground. Give yourself plenty of extra cable to allow for sag and for slight errors in the position of the anchors. Let's assume that you want to find the guy length needed for the topmost guy on a 80-foot (24-meter) tower with a 40-foot (12-meter) guy radius. The guy bracket is the middle of the topmost tower section, 70 feet above the ground (see Figure 14-6).

$$\begin{aligned} \text{Guy Length} &= \sqrt{(40^2 + 70^2)} \\ &= \sqrt{(1{,}600 + 4{,}900)} \\ &= 81 \text{ feet} \end{aligned}$$

This tower will need three lengths of cable at least 85 feet (26 meters) long for the topmost guy level. If the second guy level is at 30 feet, we'll need three guys each 55 feet (17 meters) long. As a result this tower will require a total of 420 feet (128 meters) of cable.

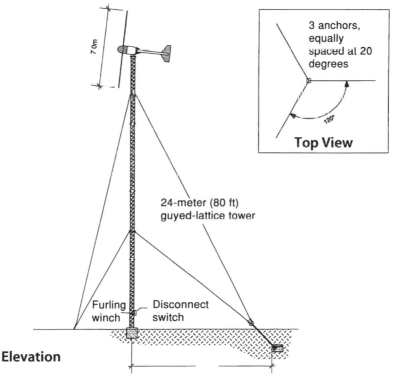

Figure 14-6. Guy cable positioning. (Bergey Windpower Co.)

Mark the length of each guy cable on the ground. Unreel the cable and cut it to length with bolt cutters (18-inch bolt cutters will handle up to $5/16$-inch cable). Next, bolt the guy brackets to the tower sections and then attach the guy cables to the guy brackets.

Cable attachment is an area, like anchors, where installation has been greatly simplified. Developments in the utility industry are being adapted for use with guyed wind systems. Guy cables can be attached with U-shaped cable clamps (known as Crosby clamps), with strand vises, and with preformed cable grips.

Past practice dictated using the cable clamps. The guy cable would be wrapped around a metal thimble and passed through two or more clamps. The clamps prevented the cable from slipping off the thimble, which distributed tension uniformly over a large radius. (Too sharp a bend in the cable severely weakens it.) Some guy brackets, particularly those in the utility industry, have built-in thimbles.

Cable clamps are awkward to use. The stiff cable resists being passed around the thimble and through the clamps. The U-bolt of the cable clamp must act on the dead end of the cable, necessitating a struggle with the cable to position the clamp properly. When finally in place, the clamps can be tightened. This whole process is awkward and time-consuming.

Strand vises simplify the task. They operate on the wedge principle. The guy cable is passed through the vise and, when tensioned, wedges the cable in the vise grip. The strand vise has a wire bale or loop, which passes through the guy bracket. No thimble is needed. Because this bale is usually fixed around the vise, strand vises are suited only for the guy bracket and not for the anchor rod. (The brackets are assembled at the site and the bale of the strand vise inserted into the bracket before it is bolted together. Not so with the anchor rod, where the eye is forged within the rod.)

Less expensive and adaptable to both ends of the guy cable is the preformed cable grip. In use by utilities since the 1950s, this works like a Chinese finger puzzle. After first passing one leg of the helically wound steel strands through the guy bracket or eye of the anchor rod, the grip is wrapped around the guy cable. Tension on the cable pulls the strands tighter together. A fine grit on the inside of the strands ensures that the grip has a firm hold on the cable.

Preformed cable grips allow quick adjustment in the guy cable during installation by being easy to remove and reapply. They are simply unwrapped, the cable pulled taut, and then rewrapped. Like cable clamps, these cable grips require the use of thimbles, but their ease of application more than compensates for the extra step.

Using a Crane or Gin Pole

Either the entire tower and wind turbine can be lifted into place with a crane, or each can be lifted separately. If you're lifting a complete assembly, the crane doesn't have to be taller than the tower. For small wind turbines the combined tower and wind turbine can be lifted at some level below the top of the tower. Because the lattice mast is somewhat frail, a nylon sling should be used to spread the lift over the entire cross section of the tower and not act on one leg alone. Lifting a complete assembly also requires that the wind machine be fully assembled. This may require lifting the tower off the ground slightly to attach the blades and tail.

If a crane is to be used, bolt all tower sections together, making sure that the section with the guy brackets and cable is in the right position. Attach the wiring conduit and thread the power cable through it as discussed in the subsequent section on wiring. While the crane holds the tower upright, position it over the base plate or pier. Rest the tower on the pier and connect the guy cables to the anchors and pull them taut. Once this is done, the crane can be removed and final adjustments made to the guy cables.

If a gin pole is used, bolt two 10-foot tower sections together and attach temporary guys (⅜-inch polypropylene rope works well). Tip the two sections up onto the pier and tie off the temporary guys to the anchors. They will hold the tower in place until the first guy level is reached. Someone must now climb the tower and bring up the gin pole.

The gin pole is a boom or davit that extends above the top of the tower (see Figure 14-7). It permits tower sections or the wind turbine to be lifted up the tower and set in place without the use of a crane. Gin poles need be nothing more than a long section of pipe strong enough to handle the expected loads and which have some means of being attached to the tower. Some gin poles are a little more sophisticated and incorporate a horizontal arm that allows the load to be centered on the tower. The gin poles used to install the old Jacobs windchargers were nothing more than an 11-foot length of 3-inch steel pipe. The gin pole used to lift the modular tower sections of Unarco Rohn's lattice mast is simply a 12-foot long aluminum pipe. This same gin pole has been used numerous times to hoist small wind turbines, though its strength limits it to machines no larger than 3 meters in diameter.

Figure 14-7. *Installing a guyed lattice tower one section at a time with a tower-mounted gin pole.*

Pulleys are used to direct the hoisting rope over the gin pole to the load. The pulley is either built into the top of the gin pole like those supplied by Unarco Rohn for use with their guyed towers, or attached separately. Never use a gin pole without first routing the hoisting rope through a pulley at the base of the tower. This pulley permits the hoisting crew to stand clear of the tower and be well away from any falling objects. It also prevents any unnecessary bending of the gin pole. With a base pulley in place, the hoisting tension acts directly on the gin pole from below. This minimizes the bending forces on the gin pole.

Art and Maxine Cook can testify to the need for a gin pole strong enough for the job at hand. Their mountaintop farm in western Pennsylvania has two wind machines: one, an Aermotor for pumping water, and the other, a wind generator for charging batteries.

Art's windcharger once developed a disturbing noise. He called the re-manufacturer who sent a technician to replace the heavy Jacobs generator. Both Art and the technician had previous experience with tower-mounted gin poles (Art has replaced more than one generator since he began using wind power in the mid-1970s). The technician scrounged a gin pole from a local scrap yard. It looked good. It was the right length, right size. The gin pole was bolted to the tower and the noisy generator removed. But raising its replacement proved more difficult. As Art strained on the rope at the base of the tower, the machine moved slowly skyward. It was only a few feet off the ground when the gin pole collapsed. The generator fell as Art scurried for cover. Fortunately, no one was hurt and the generator wasn't damaged.

Similar mishaps have occurred when the attachment of the gin pole to the tower has failed. It's paramount that the gin pole always be firmly attached to the tower and not move laterally when the load is applied. The manufacturer's recommendations for the materials used in the gin pole and its attachment to the tower should be followed religiously. Otherwise you may learn an expensive physics lesson.

> Erecting a wind turbine can be dangerous. Use extreme caution.
> When in doubt, consult the manufacturer.

With the gin pole now in place, the next tower section is brought up. The third 10-foot section will usually have the lower guy bracket attached. Once the section has been bolted down, the guy cables can be strung to the anchors. These cables are tensioned by hand until the three assembled tower sections are vertical and the tower straight. Vertical alignment can be checked with a level (held vertically) on the lowest section, with a transit, or with a plumb bob.

The gin pole is then released and moved to the top of the top-most section. A new section is hoisted up and bolted into place and so on until the tower is completed. After each set of guy cables is attached, the tower should be checked for plumb and twist. If the lower sections were aligned properly it is possible to simply sight along the tower to check the alignment of the upper levels. A transit can be used to be certain. The tower must be vertical for proper yawing of the wind turbine.

Normally, the guys on small machines can be tensioned by hand. For wind machines greater than 7 meters in diameter it may be necessary to mechanically tension the guys with a coffing hoist (come-along) pulling on a cable grip. The tension is then measured by a dynamometer in line with the hoist. Follow the manufacturer's directions for attaching the hoist, and for the amount of tension required.

After the guys are in place, you're ready to raise the wind turbine. Do so only on a calm day. Any wind at all will make it more difficult to position the turbine once it's atop the tower (see Figure 14-8).

For small wind turbines it's easier to do as much of the assembly on the ground as possible since it's awkward to work at the top of the tower. Bigger machines may require assembly on the tower. Use a machine stand near the tower if you have one. This permits attachment of the blades and other components. Put the stand in the bed of a pick-up truck if you need more clearance.

Attach the hoisting rope and one or two tag lines. The wind turbine may have an eye bolt used for lifting or it may require a special lifting jig. Whatever's used, it's important that the hoisting line lift at the wind turbine's center of gravity. If not, the wind turbine will be a lot more difficult to handle on its way up the tower, and once on top you'll have a heck of a time mounting it to the tower. The turbine may be easy to move around on the ground, but it's a lot more difficult at the top of the tower. Everything you do on the tower is harder and more dangerous than when you do it on the ground.

Use the tag lines to keep the machine from banging into the tower and tangling with the guy cables. The tag lines must be longer than the tower is high. For a lift up a 100-foot tower, the tag line will need to be at least 150 feet long or about 1½ times the tower's height. The pull on the tag lines

Figure 14-8*. Installing gin pole on truss tower.*

must be moderate (just enough to prevent the machine from hitting the tower), particularly as the turbine nears the top of the tower. It's easy to buckle the gin pole if too much force is used on the tag line.

Once the turbine is mounted atop the tower, string the conduit and fish the power cable to the generator. Torque all tower fasteners to the specified value and apply locking nuts where required.

Free-Standing Towers—Assembly and Erection

Technically truss towers can also be erected with a tower-mounted gin pole. But no one does it in practice. Only in rare cases such as remote sites in Alaska or other inaccessible areas are gin poles used to assemble truss towers. The sections are much heavier and more awkward to work with than those on a guyed tower. Most installers simply call in a crane.

Each individual member on a truss tower is so heavy that by the time the first section is bolted together you're not going to move it anywhere without heavy equipment. Ideally, you'd like the crane to simply drive up to the tower, raise it, and in one lift set the tower on its foundation and then leave. You don't want the crane to move sections of the tower around the site because you didn't thoroughly plan the assembly. This is particularly important if the wind turbine has been mounted on the tower. The more moving that's required, the greater the likelihood of damaging the turbine.

In the United States crane service is usually billed by the hour. Travel time is included. Some companies have a minimum charge; others don't. Two factors are important in crane rentals: the weight, and the height of the lift. An 80-foot (24-meter) truss tower for a 3-meter turbine will weigh less than 1500 pounds (about 700 kilograms); one for a 7-meter machine will weigh more than twice that.

The tower can be raised first, and then the wind turbine mounted on top in a second lift. For wind systems over 7 meters in diameter, this is the common practice. When using this method the boom of the crane must extend about 10 feet (3 meters) above the top of the tower to allow for the crane hook and the lifting jig.

A crane with a shorter boom can do the same job if it raises the wind turbine and tower at one time (see Figure 14-9). This works well with wind machines up to 3 meters in diameter, though Bergey Windpower also uses this method for its 7-meter turbine. When lifting the turbine and tower together, the lift should be made some distance below the top of the tower yet well above the tower's center of mass. As the tower is raised, the weight

Figure 14-9. *Erecting lattice tower with a crane. The crane lifts the turbine and tower after assembly on the ground.*

of the tower keeps the bottom sections on the ground while the upper sections move toward the vertical. If the tower was positioned correctly during assembly, the bottom will slide across the ground toward the foundation. Because the lift is being made below the top of the tower, the rotor blades are able to clear the boom as the tower nears the vertical.

With the skillful use of hand trucks, dollies, and come-alongs, even the heaviest towers can be fully assembled without power equipment. But it's wise to assemble the tower where it will make the crane operator's job as simple as possible. Begin by bolting the lowest (the heaviest) section together. Its placement will determine the location of the remaining tower sections and it won't be moved again until the crane arrives.

Truss towers are like giant Erector sets and will be puzzling until a few of the cross-girts are bolted into place. Tighten the bolts on the cross-girts snug, but not tight. As you move along the tower section you'll find that some of the pieces won't fit properly. There are always a few pieces that need some convincing before they fall into place. With the bolts just snug the members can give a little, allowing an obstinate bolt to slip through a hole where it once wasn't possible.

Erection wrenches (sometimes called *spud wrenches*) and *drift pins* are helpful in these situations (see Figure 14-10). An erection wrench has a long tapered shaft that's used by ironworkers to solve stubborn alignment problems between two pieces of metal. The shaft is inserted into the holes, and with a little muscle it's used to lever the pieces into position. A bull pin is a similar device. It too has a long tapered shaft, but instead of a wrench on one end it has a striking face. The bull pin is dropped into the holes needing a little nudge and then driven with a hammer until they're aligned. Both of these tools are well suited for aligning holes on the flange plates between each tower section.

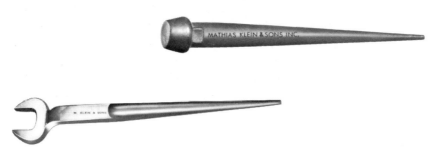

Figure 14-10. *Tower assembly tools. Left, erection wrench; right, drift or bull pin. (Klein Tools)*

On lightweight towers it's possible to assemble the next 20-foot section near the first and then bolt the two together. This proceeds until the tower is fully assembled. The wind machine is mounted on the tower and the conduit for the power cable installed. You'll find that strapping the conduit to the tower while it's on the ground and fishing through the conductors is much easier than trying to do it after the tower is erected. Once all the sections have been assembled and the tower erected, tighten the bolts to the desired torque.

Because the tower vibrates when the wind turbine is running, the nuts on all bolts must be prevented from loosening. Use self-locking nuts, or add special locking nuts after the nut and bolt have been tightened. These are easiest to install while the tower is still on the ground. Don't overlook this step. Locking nuts are specified for a reason; towers have failed without them.

The tower is now ready for the lift. Attach the lifting sling so that the stress is distributed onto tower members strong enough to take the load. Don't, for example, wrap the sling around a tower cross-girt. Instead use a tower leg; better yet, use two legs. Note that the sling must not slide along the leg as the tower is being lifted.

The crane will slowly set the tower down on the flanges or threaded bolts in the foundation. On foundations with flanges, align the holes with the drift pin and judicial use of a crowbar. Drop in the bolts once the holes are aligned. Before removing the sling, level the tower.

There are two ways to level the tower. On lighter towers, shims are forced in between the section flanges. The heavier towers use threaded bolts, between the foundation and the lower tower section, that have adjusting nuts. These nuts are used to level the tower. When the tower is level, tighten the mounting bolts and install the lock nuts. The sling can now be safely removed and the crane sent on its way.

Truss towers for wind turbines larger than 5 meters in diameter are usually erected in 20-foot sections when no heavy equipment is available to move the sections around while still on the ground. In this situation the crane places the first section on the foundation, and the tower is leveled. The following sections are then added. Where heavy equipment is available to move sections or components around the site, the entire tower can be assembled on the ground.

Tapered Tube Towers

Some free-standing tubular towers are erected a section at a time and depend on a slip fit for holding the sections together. For these towers the first section is placed on the foundation and leveled with adjusting nuts. The next section is slipped onto the first and so on. Gravity does the rest. The wiring run is then strung inside the tower and the wind machine installed.

Some towers using this approach go further and embed a short section in the foundation. When the first full tower section is raised, it's slipped into the first. This is easier than trying to align several J-bolts in the foundation with the flange on a tubular tower, but it does demand that the embedded section be perfectly level—there's no adjustment once the concrete has set.

The sections for tubular towers used on medium-sized turbines are usually bolted together. They're assembled on the ground and then hoisted onto the foundation.

Wood poles can be installed with a crane or with the boom on truck-mounted power augers. The boom on these vehicles was designed for setting utility poles and has a clamp that grasps the pole near the midpoint and tips it upright. The pole is set on a foundation and bolted down or set in an augered hole and concrete added. (Temporary guys are needed to keep the pole vertical until the concrete sets.)

Farm Windmill Towers

Installing a farm windmill tower is similar to installing a wind-electric system, though much simpler. Aermotor recommends that for windmills less than 14 feet (4 meters) in diameter, excavations for the tower's piers need be no more than 5 feet deep and 2 feet in diameter. For the 16-foot (5-meter) windmills, the holes for the piers should be 1½ foot deeper.

After excavating or augering for the legs, insert the bottom sections of each anchor post. Then erect two sections of the tower. Level and square to ensure that the windmill sits directly over the well head. Then fill the bottom 2 feet of each pier with concrete. After the concrete has set up, tamp dirt into the remainder of the hole. According to Aermotor, this technique is suitable for water-pumping windmill towers up to 50 feet tall. For more details on installing and servicing water-pumping windmills read *Windmills and Pumps of the Southwest* by Dick Hays and Bill Allen (see Appendix I).

Hinged Towers—Assembly and Erection

All towers, whether guyed or free-standing, can be hinged and tipped upright into the vertical position, eliminating the need for a crane (see Figure 14-11). Hinged towers simplify assembly because all tower and wind turbine components can be added while safely on the ground. They simplify service and repair for the same reason. Rather than climbing the tower and manhandling awkward components in the air, the tower can be lowered and the job done on the ground. Avoiding the need for a crane also reduces the cost of installation and service.

These towers can be raised by a heavy-duty industrial winch and gin pole, or by using a vehicle and gin pole. Bucket trucks or small cranes such as those for servicing highway signs can also be used to pull the tower upright on small wind turbines.

Hinged towers have their share of limitations too. The hinges add to the cost and complexity of the installation, and they introduce a potentially weak link in the tower structure. Some manufacturers refuse to install their wind turbines on hinged towers for this reason. Worse, many a hinged tower has been dropped by an installation or service crew unfamiliar with the loads involved. Some unwary crews have watched helplessly as their hoisting truck was dragged across the ground as a tower with an expensive wind turbine atop it headed inexorably toward the ground. Despite these limitations, hinged towers make sense when designed and operated properly. The advantages of hinged towers dictate that they'll continued to be used.

Tilt-up towers typically use four guy cables. Several manufacturers have marketed products for sites on the Great Plains that featured a guyed-tubular tower with accompanying gin pole. The tower is prevented from lateral

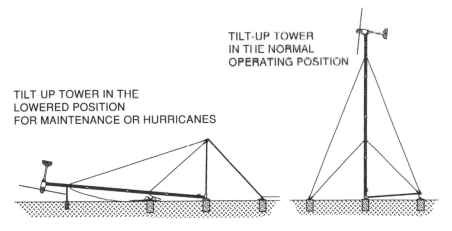

TILT-UP TOWER
IN THE NORMAL
OPERATING POSITION

TILT-UP TOWER IN THE
LOWERED POSITION
FOR MAINTENANCE OR HURRICANES

Figure 14-11. Erecting hinged tower with gin pole and winch. (Bergey Windpower Co.)

movement during erection by the two guys at right angles to the lift. The free guy cable is routed to the top of the gin pole. The tower is raised by drawing a block and tackle together between the gin pole and the free anchor.

Towers with three guy cables may also be used but there's no lateral restraint on the mast during the lift. When only three guy cables are used, the tower tends to swing from side to side until it's upright and the guys become taut. This lateral motion can be prevented by using a two-piece gin pole in the shape of an A-frame. The A-frame can be built inexpensively from four sections of lattice mast, but the bases must be hinged to allow the gin pole to move with the tower.

Truss towers have been raised in the same way. One manufacturer uses the A-frame gin pole; others use a single pole. Either the gin pole is attached to the tower's foundation with a hinge or it stands separately. When standing apart from the tower, the gin pole must be guyed to prevent it from moving laterally out from under the load.

Free-standing tube towers have also been erected with a gin pole mounted at the base of the hinged tower. In both the case of the truss tower and the tube tower, the hoisting cable is passed over the gin pole to the top of the tower. Once upright, the raising cable can be removed or it can be left in place for service calls.

In Europe several small wind turbines have been installed atop hinged, tubular towers using a powerful hydraulic cylinder.

Wiring

The bible on wiring in the United States is the National Electrical Code™. It's the rule book for what can and can't be done, but the final say in electrical matters is in the hands of your local electrical inspector, fire underwriter, or code enforcement officer.

Licensed electricians in your area will be familiar with the local application of the codes. They won't be familiar, however, with wind generators and their specific requirements, especially as they relate to interconnection with the utility. This section will be helpful to your electrician and essential if you plan to wire the wind system yourself. In either case, when unsure of your next step, check with your local electrical inspector before you begin work.

Why is the approval of the electrical inspector or fire underwriter so important? Because your fire insurance may be void without it. Also, in some communities you can't sell your home without electrical wiring that's "up to code."

The electric utility is responsible for all wiring from the nearest transformer to the service drop (where the wires are secured to your house above the kilowatt-hour meter). Your responsibility begins at this point, as does coverage by NEC.

The wiring from a wind system is much like the power lines from a small generating station. The lines can be strung above ground or buried. They enter your home in a manner similar to those from the utility. Let's first look at wiring on the tower and then consider how we're going to deliver the power to your home, barn, or business.

In all cases, the leads from the generator must be connected (*terminated* or spliced in the jargon of electricians), to the wires (*conductors*) running down the tower. These connections must be permanent and weatherproof. How this is accomplished depends on the kind of wind turbine and on the manufacturer.

When the generator is stationary within the tower, as in a vertical-axis wind turbine or in some horizontal-axis machines with right-angle drives, the generator leads can be directly connected with the conductors on the tower. On more conventional wind turbines, though, there must be a mechanism for transferring power from the moving platform of the wind machine to the stationary tower. Slip rings and brushes usually perform this task on small wind turbines.

On many larger wind turbines, however, the leads from the generator hang freely through the center of the tower. As the machine yaws, or turns to face the wind, the conductors twist. The cable is permitted to twist several revolutions before it must be unwound. This practice is common on the medium-sized wind turbines used in California.

Good connections between conductors are those that are mechanically tight, electrically insulated, and corrosion resistant. Connections may be made by using a twist-lock plug, split-bolt, wire nut, or crimp (compression) connector. Each has its merits.

Weatherproof *twist-lock plugs*, rated for the power output of the generator, have been used in the past. They're unsightly when dangling from the tower and don't offer a reliable connection. Under some building or electrical codes twist-lock plugs may be labeled an "impermanent" connection and thus may not be permitted.

Split-bolts are easy to use and come in sizes suitable for even the heaviest power cable. The stripped ends of the conductors are inserted into the jaws of the bolt and the faces tightened. These connections need an insulated boot.

Wire nuts are the most popular connector because of their ready availability and ease of use. Nearly every home handyman is familiar with them. They do have their drawbacks. They too may loosen under vibration and need to be secured with electrical tape. Moreover, they're most practical on solid rather than stranded wire and only on the smaller wire sizes. Wire nuts for heavy-gauge wire are hard to find and are more difficult to use.

Crimp or *compression connectors* are the best all-around termination. They are mechanically sound and will not loosen from vibration. Some include an insulated covering, but all can be insulated with a boot of shrink tubing slipped over the connector. Their chief disadvantage is the need for a special crimping tool, and in the larger wire sizes this tool is both expensive and awkward to use.

For small-gauge wire, crimp connectors or wire nuts are best. When joining heavy-gauge conductors, use either crimp connectors or split-bolts. Cover the connection with shrink tubing or wrap securely with good-quality electrical tape.

The connection between the leads from the slip rings and tower conductors can be made in the open or within a junction box. The conductors are then run down the tower, protected within metal or plastic conduit.

Previously, some installers have taken shortcuts and used twist-lock plugs at the top of the tower and then laced the power cable down a tower leg. They would hold the cable in place with electrical tape or with nylon cable ties. This has proved unsatisfactory. It doesn't meet building or electrical codes in many communities and presents a hazard to servicemen, to say nothing about the potential for maintenance problems.

Conductors on the tower should be protected within conduit (see Figure 14-12). Either electrical metal tubing (EMT) or plastic tubing can be used. Metal conduit is preferable because it's strong and provides some shielding against voltage transients and lightning. Each 10-foot section of EMT is joined by weatherproof compression couplers and mounted on the tower with conduit hangers. The hangers should be spaced two per section with one at each end near the coupler. If you're using PVC (polyvinyl chloride) conduit, make sure it's rated for electrical use. Because PVC flexes, use three or four conduit hangers per section.

If the connection between the leads from the slip rings and the tower conductors is made in the open, then a *weatherhead* will be needed at the top of the conduit string. The weatherhead, which prevents rain and snow from entering the conduit, is the same kind used for the service entrance at

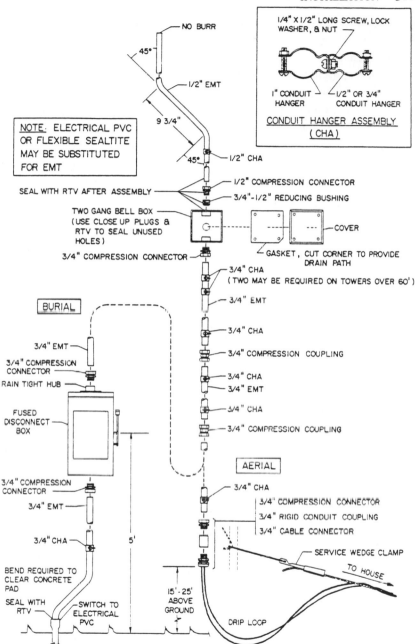

Figure 14-12. Tower conduit assembly. Power cables from the wind machine should be protected within conduit. (Bergey Windpower Co.)

your home. The connection is made, and slack in the conductors is used to make a drip loop before they are fed into the weatherhead. It's preferred, however, that all connections be made within a weather-tight enclosure or junction box. (Most codes insist on it.)

The junction box doesn't have to be elaborate. An inexpensive bell box is sufficient for small turbines. The conduit on the tower is coupled directly to the box. Similarly, the leads from the slip rings can be fed through a short section of conduit into the box. Most manufacturers provide fittings for attaching conduit to the slip ring assembly for this purpose. Once the splices have been made, the box can be sealed. The entire conductor run from the wind turbine to the base of the tower is then protected.

From the junction box to the base of the tower, conduit of ¾-inch diameter or larger is best. You can get by with ½-inch diameter on short runs, but the few pennies saved are not worth the aggravation. Strain reliefs such as a basket grip will be needed to support the conductors within the conduit. Otherwise, the weight of the conductors will pull on the splices and eventually cause them to fail.

Above-Ground and Buried Wiring

Above-ground cable runs look easier than they are. If there's any distance at all to cover, you'll need what's called a *messenger cable* to support the conductors so they won't stretch. This steel or aluminum cable is secured at the tower and at the load with a service wedge clamp, prefabricated cable grip, or wire tie. The objective in stringing cable is to keep the conductors out of reach. To achieve these clearances over long distances you may need to essentially build your own power line by setting poles and hanging insulators. You'll also need to use conductors that are rated for exposure to sunlight.

For above-ground runs from the tower to the load, a weatherhead should be used both on the tower and at the service entrance. These weatherheads contain a plastic bushing where the wires emerge. This bushing prevents the metal conduit from nicking the insulation on the conductors and grounding a circuit as the cables sway in the wind.

You'll need conduit from the weatherhead at the service entrance to a disconnect switch. The weatherhead should extend above the conductors and be at least three feet from windows and doors. It should be mounted so that there's sufficient clearance between the conductors and the roof to meet local code requirements.

Positioning of the weatherhead, or the wind generator's service entrance as it could be called, is determined by where the utility's service enters the house. In many cases you'll want to install a disconnect switch in the line between the weatherhead and the wind system's control panel (see Figure 14-13). The disconnect switch is included to permit isolation of the

Figure 14-13. Utilities often require a lockable disconnect switch on the service entrance from the wind machine. This switch must be easily accessible to their linemen and should be located as near to the utility's service drop and billing meter as practical. Note PVC conduit rising out of the ground to meet a junction box, which is attached to the disconnect switch. Output from the switch is routed through metal conduit into the basement via the windowsill.

wind system from the utility and from the service panel during emergencies. (It's also useful during maintenance of the wind system.)

The disconnect switch is a redundant safety device on interconnected wind systems because they automatically disconnect from the utility line during power outages. The switch gives a utility lineman, though, a positive mechanism for ensuring his own safety. By throwing the switch off and inserting a "lockout" (a metal tag warning that someone is working on the line) through the switch handle, the lineman ensures that no power can flow through the line while it's being serviced.

From the utility's viewpoint, the lockable disconnect switch should be located near its kilowatt-hour meter. This isn't a sinister plot by the utilities. Requirements for lockable disconnect switches accessible to repairmen (whether they're utility linemen or private electricians) are a standard safety practice in industry.

Building codes, however, may call for the disconnect switch to be near the service panel and main disconnect switch (the switch between the utility and the service panel) for aiding firefighters. One of the first steps in fighting a fire is to cut all power to the building. Ideally, the switches should be located in one place so that firefighters need not fear that an undetected line in the building is "live" (energized).

The American Wind Energy Association suggests that the wind system's service entrance and the manual disconnect switch be located at or near the utility's service entrance. They also recommend that the conductors from the wind generator not pass through any buildings before reaching the disconnect switch. In other words, the distance from the wind system's service entrance and the control panel within the building should be minimized. The conductors should run externally to the point on the building nearest the service panel. In most cases this will be where the utility's service also enters the building.

The wind generator's output is then wired to a control panel or synchronous inverter depending on the type of wind system. In no case should DC output ever be connected directly with the AC service panel. Generators producing DC must incorporate a synchronous inverter before they can be interconnected with utility-supplied AC. Those wind systems producing "dirty" AC, that is, AC of varying voltage and varying frequency, must also use a synchronous inverter.

For those wind systems intended for interconnection with the utility, conductors from the control panel or synchronous inverter are wired to a dedicated circuit in the building's existing service panel. Output from the wind system must not be plugged into a wall outlet or wired to a circuit that's already supplying a load in the building, such as a refrigerator or a series of receptacles. Instead, a circuit breaker must be "dedicated" to the wind system.

In modern buildings or those with upgraded wiring the service panel is larger than needed for existing loads. The panel has knock-outs that can be removed for the addition of a new circuit. One or more of these unused slots can be set aside for the wind system circuit. On small interconnected wind systems a 30-amp circuit breaker may be adequate. For larger systems a 100-amp breaker may be required.

Some manufacturers suggest that a special outlet or receptacle should be wired to a dedicated circuit in the service panel. This receptacle will then be used solely for the wind system, similar to the current practice for large appliances such as electric ranges and dryers. No other loads are placed on the circuit. The output from the wind generator's control panel or synchronous inverter is wired to a pig-tail with the plug at the free end.

During normal operation, the wind system is plugged into the circuit. During an electrical storm where lightning may be a hazard, the plug is pulled. This physically separates the wind system wiring from the AC wiring in the building. Those using such receptacles argue that this gives greater protection to the other circuits in the building than does a standard circuit breaker. Lightning, they say, is less likely to jump the distance between the plug and its receptacle than it is across the terminals of a circuit breaker.

Overhead lines increase the likelihood of damage from static electricity and lightning strikes. Falling trees frequently interrupt service on the utility's lines, and the same can happen with your wind system. Burying the conductors minimizes these hazards.

To bury the run from the tower to the load, dig a trench with a backhoe or power trencher. The trench should be 18 inches deep if you're planning to use PVC conduit, 2 feet deep if you're planning to use direct burial cable. Install a fused disconnect switch at the base of the tower. This minimizes the damage done if someone inadvertently digs up your cables.

Direct burial cable is easier to use than conduit. You simply lay the cable directly into the trench. However, conduit allows the conductors to be removed if defective, and there's less likelihood that the conductors will be damaged as the trench is backfilled.

When using direct burial cable, thread the conductors through several feet of PVC conduit from the disconnect switch on the tower to the bottom of the trench. This protects the cable where it's more likely to be damaged. Repeat the procedure where the cable run leaves the ground at the service entrance.

Conductors and Conductor Sizing

Wind turbine manufacturers typically specify the type and size of conductors suitable for wire runs of varying lengths. But they seldom tell you much more. They may gloss over why wire size is important, and tell you nothing at all about how to work with wires as thick as your thumb. We'll go over some of that material here.

All conductors used for wind turbines are insulated. In the United States you can identify the insulation by the multiletter code printed on its surface: R = rubber, RU = latex rubber, T = thermoplastic, H = heat resistant, HH = high heat resistant, and W = moisture resistant. The type of insulation is important because not all insulation can be used in all applications.

Most conductors are insulated with a heat- and moisture-resistant thermoplastic. These are labeled THW. Because some insulation degrades in sunlight or is susceptible to mildew, conductors intended for use above ground or for direct burial should be so rated. Conductors that can be exposed to sunlight, such as those using cross-linked polyethylene insulation, are rated SE, for service entrance. Insulation intended for direct burial is fungus and corrosion resistant and can be identified with the UF label for underground feeder, or USE for underground service entrance. Conductors not labeled SE, UF, or USE must be used within conduit where exposed to the elements.

The quality of the insulation is especially important for wind systems using synchronous inverters. Leakage through the insulation to ground can destroy an inverter in seconds. To guard against this, the insulation must be rated for 600 volts. Though the operating voltage is well below 600 volts, some wind turbines can generate voltage spikes when unloaded. This burst of high voltage will shoot through any weakness in the insulation. Nicks in the insulation from rough handling or from sharp conduit edges can be enough to cause a ground fault and destroy an inverter.

The insulation should always be tested before the conductor is put into service. One dealer suggests dipping the whole spool of cable into a tub of water and testing it with a "megger" or meg-ohm meter. The megger should register a high resistance on the insulation. A leaking or faulty insulator will register with a relatively low resistance. The insulation should be checked again after the wiring has been completed to detect any faults created during installation.

Wire size is determined by the maximum voltage drop allowable between the wind generator and the load. Voltage drop increases with increasing resistance.

Resistance in a wire is a function of its diameter, length, and material. Resistance and, therefore, voltage drop increases with decreasing diameter and with increasing length or distance. As seen in Table 14-5, copper offers less resistance than aluminum. The voltage drop is therefore greater in aluminum conductors than in copper cables of the same size.

On interconnected wind systems, the voltage drop between the wind turbine and the load is critical to the proper operation of the wind generator and its power conditioning equipment. Take the case of a vocational-technical school that installed a 6-meter wind turbine driving an induction generator. The school was responsible for wiring the system into their service panel. Students and their instructor mapped out the conduit run and laid the conductors. When all was finished the dealer flipped the switch and nothing happened. He checked the wind turbine. Everything seemed fine. He then checked the students' wiring; okay, too. As they sat and scratched their heads, some wise guy suggested they measure the voltage.

The problem? Low voltage. The wind turbine's control system sensed a voltage below its disconnect value and wouldn't engage the generator with the line. (The manufacturer incorporated this feature to detect a power outage so the generator wouldn't energize a downed line and kill a lineman.)

The dealer and the students then went back to the books. The wiring was sized according to the installation manual, or so it seemed. Their mistake was failing to take into account that the panel where the wind system was interconnected with the school's service was a long way from the utility's entrance to the school. The wind turbine's conductors were sized properly for the run from the service panel in the outlying classroom. But the long distance from the utility's service entrance to the wind turbine's control panel caused an excessive voltage drop. They remedied the situation by upgrading the conductors.

The problem with wind systems using synchronous inverters is similar, that is, low voltage disrupts optimum operation of the wind system. Synchronous inverters are used with wind generators whose voltage output varies with wind speed. As wind speed increases and more power can be produced, the inverter increases the load on the generator. The inverter knows when to do this by monitoring the generator's voltage. If the voltage drop from the generator to the inverter is too great, the inverter is fooled into thinking that less power is available than may actually be the case. The wind generator is never loaded to its full potential, and performance suffers.

Given a wire run of a certain length, you have a choice between aluminum or copper conductors in different sizes to limit voltage drop to tolerable levels. In the United States wire size is designated by its American Wire Gauge (AWG) as shown in Table 14-5. The lower the number, the larger the wire. House wiring is usually no. 12 or no. 14 gauge, with the heavier gauge wire (no. 12) being specified in some communities. The leads from the generator of a small wind turbine are heavier still: often no. 10 gauge.

Table 14-5

Resistance for Common Wire Sizes in the United States

(Ohms/1000 ft)

Wire Size (AWG)	Copper (Cu)	Aluminum (Al)
10	1.29	2.04
8	0.809	1.28
6	0.51	0.808
4	0.321	0.508
3	0.254	0.403
2	0.201	0.319
1	0.16	0.253
0	0.127	0.201
00	0.101	0.159
000	0.0797	0.126

Wire size is graded in even numbers down to no. 0 or 1/0, read "one nought." Larger diameters are then designated as no. 00 (2/0) or "double nought," no. 000 (3/0) or "three nought," to no. 0000 (4/0) or "four nought." Larger sizes are available, but the designation system changes to the number of circular mills. The largest size commonly used for wind systems is 2/0 and the smallest size is no. 8, or one size larger than the leads from the generator. The size used depends on the allowable voltage drop, the distance from the generator to the load, and whether copper or aluminum wire will be used. For runs over 500 feet, aluminum has a definite cost advantage.

From Ohm's law we can derive a simple formula for calculating the maximum wire run permitted by the manufacturer:

$$R = V/I$$

where R is resistance in ohms, V is voltage in volts, and I is current in amps. But first consider what happens when we solve for volts.

$$V = R \times I$$

The voltage drop increases in direct proportion with resistance and with current. An increase in current increases the voltage drop through the conductor. If we wanted to use a 1-kilowatt wind generator in a 12-volt system, we'd need to pump about 80 amps through the conductors. In a 120-volt system we'd need to handle only 8 amps. If we'd sized the system so it would handle 8 amps with only a 1 percent voltage drop, we'd encounter a 10 percent voltage drop at 80 amps.

To minimize resistance losses and limit voltage drop, system voltage increases with the size of the wind generator. At 12 volts a 1-kilowatt wind turbine will produce about 80 amps; a 10-kilowatt turbine, 800 amps. For example, micro wind turbines (those less than 1 kilowatt) are designed to operate at 12-24 volts. Battery-charging wind turbines in the 1-5 kilowatt range usually operate at 24-48 volts. Larger wind turbines operate at 120, 240, and 480 volts.

Now we can determine the acceptable distance to the load for various size conductors, and under various loads. The total resistance R_T in a wire run is the product of the wire's resistance per foot times the total distance the current must travel: two times the run L from the generator to the load.

$$R_T = R \times 2L$$

By substituting the equation for total resistance into Ohm's law, we can solve for L or the length of the run where V is the permissible voltage drop.

$$L = V/2IR$$

This formula was used to construct the table in Appendix G for a 1 percent voltage drop at 12, 24, 48, 36, 120, and 240 volts. An excerpt from this table for a generator voltage of 120 volts is given in Table 14-6.

In practice you will use this table backwards. The distance between the wind turbine and the load is known, and you need to find the right wire size to span it. You can use the table in Appendix G and find the appropriate wire size once you know the voltage drop permitted. Acceptable voltage drops range from 1 percent for interconnected wind systems using synchronous inverters to 3 percent for those with induction generators. Battery-charging systems can tolerate up to 5 percent. The distance to load for a 3 percent voltage drop is three times that shown in Table 14-6 or Appendix G; for a 5 percent drop, it's five times that shown in the tables.

If you don't want to use the table, you can calculate the maximum resistance and match this with the resistance of various wire sizes in Table 14-5. Let's work an example of a 120-volt wind generator producing a maximum of 25 amps (3 kilowatts) and an allowable voltage drop of only 1 percent to the synchronous inverter. The one-way distance from the generator to the service panel is 300 feet. (Note that this distance includes the tower height and the run from the wind system's service entrance to the service panel.) What size wire is needed?

$$R = V/2LI$$
$$= \frac{120 \text{ Volts x } 1\%}{2 \text{ x } 300 \text{ ft x } 25 \text{ amps}}$$
$$= \frac{1.2}{15,000}$$
$$= 0.00008 \text{ ohms per ft}$$
$$= 0.08 \text{ ohms}/1000 \text{ ft}$$

Table 14-6

One-Way Distance to Load for 1 Percent Voltage Drop for Copper and Aluminum Wire in the United States

Approx. Gen. Size (kW)	Max. Current (amps)	Wire Size (AWG)									
		10		8		6		4		3	
		(Cu)	(Al)	(Cu)	(Al)	(Cu)	(Al)	(Cu)	(Al)	(Cu)	(Al)
120 Volts											
1.2	10	48	30	74	47	118	74	187	118	236	149
1.8	15	32	20	49	31	78	50	125	79	157	99
2.4	20	24	15	37	23	59	37	93	59	118	74
3	25	19	12	30	19	47	30	75	47	94	60
3.6	30	16	10	25	16	39	25	62	39	79	50
4.2	35	14	9	21	13	34	21	53	34	67	43
4.8	40	12	8	19	12	29	19	47	30	59	37
7.2	60	8	5	12	8	20	12	31	20	39	25
10	80	6	4	9	6	15	9	23	15	30	19
12	100	5	3	7	5	12	7	19	12	24	15

Using Table 14-5 we find that a 3/0 copper conductor would be needed to span this distance with only a 1 percent voltage drop. If we used the same wind turbine in a battery-charging application limited to a 5 percent voltage, AWG no. 3 aluminum conductors are adequate.

The large conductors required for the long wire runs sometimes needed to optimally site the wind turbine are heavy, stiff, and difficult to use. The terminal blocks inside inverters, control panels, and disconnect switches were designed to accept much smaller wire than, say, the 3/0 used on a 500-foot run. Disconnect switches, for example, are rated by their current-carrying capacity and not by the size of wire they can accept. If you use a switch of the correct current rating it will be too small to make the terminations. On the other hand, if you buy a switch large enough to accept the heavy-gauge conductors, it's much more costly than that needed. You can solve the problem by using junction boxes at each end of a long run. The heavy-gauge wire is joined with a wire sized appropriately for the control panel or disconnect switch. It's important to make good splices when using junction boxes, to avoid shorts or grounds, and to keep the overall resistance of the wiring run to a minimum.

It's to your advantage to use smaller gauge wire on the tower. It's much easier, for example, to handle no. 4 copper wire on the tower and in the

Wire Size (AWG)									
2		1		0		00		000	
(Cu)	(Al)	(Cu)	(Al)	(Cu)	(Al)	(Cu)	(Al)	(Cu)	(Al)
299	188	375	237	472	299	594	377	753	476
199	125	250	158	315	199	396	252	502	317
149	94	188	119	236	149	297	189	376	238
119	75	150	95	189	119	238	151	301	190
100	63	125	79	157	100	198	126	251	159
85	54	107	68	135	85	170	108	215	136
75	47	94	59	118	75	149	94	188	119
50	31	63	40	79	50	99	63	125	79
37	24	47	30	59	37	74	47	94	60
30	19	38	24	47	30	59	38	75	48

tower's disconnect switch than the 3/0 for the span from the turbine to the load used in the previous example. The lighter gauge copper wire will also fit satisfactorily within ¾-inch conduit, as discussed in the next section. By selectively using various size wires in this way you can keep total resistance and hence voltage drop within desired limits while simplifying installation.

Small wind machines producing AC are typically single-phase. Larger machines, those in the 10-meter range and above, are often three-phase. Three-phase is more efficient at transmitting energy by reducing the size and number of conductors needed than single phase. From our earlier discussion, remember that voltage drop in a conductor is partly due to the amount of current we are trying to push through it. In a three-phase system, each phase carries only one-third the current of the generator's total output. For example, the current carried by each phase of a three-phase 10-kilowatt generator operating at 220 volts is 15 amps, whereas a single-phase generator pushes 45 amps through the conductors to deliver the same power. The wire size required for the three-phase generator can be significantly smaller than that required to carry the 45 amps of the single-phase wind turbine for the same voltage drop.

Conduit Fill

The size of the wire you use will determine the size of conduit needed for protected cable runs such as on the tower. For practical reasons there's a limit to the amount of wire that can be stuffed into conduit. Electrical codes also limit the size of conductors used inside conduit. Table 14-7 gives the maximum number of conductors that can be used with different conduit sizes in the United States.

The cross section of a conductor is based on not only the wire size but also the amount of the insulation covering it. For example, ¾-inch conduit will handle three pieces of no. 6 gauge wire with ease but can be used with no. 4 gauge wire only when covered with certain types of insulation. Electricians generally prefer larger conduit than that necessary to meet code requirements, even though it's slightly more expensive. The larger diameter eases the task of pulling the conductors through the conduit. For example, a wind turbine installer selected 2/0 THWN on one 1000-foot run. The electricians then pulled the conductors through 1.5-inch conduit rather than struggle with the 1.25-inch permitted under the electrical code.

There are several ways to get the wire inside the conduit. For short runs, the wire can simply be pushed through. On longer runs a metal fish tape is pushed through the conduit first. The fish tape has a woven basket or cable

Table 14-7

Maximum Number of Wires inside Conduit in the United States with THHN-THWN (THW) Insulation

Wire Size (AWG)	Conduit Size (Diameter in inches)				
	1/2	3/4	1	1-1/4	1-1/2
10	6				
8	3	3			
6	1	4			
4	1	2			
3	1	1			
2	1	1			
1		1	1	3	
0		1	1	3	
00		1	1	2	3
000		1	1	1	3

grip to which the conductors are attached. The tape is then used to pull the wires through the conduit.

It's always easier to fish conductors through a string of conduit strapped to a tower while the tower is on the ground. If you use a gin pole for erecting the tower, or if you have to install a large truss tower a section at a time, you'll have to assemble the conduit string after the tower is installed. This necessitates fishing the conductors through the conduit once it's on the tower.

When wiring a tower in this manner, it's best to fish a line from the bottom up. In this way the ground crew must pull the conductors back down through the conduit from below. All the person on the tower has to do is feed the wire into the conduit. For the person at the top of the tower this is far easier than pulling the cables up the conduit.

You can simplify tower wiring by using armored cable that contains the correct conductors. Bergey Windpower specifies a cable assembly that uses a flexible metal jacket to protect the conductors that are inside. Armored cable eliminates two steps: stringing conduit down the tower, and fishing the conductors through the conduit.

Surge Protection

Wind turbines, like most electrical appliances, are susceptible to damaging voltage spikes. Electrical systems are grounded to limit voltage surges due to nearby lightning strikes. Grounding also ensures the prompt operation of fuses, circuit breakers, and protective devices from other electrical faults.

Lightning is only the most obvious of several sources for voltage spikes; passing clouds and the rapid opening and closing of switches on the utility's distribution system are others. Many wind turbine owners have found that voltage surges from the utility's lines are more frequent than those from lightning itself.

Lightning can short-circuit a wind generator in less than a second. Though it occurs almost instantaneously, lightning can be of sufficient voltage to break down thick insulation and arc over insulators. We try to minimize the effects of lightning by using lightning arrestors and lightning rods.

Lightning or surge arrestors furnish a path to the ground when a greater than normal voltage exists in a conductor. This drains off the excess voltage. After the voltage has returned to normal, the flow to ground ceases. Though there's no foolproof protection against lightning, arrestors do provide some protection to power lines and other electrical components.

Equipment and buildings can also be protected by raising the effective ground level with a static line or lightning rod. The static line on the utility's distribution system and the lightning rods on farm buildings are attempts to do just that. Lightning rods, such as those at utility substations, offer a 45-degree cone of protection beneath them. When lightning strikes a lightning rod, the lightning passes directly to ground without first going through the object shielded below. Some wind generators, notably those 10 meters in diameter and larger, sport lightning rods above the nacelle for this reason. Most systems, however, rely on thorough grounding to drain off any static charge.

Reducing the buildup of static electricity during a storm minimizes the possibility of a direct hit. Lightning doesn't always strike the tallest object. There are many documented cases where telecommunications towers, because they are thoroughly grounded, have been spared a direct hit while nearby trees have been incinerated. Proper grounding lessens the possibility of a direct strike and minimizes the damage if one does occur.

Unfortunately, concrete's an insulator. Steel towers on concrete piers with concrete anchors are electrically isolated from the earth. When hit by lightning, towers with concrete foundations will sit and cook like a king-size version of the coil in your toaster. To ground the tower, drive several $5/8$-inch copper-clad ground rods at least 8 feet into the soil. Mick Sagrillo of Lake Michigan Wind & Sun suggests that when bedrock is near the surface, the grounding rods can be driven in at an angle. In this way you'll still make a good ground without the necessity of drilling into rock. You can also bury ground rods in a trench.

On guyed towers, drive a ground rod near the mast and near each of the anchors. Connect the ground rods to the tower or guy cables with solid no. 4 gauge wire. Use heavy brass or bronze ground clamps (sometimes called acorn clamps) on the ground rods and split-bolts on the guy cables or tower girts.

In areas of high lightning incidence or where the soil is dry and sandy (this soil makes a poor ground), Jim Sencenbaugh recommends installing a ground net. On guyed towers the mast and each anchor is wired to its own ground rod. Then all the ground rods are tied together with a buried ground wire. On free-standing towers there should be a ground rod for each leg of the tower. Sencenbaugh also electrically bonds each tower section together with a jumper wire to ensure a continuous path down the tower to ground. To ensure a direct path to ground, he also suggests eliminating all sharp bends from the ground wire, as high voltage prefers to travel in the shortest, most direct path.

As mentioned in a previous section, an inexpensive means to protect your wind turbine from lightning-induced surges coming from the utility line is to use a plug and receptacle. It's a cheap and reliable disconnect for lightning storms, as long as you don't forget to pull the plug.

For further protection, run all tower wiring in EMT, not plastic conduit. Then ground the conduit properly. The metal conduit not only shields the conductors inside but also furnishes a path to ground for stray voltage induced by electrical storms. Your goal is to direct the surge into the ground rather than along your conductors into your control panel or synchronous inverter.

Additional Notes on Wiring

Electrical codes in the United States require that all metal electrical enclosures such as EMT conduit, disconnect switches, control panels, and inverters be grounded so that any fault (short circuit) between a live or hot conductor and the enclosure will be conducted safely to ground. Grounding causes the circuit's protective devices to function—fuses to blow, circuit breakers to trip. This prevents the metal enclosure from becoming energized and presenting a shock hazard.

Control boxes and switches must be properly mounted and all holes or cut-outs not in use must be sealed. There must also be a minimum three feet of clearance in front of any panel for servicing. Control panels and inverters with ventilation louvers must be located so as to allow for free air circulation.

All terminals, connectors, and conductors must be compatible. Poor connections and the use of dissimilar materials such as copper and aluminum are a major cause of electrical fires. Aluminum is particularly troublesome. It oxidizes when exposed to the atmosphere and forms a highly resistant crust. This increased resistance causes the connection to heat up under heavy loads and is believed to be responsible for numerous fires. Whenever aluminum is used, terminal blocks, split-bolts, and other connectors must be rated for aluminum (Al) or for copper and aluminum (CO/AL, or Cu-Al). Aluminum connections should also be coated liberally with an antioxidizing compound. (Don't overdo it, though. These compounds are conductive and could cause a short circuit if you get sloppy.)

Once the installation is complete, you're ready for the final check of your wind system before you begin operation, the subject of the next chapter.

 **15**

Operating and Maintaining a Small Wind System

Try not to let the thrill of starting your new wind turbine get the better of you. True, small wind turbines must be ruggedly built to withstand the elements. However, they're still electrical machines. As such, they're sensitive to proper installation. You must get it right the first time. As with most electrical components, you don't get a second chance. An improperly wired inverter or control panel can lead to costly repairs before the wind system has generated its first kilowatt-hour.

Initial Startup

Take your time and look over the entire installation. Go over the manufacturer's check list: all nuts snug, cables secured, wires connected to the proper terminals, and so on. If everything meets your satisfaction, go through the suggested startup procedure. Avoid starting the machine for the first time in a high wind. If a problem develops, light wind minimizes the potential for damage and makes it easier to bring the turbine under control. It's also a good idea to keep a watchful eye on the entire wind system the first few days to see that it operates as expected.

Interconnected Wind Systems

You should notify the utility—in writing—several months before installation that you plan to interconnect a wind turbine with their lines. This gives your letter ample time to move through the utility's bureaucracy and get the proper clearances. If the utility requires any special switches or metering, it's better to find out as early as possible to minimize costly modifications once the turbine is in place. Expedite the process by calling first and finding the person responsible for *small power producers* or *qualifying facilities*. Address the letter to that person's attention.

Include in your notification the following information:

brand name and model number of wind generator;
type of wind system (whether it uses an induction generator or synchronous inverter);
maximum power output in kilowatts;
operating voltage;
number of phases;
line drawing of the proposed installation, including any disconnect switches.

The line drawing should show the location of all disconnect switches and protective relays (called contactors) relative to the service panel, the kilowatt-hour meter, and the utility's service drop (see Figure 15-1). Describe how the wind system functions under both normal and emergency conditions (such as a power outage on the utility's lines). The utility wants to know how the design of the wind system guards against energizing a downed line and endangering their linemen.

Your letter should allay the utility's fears. It should tell them that the wind turbine's controls will automatically isolate the wind turbine from the utility's lines whenever a fault on the wind turbine side of the interconnection is detected, a fault on the utility's line is detected, or abnormal operating voltage or frequency is detected.

Additional information on the wind system's power factor or VAR (volt-ampere-reactance) characteristics, current harmonics, and maximum inrush currents may be helpful to the utility but is unnecessary to ensure a safe interconnection. The turbine's manufacturer will provide this information to you if the utility insists on it. Most utilities in the United States should have enough experience with small wind turbines that these questions have already been answered.

The utility is responsible for determining whether the interconnection poses a safety hazard to their linemen and whether it will interfere with their service to other nearby customers. To do so, they will want to inspect the installation before the wind system begins operation. Don't panic; this is a reasonable action. In some cases they will accept the inspection report of the local fire underwriter or code enforcement officer and issue you an approval

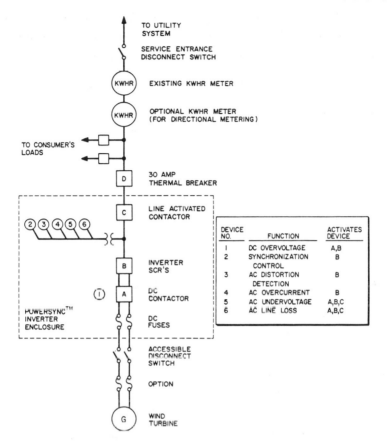

Figure 15-1. Utilities usually require line drawings such as this before they will permit an interconnection. The schematic tells them how the wind system will be interconnected with their lines and what precautions have been taken to protect their linemen. (Bergey Windpower Co.)

to begin operation until they can get out to your site themselves. But don't tolerate delays. Two weeks should be ample time for the utility to schedule an inspection if they need one.

Stand-Alone Wind Systems

Stand-alone power systems are much more complex than interconnected wind machines. There are more components (batteries, inverters, and backup generators), more wires, and more connections. There's not only a lot more room for error, there's no one to rely on but yourself. Each component demands special attention and should be carefully checked before initial startup. Consult the manufacturers' service or operation manuals for a startup check list on each component. Again, follow the checklists carefully.

Measuring Performance

The simplicity of modern small wind turbines has one drawback. When they're running properly—no frightening sounds or dangerous vibrations—it's difficult to know how well they're performing. Unlike their larger brethren, small wind turbines have little or no metering to indicate their production. You can make up for this deficiency by installing your own monitoring system and periodically checking the wind system's performance. At a minimum, every interconnected wind system should include a kilowatt-hour meter (see Figure 15-2). You may also want to install an anemometer on the tower (for more about anemometers see Chapter 3).

One simple test is to measure the wind speed when the turbine starts production. You can subsequently measure when it begins to furl or otherwise curtail generation. Begin by installing an anemometer as near the rotor

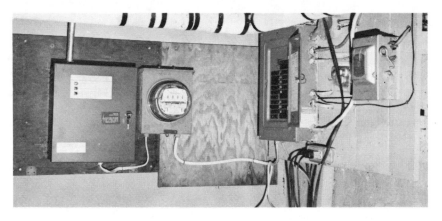

Figure 15-2. Control box and service panel. The service from the wind machine enters the top of the synchronous inverter on the left. The line-quality power leaves the bottom of the inverter in the plastic insulated cable and flows through the kilowatt-hour meter to a dedicated circuit breaker in the service panel on the right. The arrangement shown is the same on wind systems using induction generators. The control box for the induction wind machine would replace the synchronous inverter in this photo. The kilowatt-hour meter measures the amount of energy generated by the wind machine.

as possible. The measurements you're seeking are only approximations. First, the wind measured by the anemometer will never be the same as that striking the rotor, even when they're nearby. Second, wind speeds are always fluctuating. There's a lag between the time when a gust hits the rotor and when the wind turbine responds. Similarly, there's a lag when the wind ebbs and the rotor begins to slow down. Don't be alarmed if recorded wind speeds and power don't match the manufacturer's specifications the first time. You need a large number of observations to make any sense of what's happening.

Another test is to compare the power output from the wind machine at various wind speeds with the manufacturer's performance curve. The control panel on some machines includes a watt meter for this purpose, but on many small machines no meters are provided. You can improvise by using a kilowatt-hour meter on interconnected wind systems.

The kilowatt-hour meter records net energy production—the bottom line in reference to the economic value of a wind system. By measuring the rate at which a kilowatt-hour is being produced, you can calculate the average power. This is a good way to detect power-robbing problems before they show up in lower overall energy production.

Within the standard AC kilowatt-hour meter in the United States is a metal disc that spins in response to the flow of power. On the meter's face is a number labeled *kh*. This factor is in watt-hours of energy that passes through the meter per revolution of the disc. Some common values are 1, 1.8, 2, and 3. If kh = 1 the disc must revolve 1000 times before 1 kilowatt-hour registers on the meter. Where *P* equals average power:

$$P = \text{rev/min} \times kh \text{ (watt-hr/rev)} \times 60 \text{ min/h.}$$

For example, watch the meter for 1 minute and count the number of revolutions (there's a black mark on the disc for this purpose) and plug the data into the equation. You now have the 1-minute average power output. If you don't want to wait around for 1 minute, time the disc for 30 seconds and double the value (and so on). When you want to catch the output during a strong gust, measure the time it takes for a single revolution.

Now you can compare the average power output with the average wind speed for the same interval. You will quickly find that there is rarely a perfect match between manufacturer's power curve and the generator's actual performance. Sometimes it will be greater, sometimes less, depending on whether the rotor is coasting after a big gust or is just coming up to speed. You're looking for trends. If power output is consistently below that expected, try to find an explanation. For example, is it due to turbulence or to overly optimistic power ratings?

Maintenance

There's no concise answer to the question of how much maintenance is required on small wind turbines. Some wind machines are marvels of simplicity and are nearly maintenance-free. Others are more complex and the level of regular service required is higher. There are those that never seem to work right, or for very long, without a major repair. The amount of maintenance required depends on the type of wind machine, its size, and the approach of its designers.

Here are a few guidelines. Rotors with fixed-pitch blades require less maintenance than those with governors. Machines using direct drive require less maintenance than those using transmissions. Free-wheeling drive trains require less maintenance than those where the rotor must be motored up to speed. And those turbines using passive yaw to orient the rotor require less maintenance than those using active yaw drives. On the whole, small integrated wind turbines require far less maintenance than medium-sized wind machines.

If minimizing maintenance is a top priority of the designer, it's reflected in the final product. Wind machines destined for remote, stand-alone applications, where maintenance is not only infrequent but also costly, are designed from the beginning to be as maintenance-free as possible. This approach is typified by Karl Bergey's design philosophy. According to Karl, the maintenance instructions for a wind machine should simply read: "Once each year, on a windy day, walk out to the base of the tower and look at the mill. If the mill is turning, it's okay."

Many integrated small turbines, such as those built by Bergey Windpower and Northern Power Systems, are nearing such a state of hands-off operation. But we're not there yet. In an age where we're accustomed to automatic chokes and "idiot lights" on our cars, we should hardly be expected to run up and down a tower in foul weather carrying a grease gun. Yet after 100 years of development we still service our autos regularly. We shouldn't expect a machine operating in an environment as punishing as the wind to perform as maintenance-free as a refrigerator in the comfort of the kitchen.

Turbine and Tower

For modern small wind turbines, maintenance is principally a thorough inspection. However, it shouldn't be overlooked. Early detection of a problem can prevent costly damage to the wind turbine.

Maintenance may include little more than tightening loose bolts. Occasionally it may also entail cleaning the contacts on the slip rings. If your wind machine has grease fittings on any of the bearings, they will have to be greased semi-annually or quarterly. The trend, however, is toward sealed bearings and bushings that are designed to last the life of the machine. Oil in transmissions (where used) will have to be changed on a regular schedule, say every two to three years. This can be messy. But if it's required, make sure it gets done.

Inspect the turbine at least once each spring (after winter storms) and each fall (in preparation for winter). You can do this in a cursory manner from the ground on a calm day if you like by using binoculars. Check the rotor for symmetry. See if each blade looks alike. Obviously, if they don't, you have a serious problem. Watch how the turbine changes direction as the wind shifts. Note if the turbine yaws smoothly or abruptly. Erratic yawing can be due to turbulence, in which case the only solution is to install a taller tower. Erratic yawing can also occur if the tower is not vertical. (You can check the plumb of the tower with a level, plumb bob, or transit.) While still on terra firma, check that the tower is properly grounded and that the guy cables (where used) are tensioned correctly.

The remainder of the inspection must be made on the tower. Before performing any service, carefully read the next chapter on safety and take the appropriate precautions. Remember to furl the tail or apply the brake before climbing the tower.

At the top of the tower, check that all bolts on the rotor are snug and that the pitch of all the blades is the same. If any bolts or nuts are missing, replace them immediately. (Usually, if there were any missing, you would have known about it shortly thereafter.) Locking nuts are used in the rotor because of the severe vibrations. Locking nuts with a nylon insert can be applied several times before they need to be replaced. (Don't replace them with a regular nut.) Check whether the blades track each other, that is, follow the same path. You can do this by measuring the distance from a point on the tower to the tip of each blade. They should all be the same.

Pull the machine about the top of the tower and check whether it moves freely or binds in one position. If it does, you may need to grease the yaw bearings.

Change the oil, grease the bearings, and adjust the brakes as suggested in the service manual. Check that the bolts securing the tower adapter to the wind turbine and to the top of the tower are snug and that the yaw assembly is level. If not, use the leveling nuts provided on some machines or plumb the tower by adjusting the cable tension. Check that the brushes are aligned and their contacts clean. (Make sure the power is off!) Slip ring brushes can be cleaned with an alcohol-based fluid. Buff or lightly sand the contacts as required. Swing the turbine around again and ensure that the slip rings make good contact no matter which way the wind blows.

As you're coming down the tower, check that all bolts are snug. Don't overtighten as this will score the galvanized metal coating on the tower, causing rapid corrosion. Galvanizing acts somewhat like a lubricant, making it easy to overtorque a bolt.

Check for corrosion and secure connections at all wiring terminations. Use extreme caution anytime you open the control panel, synchronous inverter, or disconnect switch. If the connections look good, leave well enough alone and don't unwrap the insulation. When closing the door on any electrical enclosure, make sure you don't pinch the conductors between the door and the box.

Balance of Remote Systems

In stand-alone systems, batteries are the single most maintenance-intensive component, generally followed by the backup generator. The batteries' state of charge should be routinely monitored. If the state of charge indicates the batteries are low, remember that lead-acid batteries should never be discharged to more than 20 percent of their capacity or permanent damage can result.

Make sure that the backup generator operates properly to recharge the batteries when needed. To do so it may be necessary to start the backup generator occasionally to keep it well lubricated and to ensure that it will work when you need it.

Periodically it will be necessary to produce an equalizing charge on lead-acid batteries to balance out any voltage discrepancies between cells. Most battery chargers are capable of providing an equalizing charge. Occasionally top off battery electrolyte fluid with distilled water. Protect battery terminals from corrosion, and keep battery tops clean. Also keep the battery storage area tidy and don't store anything above the batteries. This will prevent metal objects from falling across battery terminals.

For more information on proper handling of batteries and their maintenance in a remote power system read, *The New Solar Home Book* by Joel Davidson. *Home Power* magazine also carries articles with helpful tips on how to get the most use out of batteries and inverters. (See Appendix I.)

Troubleshooting

You may be able to make some minor repairs yourself. Others will require professional help, and only you can be the judge. The following guide will give you some idea where to look when trouble develops. When you find a defect in your wind system, first determine if continued operation will cause any harm or create a hazard. If it will, shut the machine down and take corrective action. Make the needed repairs yourself or call in the dealer. If you're unsure, it's best to play it safe. You have a sizable investment to protect. If it's a minor problem, you may have the luxury of letting the machine run until you can get around to fixing it. But do fix it, or hire someone who will.

Problem: Rotor does not turn.

Check the anemometer to see that the wind speed is above the startup speed. If not, wait until the wind speed increases. If the wind speed is above startup, first check that the tail is fully extended behind the rotor on upwind machines, or that the brake is fully released on downwind machines. Second, if the tail is extended or the brake released, check whether the rotor shaft is frozen. Furl the tail or apply the brake, climb the tower, and spin the rotor by hand.

Problem: Rotor turns slowly in high winds.

On rectified alternators, check for shorted varistors, which are used for surge protection, or diodes, which rectify DC to AC. On induction generators, check for dragging brakes. On rotors with tip brakes, check that the tips are in their proper position and not deployed. Check for correct field excitation. On generators with synchronous inverters the electronic controls could be loading down the generator prematurely, preventing the rotor from getting up to speed.

Problem: Rotor vibrates or is unusually noisy.

Furl the tail or apply the brake, then check the rotor for symmetry (that each blade looks alike). Check for missing balance weights, stretched springs, or loose bolts. Shake each blade to see that it's securely attached. Check that the pitch (angle of blade to the wind) of each blade is the same. Check whether the blades are chipped or splintered and that the leading edge covers on the blades are secure. During the winter, there may be ice on the blades. This will correct itself. Check that the bolts mounting the wind turbine are snug. Check whether the blades track each other.

Problem: Turbine squeaks.

Furl the tail or apply the brake, then open the nacelle cover. Check that the bearings on the main shaft are firmly mounted. Lubricate the bearings if needed. Check that the main shaft spins true.

Problem: Wind turbine yaws erratically.

Check guy cable tension and plumb of tower. Check yaw assembly. Check for turbulence.

Problem: Tower vibrates.

Check rotor as described earlier. Check guy cable tension.

Problem: No voltage or power at the control panel.

Check that the tail or brake is released and that the wind speed is above 10 mph (5 m/s). If so, check the ammeter for current. If there's current, the voltmeter or its sensing circuit may be defective. Check the voltmeter by measuring voltage with a multimeter. If there's no current at the ammeter, check for a fault or open circuit between the generator and the control panel or between the control panel and the service panel. For example, check the slip rings. Check the field for open circuits, and check the brushes.

Problem: High voltage with no amperage.

Check for open circuit.

Problem: Power too low, voltmeter and ammeter erratic.

Furl or brake the rotor and check for loose connectors from the control panel to the wind generator. For example, check the slip rings, tighten loose connections, and clean tarnished terminals as needed.

Problem: Power output low.

On rectified alternators check for shorted diodes. Check that the turbine is fully unfurled or the brake fully released. Check that tip brakes (where used) are not deployed. Check that the transmission or main shaft is not binding.

This is only a sample of troubleshooting hints. The operations and maintenance manual or owner's manual for your turbine will offer more specific suggestions for problems characteristic of your particular wind system. Always refer to the manual, and if that fails, to the manufacturer for specific treatment of your machine's problems.

16
Safety

BOULEVARD, CA (UPI)—Terrence Mehrkam, 34, owner of a Hamburg, PA, windmill manufacturing company, was struck and killed by the blade of one of his own windmills at a "wind farm" in this San Diego County community.

The coroner's office said Mehrkam was struck by one of the blades after falling from a platform.

Contrary to popular belief, wind machines are not an entirely benign tool of a future solar society. An accident can easily kill or cripple. You could fall off the tower, get caught in the rotor, get hit by falling debris, or be electrocuted. Sadly, Terry Mehrkam wasn't the first, nor was he the last person killed working with wind energy. His accident and those of others should serve to constantly remind us of the danger inherent in working on a power plant high above the ground.

Despite their hazards, wind machines are no more dangerous than many other aspects of modern life. We have all grown to accept the hazards of the electricity and natural gas that flow through our homes. Yet accidents with these common energy sources, though not frequent, are certainly not rare. Common do-it-yourself projects, such as painting the eaves or repairing the family car, are just as dangerous as working on a wind machine. Treat wind systems with the same respect you would give any machine, and work as though your life's at stake, because it is.

Safety is both a state of mind and a way of life. This isn't a new Eastern philosophy. It's a necessity. Foremost in any discussion of safety equipment is that it must be used, and be used properly, before it's of any value. A safety belt or hard hat is no good when left on the ground.

Work on or around the tower poses the most risk: The possibility of accidents is greatest, and the severity of possible injuries is highest. You appreciate this best when hanging atop a swaying tower in a strong wind. So let's turn first to work on the tower.

Tower Safety and Climbing Gear

The simplest and most reliable way to avoid accidents is to avoid the hazard. Both wind machine designers and manufacturers should strive to eliminate working on the tower altogether. Wherever possible, for example, assembly should be completed on the ground and the wind turbine erected as an entire unit. Likewise, there should be a minimum of maintenance—preferably none—performed on the turbine once it's atop the tower. Hinged towers and integrated small wind turbines are beginning to make this goal a reality. But for most wind systems today, whether during installation or during repairs, work near the top of the tower is still very much a part of the job (see Figure 16-1).

Work and Positioning Belts

So-called safety belts serve two purposes: One is to protect against a fall, and the second is to free the hands. Many medium-sized wind turbines use a platform at the top of the tower to aid in servicing the machine. Some turbines are large enough that all work is performed inside the nacelle. On these machines the work belt is used principally to protect against a fall. Small wind turbines are far simpler, and there's seldom a work platform to stand on at the top of the tower. As a result, working on the tower of a small wind turbine demands a restraint system that frees the hands for assorted tasks.

In common parlance, the term "safety belt" is a misnomer. The safety belts used on construction sites and throughout industry are more correctly labeled positioning belts or, more simply, work belts. Work belts are made from a wide strap of leather or nylon webbing buckled around the waist. On each side of the belt are large steel D rings. A lanyard links the wearer of the belt to the tower via attachments at the two D rings and the tower. When using a work belt your legs carry all your weight.

The principal purpose of a safety belt is—as one would expect—to prevent falls. But when a fall does occur, the belt must keep you from hitting

Figure 16-1. *Servicing the fan tail on a medium-sized wind turbine (12.5-meter Aeroman) in California. Note work platform.*

the ground. And equally as important, the safety belt itself must not cause injury. Simple work belts may cause serious injury when the end of the lifeline is reached during a fall. The resulting shock is transmitted to the work belt where all the load is placed on the lower back. If the wearer is unconscious after the fall, they may even flip upside-down and slip out of the belt completely.

The popular lineman's belt, for example, isn't a safety belt or fall protection device. Originally designed for climbing utility poles, the "lineman's belt is, in fact, a positioning belt," according to William Glynn of SINCO Products, a major manufacturer of work belts. "It's only a work tool." The lineman's belt can give a false sense of security because it wasn't intended to rescue workers in a fall.

SINCO and other belt suppliers (see Appendix H) make full body harnesses designed to arrest the most severe free fall without injury. These harnesses include a work belt, leg straps, and a chest harness. The chest harness keeps the wearer upright after a fall and, with the leg straps, distributes the load from the fall uniformly over the torso.

In cooperation with belt suppliers, operators of California's wind power plants have devised their own harnesses for use with specific wind turbines. These harnesses are not directly transferable to working on small wind turbines because of the need to often work partially suspended from the tower. Standing for long periods on a narrow ladder rung or cross-girt of a small wind turbine tower is extremely tiring. Ideally, a single belt or harness could be found that safely protects you in a fall, frees up your hands to work suspended from the tower, and takes some of the load off your legs. Unfortunately, there's no one harness that does it all well.

Sit Harnesses

When examining fall protection, one casts about to find other fields where this problem is a major concern. Mountaineering quickly comes to mind. Sport climbing has developed the sit harness for direct aid or technical climbing.

Sewn sit harnesses are easy to use and, more importantly, comfortable when hanging suspended from the tower for long periods (see Figure 16-2). Sit harnesses are also superior to positioning belts during a fall (the shock is more evenly distributed among the legs, buttocks, and waist). Like positioning belts, sit harnesses allow free use of the hands. Moreover, they can be used to hang from the tower in a near sitting position. This relieves the strain on the legs.

Figure 16-2. Small wind turbines seldom offer the convenience and safety of a work platform. This necessitates hanging suspended from a work belt to free the hands.

Sit harnesses are less than ideal, and they may not meet U.S. safety standards. Alone, without a chest harness, they may also allow an unconscious person to slip out of the belt after a severe fall much like traditional work belts.

In the United States the Occupational Safety and Health Administration (OSHA) sets standards on work belts. These standards are based principally on strength. Work belts and lifelines must have a minimum breaking strength of 5400 pounds (2500 kilograms). Most sit harnesses do not meet this requirement. Depending on the style and manufacturer, sit harnesses are rated at 3000-4000 pounds (1400-1800 kilograms) breaking strength. If you must comply with OHSA, you have no choice but to stick with traditional work belts and full body harnesses designed to meet their regulations.

If neither a full body harness nor a sit harness is for you, there is a work belt that can be used to hang suspended from a positioning line in a half-sitting posture. These bosun belts (as in bosun chairs) are work belts that include a strap across the buttocks that can take some of the weight off your legs. They are more comfortable than simple work belts for extended tower work. In either case, work belts and sit harnesses are only part of a fall protection system. Without a lifeline they're not a whole lot of help.

Lifelines and Lanyards

Lanyards are short sections of rope with snap hooks attached. They are used as positioning lines or lifelines. To be used consistently and to allow freedom of movement, lanyards must be easy to release and reattach. Lanyards in the United States must "provide for a fall of no greater than six feet." In practice you should keep the lanyard as short as possible.

Both lanyards and lifelines (simply long lanyards) must have a minimum breaking strength of 5400 pounds (2500 kilograms). Like work belts, lifelines themselves should prevent the wearer from hitting the ground and should not cause any injuries. To behave this way the lifeline must "give" or stretch in a fall. Nylon rope or webbing is the material most often used for lifelines because it can stretch to minimize the shock to the body when the lifeline or lanyard reaches its limit. Steel cable, in contrast, will stop a fall abruptly, leading to serious injury. Hemp rope is little better.

Rope serves many functions during the installation of a wind system. Nylon rope, for example, could be used to lift a wind generator or restrain you in a fall. For this reason it's important to note that a safety line should be dedicated to that use and no other. Since repeated use weakens rope, make sure that lifelines or lanyards are always at their peak strength. In a fall, you get only one chance.

Lanyards must be attached to a fixed structural element. When working on small wind turbines, avoid the temptation to simply wrap the lanyard around the top of the tower like a lineman on a utility pole. Also, keep the lanyard clear of any moving parts.

Snap Hooks, Carabiners, and Slings

As the name implies, *snap hooks* are shaped like hooks and can be snapped onto a line, tower leg, cross-girt, or the D ring of a work belt. A safety catch (keeper) prevents the hook from releasing accidentally. Snap hooks are easier to attach than carabiners (snap links), and are the only form of attachment approved for work belts and harnesses used by professional wind smiths or those who maintain California's medium-sized wind turbines. Riggers and others in the construction industry also use snap hooks exclusively. A quick glance always tells you which end is up and how to apply it. This isn't the case with carabiners, which are symmetrical. Snap hooks also come in several throat sizes—up to 2-¼ inch wide for work with reinforcing bars. These rebar hooks are useful for clipping onto tower legs or cross-girts.

Carabiners still have a place. They're useful for clipping gear onto a work or tool belt. Any time a rope must be fastened and unfastened a number of times, a carabiner can make the task easier. Aluminum carabiners have a breaking strength of 4000 pounds (1800 kilograms). Steel carabiners are nearly twice as strong.

Snap hooks are designed to meet OSHA's breaking strength requirement; aluminum carabiners are not. For fall protection, use only OSHA-approved snap hooks.

Sewn loops of nylon webbing or "slings" were developed after climbers learned that a great deal of valuable rope was being lost when constructing rope harnesses. With a carabiner attached, a nylon sling can be used like a lanyard. Slings can also be used to carry equipment up the tower, tie down the rotor, and perform any number of other tasks. Wider and stronger nylon slings are used extensively in rigging. Where an erection jig or fixture is not handy, slings (appropriately rated) are a good choice for lifting heavy loads such as wind generators or a completely assembled turbine and tower.

Fall Protection Systems

One of the more dangerous activities when working on a wind system is ascending and descending the tower. It's dangerous because the common practice is simply to scale the tower, then secure yourself with a lanyard once you have reached the work station. It's even more dangerous coming down when you're tired and your timing can be off, particularly in winter when biting winds quickly sap your strength.

Several manufacturers offer a device to mitigate this hazard (see Figure 16-3). They employ a sleeve that slides along a taut cable or rope that spans the length of the tower. You attach your work belt or harness to the sleeve with a snap hook and lanyard. The sleeve rides up the cable as you climb the tower. Should you slip, it locks onto the cable, arresting the fall. (The nylon lanyard dampens the shock.)

These devices are standard equipment on all the wind turbines operating in California wind plants. They provide protection when climbing a ladder to the enclosed work platform found on most medium-sized wind turbines.

On the surface, this seems to be the ideal fall protection system. But it does have drawbacks. Chief among them is cost: about $500. Another is

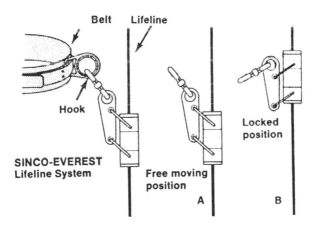

Figure 16-3. Sliding sleeve fall protection system. When the sleeve is lower than the work belt as at (A), the sleeve slides freely along the lifeline. When the sleeve is above the work belt as at (B), the sleeve locks onto the lifeline, arresting descent. (SINCO Products, Inc.)

that it seldom allows much freedom of movement, especially at the top of the tower. Wind turbines are naturally the highest point on the tower. Unfortunately, cable and sleeve systems extend only to the top of the tower to keep them from interfering with the wind turbine. When servicing the turbine this necessitates unhooking the lanyard from the sliding sleeve, then reattaching it to a sturdy anchor on the work platform.

Less costly, but more time-consuming (thus less likely to be used), is a two-lanyard system. When ascending, one lanyard is attached above as far as possible. When it is reached, the second lanyard is attached above and then the first removed, and so on. This ensures that you are always tied to the tower, even when reattaching a lanyard. The two-lanyard technique illustrates a good overall safety practice—always keep one lanyard or safety line firmly attached. This is particularly helpful when you need to reposition yourself at the top of the tower.

Although the cable and sleeve system provides acceptable fall protection on the tower, it doesn't solve the positioning problem when servicing a small wind turbine atop its tower. You will still need to use your lanyards for both fall protection and positioning. For example, assume you need to remove the nacelle cowling on a small wind machine. There might be an attachment point on the chassis useful for positioning, but you've got to get there first. A lanyard clipped at your last position should give you enough slack to reach the turbine while still protecting you in a fall. When you reach the new position you can attach a second lanyard.

More Tower Tips

Besides work belts and harnesses, there is some other gear that can make tower work both safer and more comfortable. Boots are one, gloves another. Always wear boots with firm, nonslip soles. Your feet tire less and are less likely to slip from a girt or ladder rung than with street shoes. Gloves do more than protect the hands—they help you get a better grip, and a good grip is paramount. Leather is best. The galvanizing used on towers forms droplets on the steel before it cools. These droplets can be sharp as a knife, cutting through cloth gloves with ease.

Hard hats are also essential attire. Admittedly they are uncomfortable, particularly in winter, and they're difficult to wear in a high wind unless fitted tightly or used with a chin strap. However, their value becomes apparent when working around small wind turbines that lack parking brakes, or after a wrench whizzes by your head.

Most small wind machines lack parking brakes because of their added cost and complexity. Even when furled, the rotor may still spin. Those blades may not look like much, but they can easily kill you. The drive train contains a lot of inertia when the rotor is turning, and this inertia can drive a lightweight blade with damaging force.

A similar problem is unexpected yawing of the turbine in gusty winds. Just when you think you're clear of all that machinery, the wind will change direction and bring everything swinging your way. It's then that a hard hat and a safety line are truly appreciated. Larger turbines, like those on wind farms in California, feature both a parking brake for the rotor and a brake to prevent the nacelle from yawing unexpectedly.

This brings us to a cardinal rule: never climb the tower while the turbine is running. Always apply the brake, furl the rotor, or do whatever else is required to shut the unit down first.

Never work alone. Always have a neighbor or a friend nearby, who can go for help if you need it. Hand-held radios are a useful tool for talking with your ground crew. Even with only a slight breeze it's hard to hear commands from the top of a 100-foot (30-meter) tower.

If the tower doesn't have a cable and sleeve device, climb the tower on the windward side whenever possible. The wind will force you into the tower, not off it. Never climb the tower in high winds. In California wind plants, for example, no work is performed in winds above 25-30 mph (about 12 m/s)

Keep the base of the tower clear in case a tool or some lost parts come hurtling to earth. No one should work at the base of the tower while someone is working above. A worker in California—who wasn't wearing a hard hat—received a serious head injury when his companion dropped a bolt from the top of the tower.

As for tools, always carry them in a tool belt; don't carry them in your hand. All other items should be hoisted up with a rope once you are safely in place. Take a hand line up with you on your first trip. Then use a nylon or canvas bucket to ferry small parts up and down the tower or to hold parts while you're working.

When around rotating machinery, whether it's a wind machine or a bicycle, don't wear rings, watches, loose clothing, or long hair (tuck it under your hard hat if need be).

Stay clear of the tower during ice storms or freezing rain. If operating, the rotor will shed ice by throwing it to the ground. Ice projectiles will strike directly below the wind turbine and can be a hazard in the immediate vicinity of the tower. It's also extremely dangerous to climb the tower in freezing rain or when the tower's covered by ice.

Electrical Safety

Wind generators produce high voltages. Use extreme caution anytime you open the control box or the nacelle cowling, or work around the slip rings. Always turn off the power before working around any of the electrical components, and use insulated tools whenever possible. Remember, electric shock can kill. But if you're working on the tower, the shock itself may not be the greatest danger. Electric shock can cause you to lose your grip. If the shock doesn't kill you, the fall will!

Avoid an electric shock by checking the circuits with a multimeter before you begin work. You could have thrown the wrong switch by mistake. Take your time and think about what you're doing. Never wear metal jewelry when working around electricity (even with gloves).

Avoid constructing a tower near utility lines. If you have any doubts about clearance between the tower or the boom of a crane and a power line, call the utility company before you start to erect the tower.

Stay clear of the tower if a storm is threatening, especially an electrical storm. A lightning strike anywhere near the tower will energize all metal components. Static buildup before a storm can produce the same effect.

Use extreme care when working around or servicing batteries in a stand-alone power system. Ventilate the battery storage area to avoid hydrogen accumulation and thoroughly fuse battery cables to prevent unintended discharge. Wear goggles whenever working around batteries.

Rope and Cable

Rope is of the most essential tools for installing and servicing a wind machine. It can be a woven nylon rope used for a lifeline or a cable for hoisting the turbine. Because of its importance and its many uses, rope has continued to develop over the years until it has reached a state of complexity that's almost baffling. To pick the right rope or cable for the job, an understanding of a rope's inner workings is helpful.

Rope is made from fibers. These can be natural or synthetic. Hemp, manila, cotton, and sisal are examples of natural rope. (Manila is the most common.) Synthetic rope is made of nylon, polyethylene, polypropylene, or polyester. The fibers are twisted to form yarns, and strands are wound from the yarn. Several strands are then combined to form rope.

As opposed to rope, the strands of steel cable are wound from wires instead of yarn. The wire is made from various grades of steel, depending on their use. For example, stainless steel may be used when corrosion protection is desired.

Europeans have developed a jacketed rope that has become a climber's standard accessory. In the kernmantle construction a core (kern) of braided or twisted strands is covered by a protective, braided sheath (mantle). No strands are exposed at the surface. Consequently, there's less opportunity for debris to work between the strands and damage the rope. The sheath can take a lot of abuse while still preserving the integrity of the core.

Manufacturers indicate the breaking strength of rope or cable on the spool. You should look for this when you purchase rope and cable because it will partially determine the rope you choose for a particular job. But breaking strength is not what you might think it is. The breaking strength (or tensile strength) of rope is the linear tension at which a rope breaks. So far, so good. A rope is considered to break, however, when one segment of the rope breaks—not the entire rope. And breaking strength applies only to new ropes; it decreases with age and use. Breaking strengths, furthermore, may vary between manufacturers because of the various testing methods employed, so always check the figures closely.

Breaking strengths also vary between American and European manufacturers. American manufacturers list the average breaking strength; European manufacturers list the minimum breaking strength which can be 10 percent to 15 percent less than the average breaking strength. Table 16-1 is a general guide to the breaking strength of new rope in the United States.

When choosing the type and size of rope for a specific job, buy it according to its working strength not its breaking strength. Working strength allows a reserve or margin of safety. This will account for any reduction in strength due to knots or unexpected loads on the rope.

The force acting on a rope can greatly exceed the simple weight of a load when the lift is uneven or the load drops suddenly. Knots, on the other hand, reduce the breaking strength of rope 45-75 percent. As a result of these effects, the working strength of rope is only one-fifth of its breaking strength. For example, if you are planning to lift a 500-pound (230 kilogram) wind machine, a rope with a breaking strength of 2500 pounds (1150 kilograms) should be used. Though a ½-inch (13-millimeter) manila rope would be adequate for the task, a ⅝-inch (16-millimeter) manila rope

Table 16-1

Breaking Strength (Pounds) of Rope in the United States

Size (in)	Manila	Nylon	Polypropylene	Polyester
3/8	1,350	3,725	2,350	2,700
1/2	2,650	6,080	3,990	6,080
9/16	3,450	7,600	4,845	7,600
5/8	4,400	9,800	5,890	9,500
3/4	5,400	13,490	8,075	11,875
13/16	6,500	16,150	9,405	14,725
7/8	7,700	19,000	10,925	17,100
1	9,000	23,750	13,300	20,900

would be a better choice. The heavier rope not only provides a wide margin of safety, but it's also thick enough that you can get a good grip on it. Used and worn rope may have a working strength only one-tenth of its breaking strength.

In general, nylon has twice the strength of manila rope. Nylon stretches; manila does not. Polypropylene also stretches, but is made from a coarser fiber than nylon and is more resistant to abrasion. Polyester is more resistant to sunlight than the other synthetics, but like manila, it doesn't stretch. These features determine which rope is better suited for the job. Nylon is preferred, for example, over manila or polyester for lifelines because it stretches. Manila is superior for hoisting.

Like rope, not all steel cable is created equal. Guy strand is stiff and heavy. Winch cable, in contrast, is flexible so it can be wound around the spool of a winch or threaded over a pulley. Guy strand comes in several different grades, from common or utility grade to extra-high-strength (see Table 16-2). Guy strand is also galvanized. Other cable may be of stainless steel as in aircraft cable, or simply untreated steel.

The construction of steel cable is given by the number of strands and the wires per strand. Almost all cable has at least six to eight strands. For example, a 6 x 7 cable has six strands and seven wires per strand. What you use depends on what you want to do with the cable. There's no sense in using the more expensive winch cable for tower guys; guy strand works better at lower cost. For a hoisting line, though, winch cable is best.

Inspect rope regularly for signs of abrasion. Natural ropes usually show clear signs of fatigue and wear. They become limp, surface fibers become soft and fuzzy, and the inner fibers rot. Synthetic ropes become limp and

Table 16-2

Breaking Strength (Pounds) of Steel Cable in the United States

| Size (in) | Winch Cable | Guy Strand | |
		Common	EHS
1/8	1,760		
7/32	5,600		
1/4	7,000	1,900	6,650
5/16	9,800	3,200	11,200
3/8	14,800	4,350	15,400
7/16		5,700	20,800

soft with frayed surface fibers. No rope should be used for more than 5 years even when properly treated.

When using steel cable for the hoisting line, check for nicks or kinks that may weaken the cable. You can keep the cable from fraying by wrapping tape around the ends. When cutting the cable, keep a grip on both ends so it does not recoil and whip around. The cable should not be pulled over a sharp radius or tied in a knot like a rope. Cable should be looped over a thimble and secured with Crosby clamps wherever it is attached. Remember that the U-bolts of the Crosby clamps bear on the dead end (the free end) of the cable.

Pulleys

Pulleys serve two purposes: to change the direction of the force applied to a hoisting rope and to provide a mechanical advantage. They make the job of hoisting wind turbine components easier and safer. Pulleys allow you to lift a load from the security of ground level, rather than putting yourself in the absurd and dangerous position of carrying up tower and wind turbine components by hand. Carrying a generator for a small wind turbine up a tower may seem ludicrous, but it's been done. No one does it more than once.

By fixing a pulley on a gin pole at the top of a tower and a snatch block (pulley with a hook) at the base, you can hoist components while standing a safe distance from the tower. Fixed pulleys alone provide no mechanical advantage. To gain a mechanical advantage there must be a pulley that moves with the load. The number of lines suspending the movable pulley or block determines how much advantage is gained. If the movable pulley hangs from two lines, the force needed to lift the load is reduced by one-half. To gain this advantage, though, twice as much rope must be used. Work equals force times distance. To perform the same amount of work as before when we reduce the force by half, we must double the distance through which it acts.

In practice, hoisting a small wind machine by hand requires at least one pulley in the movable block. This block is attached to the lifting ring or jig on the wind turbine. The hoisting rope is routed through a pulley at the base of the tower, over a pulley on the gin pole, and through the movable block. The fixed end of the hoisting rope is attached near the top of the gin pole. With this arrangement, a 75-pound force will be needed to lift a 150-pound wind machine, and a minimum of 300 feet of rope will be required on a 100-foot tower. (For this 2:1 advantage, you'll have three lines draped down the tower when the load is on the ground.) A heavier load requires either more force or another pulley in the fixed block on the gin pole. When the fixed end of the hoisting line is routed through the second pulley in the fixed block and then attached to the movable block, a 3:1 advantage is gained, but another 100 feet of hoisting rope will be needed.

For a thorough discussion of this topic read Michael Hackleman's *The Homebuilt, Wind-Generated Electricity Handbook.* (See Appendix I.)

Winches and Hoists

Heavy loads, such as complete tower assemblies and wind machines weighing several hundred pounds, can be hoisted up the tower using either a vehicle or a winch. However, using a truck to hoist a wind turbine is ticklish. Whenever a vehicle is used, you lose a degree of control over the lift no matter how well the driver feathers the clutch. Winches are preferred over vehicles for raising wind turbines and heavy tower components, but most winches have limited spool capacity. They can hold only 50-100 feet of cable (truck-mounted winches usually have less than 25 feet of cable). For 100-foot towers, the winch must have at least a 100-foot spool capacity, more if a block and tackle are used. These winches are found in industrial supply catalogs along with the handy capstan winch.

The capstan winch drives a spool or capstan. The hoisting rope is looped over the spool a few turns, but instead of winding the rope onto the spool directly, the spool merely aids in pulling the rope along. You have to take up the slack yourself.

Winches should have a remote control switch so you can stand clear of the winch when it's in use. They should also have a brake that locks the spool in either direction, and the brake should automatically engage if power is lost. You can mount the winch on a special stand for heavy lifts or on the bed of a pickup truck for light lifts.

Loss Prevention

As soon as possible after the tower has been erected, prominently post warning signs at eye level on the tower. These signs should proclaim *Danger: High Voltage* or *Danger: Authorized Personnel Only.*

If there is the possibility that the tower will become an "attractive nuisance" and be scaled by thrill seekers, children, or vandals, install anticlimb guards. These can be purchased from the tower supplier, or you can improvise. Protect manual controls at the base of the tower by removing winch handles or chaining them down.

Because the attachments of the guy cables on guyed towers are so tempting to vandals, mushroom the threads of bolts to prevent the nuts from being removed. Also install a safety cable through the turnbuckles. This prevents both vandals and normal vibrations from loosening the turnbuckles and releasing the guy cable.

Avoid placing guy anchors in pathways, but where you must, consider planting low shrubs or bushes around the anchors. People tend to detour around hedgerows rather than go charging through them. Shrubs can also soften the line between the tower and the anchor. Slip fluorescent guy guards over the guy cables to make them more visible.

Fence out horses and cattle. They love to scratch their backs on the edges of the tower. Don't give them the chance to damage any disconnect switches or tower wiring at the base of the tower.

This chapter has given general guidelines for working safely around wind systems. The material given is by no means an exhaustive prescription for protecting life and limb. If you're prudent and cautious you should be able to install, operate, and maintain your small wind turbine in relative safety. As mentioned in the beginning of this chapter, however, there's no better way to avoid accidents than by avoiding work on the wind turbine, its tower, or other components altogether. If you have any doubts about your ability to perform the tasks needed to install or service your wind turbine, don't hesitate to hire a professional.

 **17**

Looking to the Future

As you've seen in the preceding chapters, wind energy works. It may not be as simple or as straightforward as we would like, but it works, and works reliably. No miracles or media-grabbing breakthroughs are necessary before we put small wind turbines to productive use pumping water, powering remote homesteads, or generating electricity in parallel with that from the utility.

With rotor efficiencies now nearing 40 percent (about 70 percent of the theoretical limit), no one within the wind industry expects any startling technological leaps that will revolutionize small wind turbines. In keeping with the steady progress of the past decade, analysts expect continuing incremental improvements. Advances in aerodynamics have already resulted in quieter, more efficient airfoils. Small wind turbine veterans Mick Sagrillo and Mike Bergey both expect spinoffs from electric vehicle research to produce similar improvements in batteries, generators, and electronic inverters.

Recent progress by the auto industry has led to new rare-earth magnets that could find widespread application in small wind turbines. By adapting these magnets with their greater flux density to new generators, Mike Bergey believes designers will eventually squeeze more power out of low wind speeds. The first to do so was Wind Baron which introduced its Neo 500 in 1992. The successor to Wind Baron's popular micro turbine, the Windseeker, the Neo 500 uses recently developed neodymium-iron-boron magnets that are more powerful than the ferrite magnets commonly used today. If Wind Baron is successful, analysts expect other manufacturers to follow suit with new generators of their own.

Power electronics are also advancing rapidly. New electronic switching devices are driving a revival of static and synchronous inverter technology. The effects have already been seen in cheaper, more reliable inverters for stand-alone power systems introduced during the late 1980s. Improvements are just ahead in interconnected wind systems as well.

In conjunction with the U.S. Department of Energy's National Renewable Energy Laboratory (NREL), Bergey Windpower is developing a new synchronous inverter that will not only deliver improved power quality at lower cost but may also be used as an uninterruptible power supply. Today's synchronous inverters do their jobs well, but they're of little use when the utility's lines go down. Owners of contemporary interconnected wind systems find that they're just as vulnerable to power outages as any other utility customer. The Bergey-NREL inverter would, for the first time, enable the wind turbine to provide an independent power supply during utility outages. The inverter would automatically isolate the wind system from the utility and begin powering essential loads in the home. This new inverter would seemingly combine the best of a stand-alone power system with that of a wind system interconnected with the utility. If the technology proves successful, the inverter should be available by the mid-1990s.

Another development that could broaden the appeal of small wind turbines is further integration of the components (turbine, tower, and inverter) with installation. "The market is moving toward turn-key systems," says Sagrillo, where wind turbines would be sold and installed like any other major household appliance. Mike Bergey agrees. He sees continued refinement of installation with screw anchors, simple inexpensive guyed towers, and precast piers. Collectively, these refinements will permit erection of small turbines in a matter of hours, not days.

Altogether these enhancements will make small wind turbines more accessible to a greater number of users, and more cost-effective. Mike Bergey estimates that manufacturers could potentially reduce the overall cost of electricity from small wind machines 30-40 percent during the next 5-10 years. But much of the cost reduction, says Bergey, will come only through increased demand.

The future success of small wind turbines in North America lies not so much with new technology, although that's always helpful, but with creating a market that will sustain larger production runs. Volume production to a small wind turbine company means something quite different than it does to General Motors. Bergey Windpower, for example, projects marked cost reductions from a modest doubling of U.S. production from 400 turbines per year in the early 1990s to about 1000 units per year by mid-decade. This now appears possible.

During the 1980s U.S. manufacturers of small turbines sustained themselves principally on export sales. Modest growth in the international market, coupled with policy developments in the United States, could push demand and hence production up to the level Bergey Windpower believes necessary to gain economies in manufacturing.

Many think the market is there. The European Wind Energy Association estimates that there's a potential worldwide demand for as many as 15,000 small wind turbines per year during the 1990s. Mike Bergey sees a market for 5000-10,000 small turbines per year through the end of the century. The challenge is to get manufacturers started on the road to greater volume. One way could be to tap the financial and organizational clout of electric utilities.

Small Turbines at the End of the Line

The near-term future of small wind systems in both industrialized and Third World nations may be at the end of the line—the end of the utility's distribution line. Small wind turbines could conceivably become useful to utilities as a tool for strengthening weak distribution systems. They may even offer a means for the utilities themselves to provide power beyond the reach of their own lines (see Figure 17-1).

Utilities, such as Pacific Gas & Electric, have found that by reducing the loads at the end of heavily used lines they can avoid construction of expensive new transmission capacity. Loads can be reduced through conservation, or by installing modular sources of generation such as photovoltaics and small wind turbines. Dispersed sources of generation help the utility meet growing demand at lower cost than by traditional methods of expanding the distribution system.

This is one new market for interconnected wind turbines that's only now being examined. To test the concept, NREL is sponsoring a million-dollar project where one or more utilities will install small wind turbines at the end of their distribution systems. Robert Thresher, head of NREL's wind program, expects results from the study by the mid-1990s. By then,

WIND POWER
AT THE END OF THE LINE

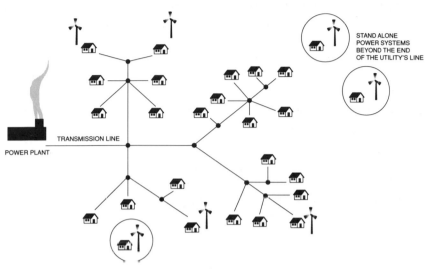

Figure 17-1. Wind turbines can be beneficial to utilities by strengthening weak distribution systems when placed at the end of long transmission lines. Wind turbines also enable utilities to serve customers outside the economic reach of existing transmission lines as part of stand-alone power systems.

utilities around the world should know whether they want to take wind not only to the end of the line, but beyond it as well.

Utilities are obligated to provide a service not necessarily to deliver central-station power to all sites. If that service can be met at lower cost by installing a remote power system, then there's little reason to extend the distribution system to remote customers just because that's the way it's always been done before.

The government of New South Wales now subsidizes stand-alone power systems for remote cattle stations in Australia's outback in lieu of providing central-station power from the provincial utility. Even a utility as conservative as Electricité de France has found this concept makes good economic sense for rural areas of France and its overseas territories.

It's likely that U.S. utilities will eventually follow suit and provide remote power systems for potential customers beyond the reach of their lines. Pacific Gas & Electric toyed with the idea. After examining the growing market for remote power systems in northern California, Pacific Gas & Electric refused to rule out their possible future entry into the market.

In some cases the utility may even find that it makes sense to take some small loads off the distribution system entirely. The Electric Power Research Institute (EPRI), in a study to identify cost-effective applications of photovoltaics, discovered that some midwestern utilities spend considerable sums serving extremely small loads, some of which were at one time served by water-pumping windmills. In one case EPRI found that a Kansas utility faced costly repairs of remote distribution lines from storm damage about every 4 years. The loads served by this Kansas utility, particularly remote water pumping, could be less expensively met by a stand-alone wind system.

These developments portend a future unimaginable a decade ago, one where the utility itself disconnects its own lines, installing a small wind system to serve the customer instead, one where the utility markets small wind systems to remote customers as part of its obligation to provide electrical service.

NREL's Thresher sees greater utility use of small wind turbines as inevitable. Even some in the U.S. utility industry see their companies owning and operating small, dispersed wind machines the way they now own and operate distribution transformers. These small power systems would simply become an extended, and sometimes discontinuous, portion of the utility's network for generating and distributing electricity.

The utility's development of dispersed small wind systems could take several forms. They could own the equipment outright, they could lease the wind system to the customer, or they could sell the wind system to the customer at a discount through a rebate program in much the same way U.S. utilities today give away energy-efficient light bulbs or offer rebates for installing energy-efficient appliances.

The utility and its ratepayers benefit by beefing up the distribution system and serving new customers at lower cost. The utility profits by expanding its rate base with new generating capacity, albeit in smaller units than they're accustomed too. And society gains a nonpolluting, renewable resource. Everyone wins, including small wind turbine manufacturers.

The buying power of the utilities could push production volumes higher, thereby helping manufacturers achieve the expected manufacturing economies. At the same time, the utility industry's admirable penchant for reliability will force wind turbine manufacturers to maintain the highest standards possible. Utility procurement practices could also ensure that only qualified products reach the field, deterring deployment of question-able wind turbine designs.

This is but one scenario of how a potentially large market for small machines may develop: the discovery of a niche where wind turbines add value to the utility over and above the avoided cost of the electricity the wind machines produce.

Fair Price for a High-Value Product

Consumers and manufacturers alike should welcome the greater utility involvement that end-of-the-line applications entail. Only with direct utility participation will many of the frivolous barriers still standing in the way of small wind machines be surmounted.

Without utility participation, the greatest single road block to develop-ment of dispersed interconnected wind turbines will remain: payment for its true value. When utilities are first permitted to include small wind tur-bines in their rate base, they will become ardent promoters of wind energy. If the utilities themselves are able to incorporate small wind turbines into their systems, either at the end of the line or beyond, then they will quickly uncover the value of wind energy's other attributes. It will be in their in-terest to do so.

With utilities at their side, renewable energy advocates will find it much easier to explain why it makes sense to pay more for wind-generated elec-tricity than for electricity from coal, oil, or even natural gas. It's an old idea, but one only now being implemented, mostly in Europe. Because wind turbines produce electricity cleanly without air or water pollution, use a renewable resource, provide generating diversity, and can be dispersed throughout a utility's distribution system, the electricity they produce is worth more than that from conventional power plants with none of these attributes. Studies in both Europe and the United States estimate that the environmental benefits alone are worth 3-5 cents per kilowatt-hour.

If this "environmental premium" were added to the current value of wind-generated electricity, the use of wind energy, whether through small or medium-sized wind turbines, would expand rapidly. Introducing this premium into the tariff for wind energy through the regulatory process in the United States may be difficult and take years of struggle, even with utility support. It's problematic without them.

A simpler approach to bringing payment for wind-generated electricity into line with its value, one adopted in Germany and Denmark, is to pay a fixed percentage of the retail rate. In Germany it's straightforward: the price paid for wind-generated electricity is 90 percent of the retail rate; in Denmark it's 85 percent. The same could be done in North America. This one act alone could do more to expedite wind energy's wide-scale use in the United States than all other incentives combined. Unlike the current determination of avoided cost in the United States, this approach would be simple and easy to administer. It wouldn't take an army of attorneys wrangling for a decade to implement such a proposal.

These environmental tariffs, whether as a simple percentage of the retail rate or as separate environmental "adders" to avoided cost, could quickly level the playing field between polluting fossil fuels and wind energy. The possibility exists. It happened in northern Europe. It could happen in North America as well.

Concerns about global warming and air quality have not waned since Germany and Denmark pioneered environmental rates. Now other European countries may follow suit. With similar tariffs on this side of the Atlantic, small wind systems could become as commonplace in North America as the water-pumping windmill once was.

Mick Sagrillo, whose dream of "a wind generator in everybody's backyard" was dashed during the 1980s, is hopeful once again. "It's no longer just tinkerers," he says, who are interested in wind energy. Sagrillo and others are convinced that environmental issues will repower the small wind turbine market during the 1990s. They believe small wind turbines will find their place in the sun.

Appendixes

A. Conversions

Speed
1 m/s	=	2.24 mph
1 mph	=	0.446 m/s
1 knot	=	1.15 mph
1 mph	=	0.870 knots

Length
1 meter	=	3.28 feet
1 foot	=	0.305 meters
1 kilometer	=	0.620 miles
1 mile	=	1.61 kilometers

Area
1 square kilometer	=	0.386 square miles
1 square kilometer	=	1,000,000 square meters
1 square kilometer	=	100 hectares
1 square mile	–	2.59 square kilometers
1 square foot	=	0.093 square meters
1 square meter	=	10.76 square feet
1 hectare	=	10,000 square meters
1 hectare	=	2.47 acres
1 acre	=	0.405 hectares
1 acre	=	4049 square meters

Volume

1 cubic meter	=	35.3 cubic feet
1 cubic foot	=	0.028 cubic meters
1 liter	=	0.264 gallons
1 gallon	=	3.78 liters
1 cubic meter	=	1000 liters
1 cubic meter	=	264 gallons
1 gallon	=	0.0038 cubic meters

Flow Rate

1 liter/second	=	15.8510 U.S. gallons/minute
1 cubic meter/minute	=	264 U.S. gallons/minute
1 gallon/minute	=	0.63 liters/second
1 gallon/minute	=	0.0038 cubic meters/minute

Weight

1 metric ton	=	1.10 tons
1 kilogram	=	2.20 pounds
1 pound	=	0.454 kilograms

Energy Equivalency of Common Fuels

1 kWh = 3413 BTU
= 3.41 ft^3 of natural gas
= 0.034 gallon of oil
= 0.00017 cord of wood

1 Therm = 10^5 BTU
= 100 ft^3 of natural gas
= 1 gallon of oil
= 29.3 kWh of electricity
= 0.005 cord of wood

1 gallon of oil = 1 x 10^5 BTU

1 cord of wood = 2 x 10^7 BTU

1000 ft^3 (Mcf) natural gas = 1 x 10^6 BTU

Approximate Primary Energy Offset by Direct Generation of Electricity

1 kWh	=	10,000 Btu
600 kWh	=	1 barrel of oil

B. Air Density Corrections for Temperature

Relative Change in Air Density with Temperature*

Temperature Celsius	Relative Change	Temperature Fahrenheit	Relative Change
−20	1.138	0	1.131
−15	1.116	10	1.106
−10	1.095	20	1.083
−5	1.075	30	1.061
0	1.055	40	1.040
5	1.036	50	1.020
10	1.018	60	1.000
15	1.000	70	0.981
20	0.983	80	0.963
25	0.966	90	0.945
30	0.951	100	0.929
35	0.935	110	0.912
40	0.920	120	0.896
45	0.906		
50	0.892		

*Change in air density relative to standard temperature of 15°C
or 60°F. Battelle's atlas of wind power density includes average temperature.

C. Rayleigh Wind Speed Distribution

You need only the average wind speed to determine the Rayleigh speed distribution. Use Table C-1 for average speeds in m/s, Table C-2 for speeds in mph. The average speed is in the top row. The *distribution* is the column below the average speed. The *wind speed bin* corresponds to the horizontal axis on graphs of wind turbine power curves. The *probability of occurrence* is the percentage of time the wind occurs within that wind speed bin.

Table C-1

*Rayleigh Wind Speed Distribution
for Annual Average Wind Speed in m/s*

Wind Speed Bin		Probability of Occurrence at Annual Average Wind Speeds in m/s					
(m/s)	(mph)	3	3.5	4	4.5	5	5.5
4	9.0	0.1728	0.1839	0.179	0.1668	0.152	0.1371
5	11.2	0.0985	0.1291	0.1439	0.1471	0.1432	0.1357
6	13.4	0.0453	0.0765	0.1006	0.1152	0.1217	0.1224
7	15.7	0.0170	0.0388	0.062	0.0812	0.0943	0.1019
8	17.9	0.0052	0.0169	0.0339	0.0519	0.0673	0.0789
9	20.2	0.0013	0.0064	0.0166	0.0302	0.0444	0.0571
10	22.4	0.0003	0.0021	0.0072	0.016	0.0272	0.0387
11	24.6	0	0.0006	0.0028	0.0078	0.0154	0.0247
12	26.9	0	0.0002	0.001	0.0035	0.0082	0.0148
13	29.1	0	0	0.0003	0.0014	0.004	0.0084
14	31.4	0	0	0.0001	0.0005	0.0019	0.0045
15	33.6	0	0	0	0.0002	0.0008	0.0023
16	35.8	0	0	0	0.0001	0.0003	0.0011
17	38.1	0	0	0	0	0.0001	0.0005
18	40.3	0	0	0	0	0	0.0002
19	42.6	0	0	0	0	0	0.0001
20	44.8	0	0	0	0	0	0
21	47.0	0	0	0	0	0	0
22	49.3	0	0	0	0	0	0
23	51.5	0	0	0	0	0	0
24	53.8	0	0	0	0	0	0
25	56.0	0	0	0	0	0	0
26	58.2	0	0	0	0	0	0
27	60.5	0	0	0	0	0	0
28	62.7	0	0	0	0	0	0

6	6.5	7	7.5	8	8.5	9
0.1231	0.1105	0.0992	0.0893	0.0807	0.0731	0.0664
0.1264	0.1168	0.1074	0.0985	0.0903	0.0828	0.0761
0.1194	0.1142	0.108	0.1014	0.0947	0.0882	0.0821
0.1049	0.1047	0.1023	0.0986	0.0942	0.0893	0.0844
0.0864	0.0905	0.0919	0.0914	0.0895	0.0867	0.0834
0.0671	0.0742	0.0788	0.0811	0.0817	0.0811	0.0796
0.0492	0.0579	0.0645	0.0691	0.0719	0.0733	0.0735
0.0343	0.0431	0.0507	0.0567	0.0612	0.0642	0.066
0.0226	0.0307	0.0383	0.0449	0.0503	0.0545	0.0576
0.0142	0.0209	0.0278	0.0343	0.0401	0.045	0.049
0.0085	0.0136	0.0194	0.0253	0.031	0.0361	0.0406
0.0048	0.0085	0.0131	0.0181	0.0233	0.0283	0.0328
0.0026	0.0051	0.0085	0.0125	0.017	0.0215	0.0259
0.0014	0.0029	0.0053	0.0084	0.012	0.016	0.02
0.0007	0.0016	0.0032	0.0055	0.0083	0.0116	0.0151
0.0003	0.0009	0.0019	0.0034	0.0056	0.0082	0.0111
0.0001	0.0004	0.0011	0.0021	0.0036	0.0056	0.008
0.0001	0.0002	0.0006	0.0012	0.0023	0.0038	0.0057
0	0.0001	0.0003	0.0007	0.0014	0.0025	0.0039
0	0	0.0002	0.0004	0.0009	0.0016	0.0026
0	0	0.0001	0.0002	0.0005	0.001	0.0017
0	0	0	0.0001	0.0003	0.0006	0.0011
0	0	0	0.0001	0.0002	0.0004	0.0007
0	0	0	0	0.0001	0.0002	0.0004
0	0	0	0	0	0.0001	0.0003

Table C-2

*Rayleigh Wind Speed Distribution
for Average Annual Wind Speed in mph*

Wind Speed Bin (mph)	(m/s)	Probability of Occurrence at Annual Average Wind Speeds in mph					
		7	8	9	10	11	12
8	3.6	0.0919	0.0895	0.0834	0.076	0.0686	0.0616
10	4.5	0.0645	0.0719	0.0735	0.0716	0.0678	0.0632
12	5.4	0.0383	0.0503	0.0576	0.0608	0.0612	0.0597
14	6.2	0.0194	0.0310	0.0406	0.0472	0.0509	0.0524
16	7.1	0.0085	0.0170	0.0259	0.0337	0.0394	0.0432
18	8.0	0.0032	0.0083	0.0151	0.0222	0.0285	0.0335
20	8.9	0.0011	0.0036	0.008	0.0136	0.0194	0.0246
22	9.8	0.0003	0.0014	0.0039	0.0077	0.0123	0.0171
24	10.7	0.0001	0.0005	0.0017	0.0041	0.0074	0.0113
26	11.6	0	0.0002	0.0007	0.002	0.0042	0.0071
28	12.5	0	0	0.0003	0.0009	0.0022	0.0042
30	13.4	0	0	0.0001	0.0004	0.0011	0.0024
32	14.3	0	0	0	0.0002	0.0005	0.0013
34	15.2	0	0	0	0.0001	0.0002	0.0007
36	16.1	0	0	0	0	0.0001	0.0003
38	17.0	0	0	0	0	0	0.0002
40	17.9	0	0	0	0	0	0.0001
42	18.8	0	0	0	0	0	0
44	19.6	0	0	0	0	0	0
46	20.5	0	0	0	0	0	0
48	21.4	0	0	0	0	0	0
50	22.3	0	0	0	0	0	0
52	23.2	0	0	0	0	0	0
54	24.1	0	0	0	0	0	0
56	25.0	0	0	0	0	0	0
58	25.9	0	0	0	0	0	0
60	26.8	0	0	0	0	0	0
62	27.7	0	0	0	0	0	0

13	14	15	16	17	18	19	20
0.0552	0.0496	0.0447	0.0403	0.0365	0.0332	0.0303	0.0277
0.0584	0.0537	0.0492	0.0451	0.0414	0.038	0.035	0.0323
0.0571	0.054	0.0507	0.0473	0.0441	0.041	0.0382	0.0355
0.0523	0.0512	0.0493	0.0471	0.0447	0.0422	0.0398	0.0374
0.0453	0.046	0.0457	0.0448	0.0434	0.0417	0.0399	0.038
0.0371	0.0394	0.0406	0.0409	0.0406	0.0398	0.0387	0.0374
0.029	0.0323	0.0346	0.036	0.0367	0.0368	0.0365	0.0358
0.0216	0.0254	0.0284	0.0306	0.0321	0.033	0.0334	0.0334
0.0153	0.0191	0.0224	0.0252	0.0273	0.0288	0.0298	0.0304
0.0104	0.0139	0.0171	0.0201	0.0225	0.0245	0.026	0.0271
0.0068	0.0097	0.0127	0.0155	0.0181	0.0203	0.0221	0.0236
0.0043	0.0065	0.0091	0.0116	0.0141	0.0164	0.0184	0.0201
0.0026	0.0042	0.0063	0.0085	0.0108	0.013	0.015	0.0168
0.0015	0.0027	0.0042	0.006	0.008	0.01	0.012	0.0138
0.0008	0.0016	0.0027	0.0041	0.0058	0.0075	0.0093	0.0111
0.0004	0.0009	0.0017	0.0028	0.0041	0.0056	0.0071	0.0088
0.0002	0.0005	0.001	0.0018	0.0028	0.004	0.0054	0.0068
0.0001	0.0003	0.0006	0.0012	0.0019	0.0028	0.0039	0.0052
0.0001	0.0002	0.0004	0.0007	0.0012	0.002	0.0028	0.0039
0	0.0001	0.0002	0.0004	0.0008	0.0013	0.002	0.0028
0	0	0.0001	0.0003	0.0005	0.0009	0.0014	0.002
0	0	0.0001	0.0001	0.0003	0.0006	0.0009	0.0014
0	0	0	0.0001	0.0002	0.0004	0.0006	0.001
0	0	0	0	0.0001	0.0002	0.0004	0.0007
0	0	0	0	0.0001	0.0001	0.0003	0.0005
0	0	0	0	0	0.0001	0.0002	0.0003
0	0	0	0	0	0	0.0001	0.0002
0	0	0	0	0	0	0.0001	0.0001

D. Maps of United States and World Wind Power

The accompanying map of the United States Annual Average Wind Power is taken from *Wind Energy Resource Atlas of the United States.* The map was published in 1987 by Battelle's Pacific Northwest Laboratory for the U.S. Department of Energy. This map and its companion state maps incorporate the most up-to-date information available to the general public. Unlike the previous survey published in the early 1980s this update includes data from both private and public sources never before assimilated into one comprehensive assessment of wind energy in the United States.

The Battelle wind atlas is the first place to turn for an overview of wind energy's potential anywhere in the United States. The 1987 atlas contains valuable background information on the power in the wind, terrain features, and meteorological conditions that affect the wind resources within each region of the country. The complete atlas can be obtained by writing

Table D

Battelle Classes of Wind Power Density for the United States*

Class	Wind Speed and Power at 10 m			Wind Speed and Power at 30 m			Wind Speed and Power at 50 m		
	Power Density (W/m²)	Speed (m/s)	Speed (mph)	Power Density (W/m²)	Speed (m/s)	Speed (mph)	Power Density (W/m²)	Speed (m/s)	Speed (mph)
1	50	3.5	7.8	80	4.1	9.2	100	4.4	9.9
2	100	4.4	9.9	160	5.1	11.4	200	5.5	12.3
3	150	5.0	11.5	240	5.9	13.2	300	6.3	14.1
4	200	5.5	12.3	320	6.5	14.6	400	7.0	15.7
5	250	6.0	13.4	400	7.0	5.7	500	7.5	16.8
6	300	6.3	14.1	480	7.4	16.6	600	8.0	17.9
7	400	7.0	15.7	640	8.2	18.4	800	8.8	19.7
	1000	9.5	21.3	1600	11.1	24.7	2000	11.9	26.7

*Increase in speed and power with height assumes one-seventh power law.

the American Wind Energy Association or the National Technical Information Service. See the Bibliography for details.

Battelle's maps don't show wind power density directly. Instead they present a range of possible values as *wind power classes* as shown in Table D. Residential wind systems are suitable in regions with wind power Class 2 or greater. Small wind turbines begin to look more economically attractive in Class 3 areas and above. Commercial wind power plants have been developed in Class 5 resources, though the most successful are sited in areas with Class 6 or better.

Following the map of the United States are six maps of wind resources around the world prepared by Battelle Pacific Northwest Laboratory. Battelle's power classes on these maps differ slightly from those used in the United States; the world maps show three power classes above the seven found on the U.S. map to provide more detail in windy regions. Classes 1 through 6 correspond directly on all maps.

Battelle Classes of Wind Energy for the World Wind Resource Maps

CLASSES OF WIND ENERGY FLUX (WEF)

WIND ENERGY CLASS	10m (33 ft)			50m (164 ft)		
	WEF W/m²	SPEED m/s	mph	WEF W/m²	SPEED m/s	mph
1	0	0	0	0	0	0
2	100	4.4	9.8	200	5.6	12.5
3	150	5.1	11.5	300	6.4	14.3
4	200	5.6	12.5	400	7.0	15.7
5	250	6.0	13.4	500	7.5	16.8
6	300	6.4	14.3	600	8.0	17.9
7	400	7.0	15.7	800	8.8	19.7
8	800	8.8	19.7	1600	10.1	22.6
9	1200	10.1	22.6	2400	12.7	28.4
10	1600	11.1	24.9	3200	14.0	31.3
	>1600	>11.1	>24.9	>3200	>14.0	>31.3

RIDGE CREST ESTIMATES (LOCAL RELIEF >1500 m)

SPEED ASSUMES A RAYLEIGH DISTRIBUTION

Pacific Northwest Laboratory
Operated for the U.S. Department of Energy
by Battelle Memorial Institute

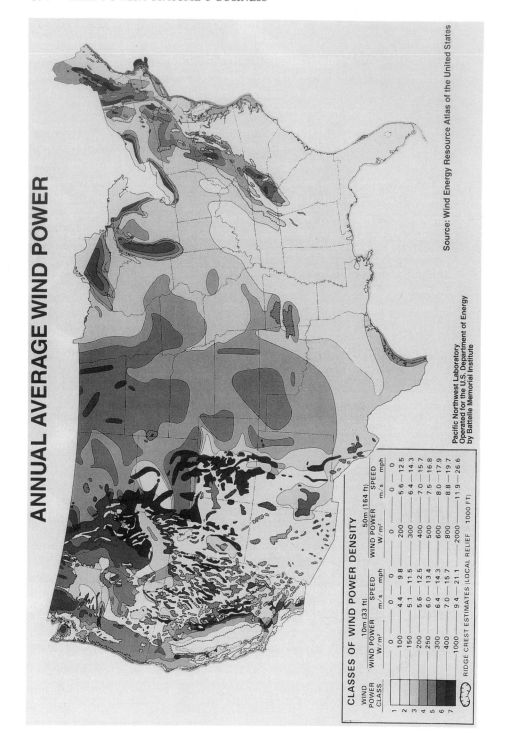

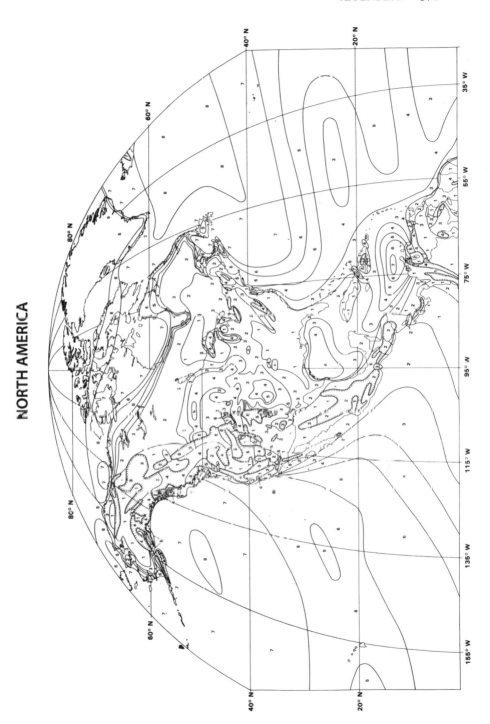

NORTH AMERICA

SOUTH AMERICA

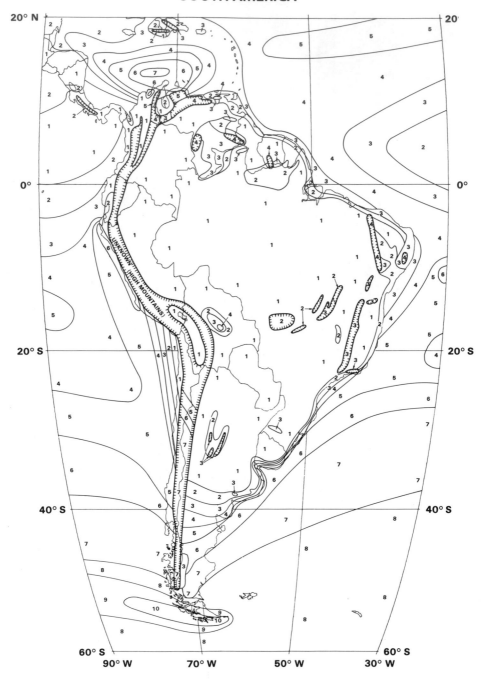

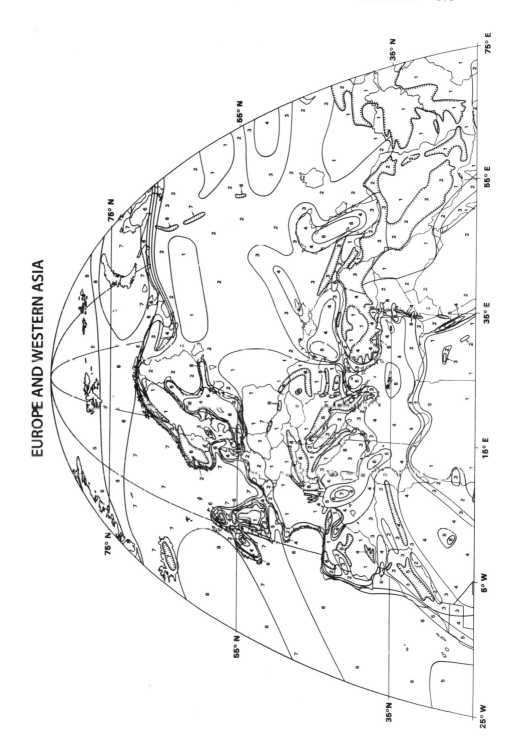

EUROPE AND WESTERN ASIA

AFRICA

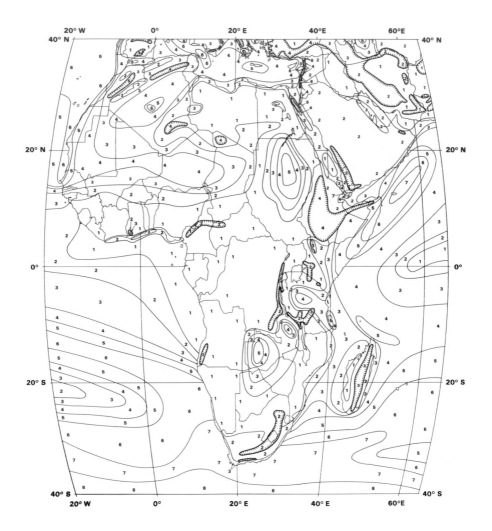

EASTERN ASIA

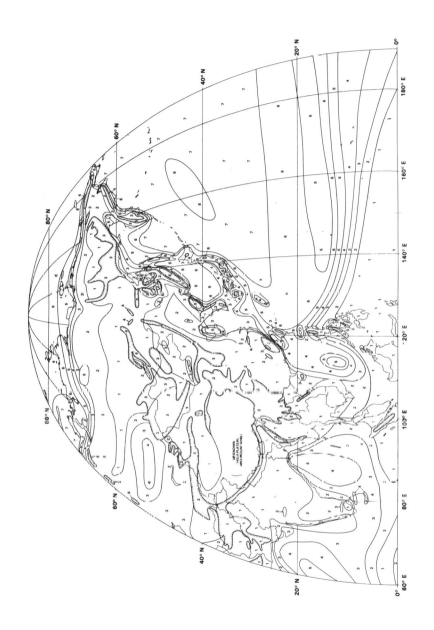

AUSTRALASIA

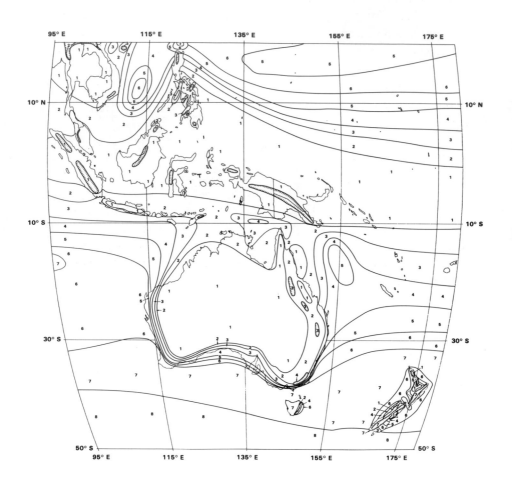

E. Estimates of Annual Energy Output

The following tables estimate the amount of energy a wind turbine of a given size will produce in a given wind regime using the swept-area method. The table assumes a Rayleigh distribution of wind speeds. The overall conversion efficiency assumed is given in the column labeled "Total Effic." The assumed efficiencies have been derived from a survey of product literature from wind turbine manufacturers worldwide. The results are approximations and won't necessarily correspond to estimates by the manufacturer for the same conditions using a power curve and a specific wind speed distribution.

The first table can be used once the average wind speed is known. The second table can be used with Battelle's wind power maps as described in the text.

To estimate the potential generation from a 7-meter wind turbine installed on the Great Plains where the average annual wind speed at hub height is 6 m/s (13.4 mph), follow these steps:

Find row 6.0 in the first column. This is the wind speed in m/s. Then move along the row until it intersects with the last column, labeled 7. This is the wind turbine's rotor diameter in meters. The value where row 6.0 and column 7 intersect, 18, is the estimated generation in thousand kWh per year. Under these conditions a 7-meter turbine will produce approximately 18,000 kWh per year.

For sites found on the world resource maps with wind power Class 8 and above, simply adjust the AEO by a ratio between the class on the world map and the class on U.S. map. For example, in Table E-2 a 1-meter wind turbine should produce 1000 kilowatt-hours per year at a Class 7 site with 1000 W/m². At a Class 8 site with 1200 W/m² (1200/1000 = 1.2) the 1-meter turbine would generate 1200 kilowatt-hours per year.

Table E-1

Estimated Annual Energy Output

at Hub Height in Thousand kWh/yr

		Power									
Average Speed		Density	Total	1	1.5	2	3	4	5	6	7
(m/s)	(mph)	(W/m²)	Effic.	(3.3)	(4.9)	(6.6)	(9.8)	(13.1)	(16.4)	(19.7)	(23)
4.0	9.0	75	0.28	0.1	0.3	0.6	1.3	2.3	3.6	5.2	7.1
4.5	10.1	110	0.28	0.2	0.5	0.8	1.9	3.4	5.3	7.6	10
5.0	11.2	150	0.25	0.3	0.6	1.0	2.3	4.1	6.5	9.3	13
5.5	12.3	190	0.25	0.3	0.7	1.3	2.9	5.2	8.2	12	16
6.0	13.4	250	0.21	0.4	0.8	1.4	3.3	5.8	9.0	13	18
6.5	14.6	320	0.19	0.4	0.9	1.7	3.8	6.7	10	15	20
7.0	15.7	400	0.16	0.4	1.0	1.8	4.0	7.0	11	16	22
7.5	16.8	490	0.15	0.5	1.1	2.0	4.6	8.1	13	18	25
8	17.9	600	0.12	0.5	1.1	2.0	4.5	7.9	12	18	24
8.5	19.0	720	0.12	0.6	1.3	2.4	5.3	9.5	15	21	29
9	20.2	850	0.12	0.7	1.6	2.8	6.3	11	18	25	34

*Assumed efficiency based on published data.

Table E-1.1

Estimated Annual Energy Output

at Hub Height in Thousand kWh/yr

		Power		Rotor Diameter in meters (feet)									
Average Speed		Density	Total	10	11	12	13	15	17	18	19	20	21
(m/s)	(mph)	(W/m²)	Effic.	(33)	(36)	(39)	(43)	(49)	(56)	(59)	(62)	(66)	(69)
4.0	9.0	75	0.25	13	16	19	22	29	37	42	46	52	57
4.5	10.1	107	0.25	19	23	27	32	43	55	61	68	73	81
5.0	11.2	146	0.3	31	37	45	52	70	89	100	110	120	130
5.5	12.3	195	0.3	39	47	56	66	88	110	130	140	160	180
6.0	13.4	253	0.3	52	62	74	87	120	150	170	190	210	230
6.5	14.6	321	0.3	66	80	100	110	150	190	210	240	270	290
7.0	15.7	401	0.28	80	90	110	130	170	220	250	280	310	340
7.5	16.8	494	0.28	90	110	140	160	210	270	310	340	380	420
8	17.9	599	0.25	100	120	150	170	230	300	330	370	410	450
8.5	19.0	718	0.25	120	150	180	210	280	360	400	450	490	540
9	20.2	853	0.22	130	160	190	220	290	370	420	460	520	570

*Assumed efficiency based on published data.

22 (72)	23 (75)	24 (79)	25 (82)	26 (85)	27 (89)	28 (92)	30 (98)	33 (108)	34 (112)	35 (115)	39 (128)	41 (134)	43 (141)
62	68	74	80	87	94	100	120	140	150	160	200	220	240
90	100	110	110	120	130	140	170	200	210	220	280	310	340
150	160	170	190	200	220	240	270	330	350	370	460	510	560
190	210	230	250	270	290	310	360	440	460	490	610	680	740
250	280	300	330	350	380	410	470	570	600	640	790	880	960
320	350	380	410	450	480	520	600	720	770	810	1000	1100	1200
370	410	450	480	520	560	610	700	840	890	950	1200	1300	1400
460	500	550	590	640	690	750	860	1000	1100	1200	1400	1600	1800
500	550	590	640	700	750	810	930	1100	1200	1300	1600	1700	1900
600	650	710	770	840	900	1000	1100	1300	1400	1500	1900	2100	2300
620	680	740	810	870	940	1000	1200	1400	1500	1600	2000	2200	2400

Table E-2

Estimated Annual Energy Output for Batelle Wind Power Classes

at 30 m (98 ft) Hub Height in Thousand kWh/yr *

	Battelle Power Class at 10 m		Wind Speed and Power at 30 m hub height			Rotor Diameter in meters (feet)							
Class	Power Density (W/m²)	Speed (m/s)	Power Density (W/m²)	Speed (m/s)	Total Effic.	1 (3.3)	1.5 (4.9)	2 (6.6)	3 (9.8)	4 (13.1)	5 (16.4)	6 (19.7)	7 (23)
1	50	3.5	80	4.1	0.28	0.2	0.3	0.6	1.4	2.5	3.9	5.6	7.6
2	100	4.4	160	5.1	0.25	0.3	0.6	1.1	2.5	4.4	6.9	9.9	13
3	150	5.0	240	5.9	0.21	0.3	0.8	1.4	3.1	5.6	8.7	12	17
4	200	5.5	320	6.5	0.19	0.4	0.9	1.7	3.8	6.7	10	15	21
5	250	6.0	400	7.0	0.16	0.4	1.0	1.8	4.0	7.1	11	16	22
6	300	6.3	480	7.4	0.15	0.5	1.1	2.0	4.5	7.9	12	18	24
7	400	7.0	640	8.2	0.14	0.6	1.4	2.5	5.5	9.9	15	22	30
	1000	9.5	1600	11.1	0.12	1	3	5	12	21	33	48	65

Table E-2.1

Estimated Annual Energy Output for Batelle Wind Power Classes

at 30 m (98 ft) Hub Height in Thousand kWh/yr *

	Battelle Power Class at 10 m		Wind Speed and Power at 30 m hub height			Rotor Diameter in meters (feet)								
Class	Power Density (W/m²)	Speed (m/s)	Power Density (W/m²)	Speed (m/s)	Total Effic.	10 (33)	11 (36)	12 (39)	13 (43)	15 (49)	17 (56)	18 (59)	19 (62)	20 (66)
1	50	3.5	80	4.1	0.25	14	17	20	23	31	40	45	50	55
2	100	4.4	160	5.1	0.3	33	40	48	56	74	96	110	120	130
3	150	5.0	240	5.9	0.3	50	60	71	84	110	140	160	180	200
4	200	5.5	320	6.5	0.3	66	80	95	110	150	190	210	240	260
5	250	6.0	400	7.0	0.28	80	90	110	130	170	220	250	280	310
6	300	6.3	480	7.4	0.28	90	110	130	160	210	270	300	330	370
7	400	7.0	640	8.2	0.25	110	130	160	190	250	320	360	400	440
	1000	9.5	1600	11.1	0.22	240	290	350	410	540	700	780	870	970

*Assumed efficiency based on published data.

21	22	23	24	25	26	27	28	30	33	34	35	39	41	43
(69)	(72)	(75)	(79)	(82)	(85)	(89)	(92)	(98)	(108)	(112)	(115)	(128)	(134)	(141)
61	67	73	79	86	90	100	110	120	150	160	170	210	230	250
150	160	170	190	210	220	240	260	300	360	380	400	500	560	610
220	240	260	290	310	340	360	390	450	540	570	610	750	830	920
290	320	350	380	410	450	480	520	590	700	800	800	1000	1100	1200
340	370	410	440	480	520	560	600	690	800	900	900	1200	1300	1400
410	450	490	530	580	630	670	700	800	1000	1100	1100	1400	1600	1700
490	530	580	630	690	740	800	900	1000	1200	1300	1300	1700	1900	2000
1100	1200	1300	1400	1500	1600	1800	1900	2200	2600	2800	3000	3700	4100	4500

Increase in speed and power with height assumes one-seventh power law.

F. Estimates of Water Pumping Capacity of Farm Windmills

The following tables estimate the amount of water a traditional American windmill of a given diameter will pump daily from a given depth within different wind regimes. The tables assume that the overall efficiency of the windmill is 5% in a Rayleigh wind speed distribution. Actual performance may vary depending on the windmill and whether it's properly matched to the well pump.

To use the tables, first find the total dynamic head in the left most column. Then find the average annual wind speed at hub height. For example, if the total pumping head is about 100 feet (30 meters) at a site with an 11 mph (5 m/s) average wind speed, a farm windmill with an 8-foot windwheel will pump about 2600 gallons (10 m³) per day.

Table F-1

Approximate Daily Output, American Farm Windmill

8-foot (2.4-meter) Diameter Rotor, in cubic meters/day and gallons/day

		\multicolumn{10}{c}{Average Annual Wind Speed, m/s (approximate mph)}									
		3 (7)		4 (9)		5 (11)		6 (13)		7 (16)	
\multicolumn{2}{l}{Pumping Head}										
(m)	(ft)	(m³)	(gals.)	(m³)	(gals.)	(m³)	(gals.)	(m³)	(gals.)	(m³)	(gals.)
10	30	6	1700	15	4000	30	7900	51	13,600	82	21,600
20	70	3	800	8	2000	15	3900	26	6800	41	10,800
30	100	2	600	5	1300	10	2600	17	4500	27	7200
40	130	2	400	4	1000	7	2000	13	3400	20	5400

Source: Center for International Development, Research Triangle Institute.

Table F-2

Approximate Daily Output, American Farm Windmill

10-foot (3.05-meter) Diameter Rotor, in cubic meters/day and gallons/day

		\multicolumn{10}{c}{Average Annual Wind Speed, m/s (approximate mph)}									
		3 (7)		4 (9)		5 (11)		6 (13)		7 (16)	
\multicolumn{2}{l}{Pumping Head}										
(m)	(ft)	(m³)	(gals.)	(m³)	(gals.)	(m³)	(gals.)	(m³)	(gals.)	(m³)	(gals.)
10	30	10	2700	24	6300	47	12,300	80	21,200	128	33,700
20	70	5	1300	12	3100	23	6100	40	10,600	64	16,800
30	100	3	900	8	2100	16	4100	27	7100	43	11,200
40	130	3	700	6	1600	12	3100	20	5300	32	8400

Source: Center for International Development, Research Triangle Institute.

Table F-3

Approximate Daily Output, American Farm Windmill

12-foot (3.7-meter) Diameter Rotor, in cubic meters/day and gallons/day

		Average Annual Wind Speed, m/s (approximate mph)								
Pumping Head		3 (7)		4 (9)		5 (11)		6 (13)		7 (16)	
(m)	(ft)	(m³)	(gals.)	(m³)	(gals.)	(m³)	(gals.)	(m³)	(gals.)	(m³)	(gals.)
10	30	14	3800	34	9100	67	17,700	116	30,600	184	48,500
20	70	7	1900	17	4500	33	8800	58	15,300	92	24,300
30	100	5	1300	11	3000	22	5900	39	10,200	61	16,200
40	130	4	1000	9	2300	17	4400	29	7600	46	12,100

Source: Center for International Development, Research Triangle Institute.

Table F-4

Approximate Daily Output, American Farm Windmill

14-foot (4.3-meter) Diameter Rotor, in cubic meters/day and gallons/day

		Average Annual Wind Speed, m/s (approximate mph)								
Pumping Head		3 (7)		4 (9)		5 (11)		6 (13)		7 (16)	
(m)	(ft)	(m³)	(gals.)	(m³)	(gals.)	(m³)	(gals.)	(m³)	(gals.)	(m³)	(gals.)
10	30	20	5200	47	12,300	91	24,100	158	41,600	250	66,000
20	70	10	2600	23	6200	46	12,000	79	20,800	125	33,000
30	100	7	1700	16	4100	30	8000	53	13,900	83	22,000
40	130	5	1300	12	3100	23	6000	39	10,400	63	16,500

Source: Center for International Development, Research Triangle Institute.

Table F-5

Approximate Daily Output, American Farm Windmill

16-foot (4.9-meter) Diameter Rotor, in cubic meters/day and gallons/day

		Average Annual Wind Speed, m/s (approximate mph)								
Pumping Head		3 (7)		4 (9)		5 (11)		6 (13)		7 (16)	
(m)	(ft)	(m³)	(gals.)	(m³)	(gals.)	(m³)	(gals.)	(m³)	(gals.)	(m³)	(gals.)
10	30	26	6800	61	16,100	119	31,400	206	54,300	327	86,300
20	70	13	3400	30	8000	60	15,700	103	27,200	163	43,100
30	100	9	2300	20	5400	40	10,500	69	18,100	109	28,800
40	130	6	1700	15	4000	30	7900	51	13,600	82	21,600

Source: Center for International Development, Research Triangle Institute.

G. Voltage Drop in Conductors

Where the permissible voltage drop is greater than 1 percent, multiply the distance values below by the total percentage of voltage drop. For example, if a 2 percent voltage drop is acceptable, double the distance values in the following table.

Table G

One-Way Distance (Feet) to Load for 1 Percent Voltage Drop in Copper and Aluminum Wire in the United States

Approx. Gen. Size (kW)	Max. Current (amps)	Wire Size (AWG) 10		8		6		4	
		(Cu)	(Al)	(Cu)	(Al)	(Cu)	(Al)	(Cu)	(Al)
12 volts									
0.01	1	48	30	74	47	118	74	187	118
0.06	5	10	6	15	9	24	15	37	24
0.12	10	5	3	7	5	12	7	19	12
0.24	20	2	2	4	2	6	4	9	6
0.48	40	1	1	2	1	3	2	5	3
24 volts									
0.24	10	10	6	15	9	24	15	37	24
0.48	20	5	3	7	5	12	7	19	12
0.7	30	3	2	5	3	8	5	12	8
1	40	2	2	4	2	6	4	9	6
1.4	60	2	1	2	2	4	2	6	4
1.9	80	1	1	2	1	3	2	5	3
2.4	100	1	1	1	1	2	1	4	2
2.9	120	1	1	1	1	2	1	3	2
3.4	140	1	0	1	1	2	1	3	2
36 volts									
0.7	20	7	5	11	7	18	11	28	18
1.4	40	4	2	6	4	9	6	14	9
2.2	60	2	2	4	2	6	4	9	6
2.9	80	2	1	3	2	4	3	7	4
3.6	100	1	1	2	1	4	2	6	4
48 volts									
1	20	10	6	15	9	24	15	37	24
1.4	30	6	4	10	6	16	10	25	16
1.9	40	5	3	7	5	12	7	19	12
2.4	50	4	2	6	4	9	6	15	9
2.9	60	3	2	5	3	8	5	12	8
3.4	70	3	2	4	3	7	4	11	7

3		2		1		0		00		000	
(Cu)	(Al)	(Cu)	(Al)	(Cu)	(Al)	(Cu)	(Al)	(Cu)	(Al)	(Cu)	(Al)
236	149	299	188	375	237	472	299	594	377	753	476
47	30	60	38	75	47	94	60	119	75	151	95
24	15	30	19	38	24	47	30	59	38	75	48
12	7	15	9	19	12	24	15	30	19	38	24
6	4	7	5	9	6	12	7	15	9	19	12
47	30	60	38	75	47	94	60	119	75	151	95
24	15	30	19	38	24	47	30	59	38	75	48
16	10	20	13	25	16	31	20	40	25	50	32
12	7	15	9	19	12	24	15	30	19	38	24
8	5	10	6	13	8	16	10	20	13	25	16
6	4	7	5	9	6	12	7	15	9	19	12
5	3	6	4	8	5	9	6	12	8	15	10
4	2	5	3	6	4	8	5	10	6	13	8
3	2	4	3	5	3	7	4	8	5	11	7
35	22	45	28	56	36	71	45	89	57	113	71
18	11	22	14	28	18	35	22	45	28	56	36
12	7	15	9	19	12	24	15	30	19	38	24
9	6	11	7	14	9	18	11	22	14	28	18
7	4	9	6	11	7	14	9	18	11	23	14
47	30	60	38	75	47	94	60	119	75	151	95
31	20	40	25	50	32	63	40	79	50	100	63
24	15	30	19	38	24	47	30	59	38	75	48
19	12	24	15	30	19	38	24	48	30	60	38
16	10	20	13	25	16	31	20	40	25	50	32
13	9	17	11	21	14	27	17	34	22	43	27

One-Way Distance (Feet) to Load for 1 Percent Voltage Drop in Copper and Aluminum Wire, con't.

Approx. Gen. Size (kW)	Max. Current (amps)	10 (Cu)	10 (Al)	8 (Cu)	8 (Al)	6 (Cu)	6 (Al)	4 (Cu)	4 (Al)	
120 volts										
1.2	10	48	30	74	47	118	74	187	118	
1.8	15	32	20	49	31	78	50	125	79	
2.4	20	24	15	37	23	59	37	93	59	
3	25	19	12	30	19	47	30	75	47	
3.6	30	16	10	25	16	39	25	62	39	
4.2	35	14	9	21	13	34	21	53	34	
4.8	40	12	8	19	12	29	19	47	30	
7.2	60	8	5	12	8	20	12	31	20	
10	80	6	4	9	6	15	9	23	15	
12	100	5	3	7	5	12	7	19	12	
240 volts										
2	10	95	60	148	94	235	149	374	236	
5	20	48	30	74	47	118	74	187	118	
10	40	24	15	37	23	59	37	93	59	
14	60	16	10	25	16	39	25	62	39	79
19	80	12	8	19	12	29	19	47	30	59
24	100	10	6	15	9	24	15	37	24	47

3		2		1		0		00		000	
(Cu)	(Al)	(Cu)	(Al)	(Cu)	(Al)	(Cu)	(Al)	(Cu)	(Al)	(Cu)	(Al)
236	149	299	188	375	237	472	299	594	377	753	476
157	99	199	125	250	158	315	199	396	252	502	317
118	74	149	94	188	119	236	149	297	189	376	238
94	60	119	75	150	95	189	119	238	151	301	190
79	50	100	63	125	79	157	100	198	126	251	159
67	43	85	54	107	68	135	85	170	108	215	136
59	37	75	47	94	59	118	75	149	94	188	119
39	25	50	31	63	40	79	50	99	63	125	79
30	19	37	24	47	30	59	37	74	47	94	60
24	15	30	19	38	24	47	30	59	38	75	48
472	298	597	376	750	474	945	597	1188	755	1506	952
236	149	299	188	375	237	472	299	594	377	753	476
118	74	149	94	188	119	236	149	297	189	376	238
50	100	63	125	79	157	100	198	126	251	159	
37	75	47	94	59	118	75	149	94	188	119	
30	60	38	75	47	94	60	119	75	151	95	

H. Further Sources of Information

Non-Governmental Organizations

The following organizations can be contacted for further information. The Alternative Energy Institute conducts field trials of small wind turbines at its Wind Turbine Test Center and is a good source of practical information on small machines. Denmark's Folkcenter for Renewable Energy also designs and tests small wind turbines. The American, British, Canadian, German, and European Wind Energy Associations hold annual conferences on the latest developments in wind technology. The Center for International Development deals specifically with water pumping in developing countries.

Alternative Energy Institute
West Texas State University
Box 248
Canyon, TX 79016
phone: 806 656 2295
fax: 806 656 2071
email: aei@wtamu.edu

American Wind Energy Association
122 C St., NW, 4th Floor
Washington, DC 20001
phone: 202 383 2500
fax: 202 383 2505
email: windmail@mcimail.com
http: www.igc.apc.org/awea

Bundesverbandes WindEnergie (BWE)
German Wind Energy Association
Natruper Strasse 70
D 49090 Osnabruck Germany
phone: +49 541 96 19 185
fax: +49 541 96 19 186
email: iwb_bwe@t-online.de

Canadian Wind Energy Association
100, 3553 31st NW
Calgary, Alberta T2L 2K7 Canada
phone: 800 9-CanWEA or 403 289 7713
fax: 403 282 1238
email: canwea@cadvision.com

Center for International Development
Research Triangle Institute
P.O. Box 12194
Research Triangle Park, NC 27709-2194
phone: 919 541 6485
fax: 919 541 6621

Centre for Alternative Technology
Llwyngwern Quarry
Machynlleth, Powys, Wales SY20 9AZ
United Kingdom
phone: +44 1654 702 400
fax: +44 1654 702 782
email: cat@gn.apc.org

Danmarks Vindmølleforening
Danish Windmill Association
Klostervej 89
DK 8680 Ry
Denmark
phone: +45 86 89 36 36
fax: +45 86 89 36 30

European Wind Energy Association
26 Spring St.
London W2 1JA
United Kingdom
phone: +44 171 402 7122
fax: +44 171 402 7125
email: 101715,1101@compuserv.com

The Folkcenter for Renewable Energy
(Nordvestjysk Folkcenter for Vedvarende
   Energi)
Kammersgaardvej 16, Sdr. Ydby
DK 7760 Hurup, Thy
Denmark
phone: +45 97 95 65 55
fax: +45 97 95 65 65
email: fcenergy@inet.uni-c.dk

Kern Wind Energy Association
P.O. Box 277
Tehachapi, CA 93581-0277
phone: 805 822 7956
fax: 805 822 8452
email: 3062869@mcimail.com

Landelijk Bureau Windenergie
Postbus 233
Zeist, Utrecht NL3700 AE
The Netherlands
phone: +31 30 691 30 19
fax: +31 30 692 36 63

Natta-Network for Alternative
   Technology & Technology
   Assessment
Energy & Environment Research Unit
Milton Keynes, Bucks MK7 6AA
United Kingdom
phone: +44 190 865 4638
fax: +44 190 865 3744

Renewable Energy Research Laboratory
University of Massachusetts
Engineering Laboratory
Amherst, MA 01003
phone: 413 545 4359
fax: 413 545 0724
email: rerl@igc.apc.org

Rutherford Appleton Laboratory
Energy Research Unit
Chilton, Didcot, Oxon OX12 0BP
United Kingdom
phone: +44 1235 44 5559
fax: +44 1235 44 6863

Government-Sponsored Laboratories

The National Renewable Energy Laboratory performs technical re-
search for the U.S. Department of Energy. Risø, ECN (Energie Centrum
Nederlands), and the Energy Technology Support Unit perform a similar
function for Denmark, the Netherlands, and the United Kingdom. The
German wind energy institute (DEWI) is financed by the state of Lower
Saxony. Sandia National Laboratories concentrates on vertical axis wind
turbines, USDA on water pumping and other rural uses. The Atlantic
Wind Test Site, USDA, NREL, Risø, ECN, Windtest Kaiser-Wilhelm-
Koog, the German Wind Energy Institute, and the U.K.'s National Engi-
neering Laboratory operate test centers where they measure wind turbine
performance and reliability.

Atlantic Wind Test Site
Rural Route 4
Tignish COB 2BO
Prince Edward Island, Canada
phone: 902 882 2746
fax: 902 882 3823

Chinese Wind Energy Development
   Center
Huayuan Road 3
Beijing, 100083, China
phone: +86 1 20 20 108
fax: +86 1 20 12 880

Deutsches Windenergie Institut (DEWI)
Eberstrasse 96
D 26382 Wilhelmshaven
Germany
phone: +49 44 21 48 080
fax: +49 44 21 48 08 43

ECN (Energie Centrum Nederlands)
Postbus 1
1755 ZG Petten
The Netherlands
phone: +31 22 46 41 15
fax: +31 22 46 32 14

Energy Technology Support Unit (ETSU)
B156 Harwell Laboratory
Didcot, Oxfordshire OX11 0RA
United Kingdom
phone: +44 12 35 43 35 17
fax: +44 12 35 43 29 23

National Renewable Energy Laboratory
National Wind Technology Center
1617 Cole Boulevard
Golden, CO 80401-3393
phone: 303 384 6900
fax: 303 384 6999
Internet: nrel@igc.apc.org

National Wind Turbine Test Site (NEL)
East Kilbride G75 0QU, UK
phone: +44 13 55 22 02 22
fax: +44 13 55 27 23 33

Risø National Laboratory
Test Station for Wind Turbines
P.O. Box 49
DK 4000 Roskilde, Denmark
phone: +45 42 37 12 12
fax: +45 42 37 29 65

Sandia National Laboratories
P.O. Box 5800
Albuquerque, NM 87185
phone: 505 844 1030
fax: 505 844 9500
Internet: snlwind@igc.apc.org

U.S. Department of Agriculture
Agricultural Research Station
P.O. Box 10
Bushland, TX 79012
phone: 806 356 5734
fax: 806 356 5750

Windtest Kaiser-Wilhelm-Koog
Sommerdeich 14b
D 2222 Kaiser-Wilhelm-Koog
Germany
phone: +49 48 56 511
fax: +49 48 56 12 70

Electronic Networks

There are several electronic news groups on the subject of energy. These electronic conferences often contain information about wind energy and are accessible by subscribers around the world. There are two conferences on EcoNet that cover wind energy topics: awea.windnet for discussions relating to medium-sized wind turbines and awea.wind.home for discussions about residential applications. For those who are not subscribers to EcoNet and wish to receive either awea.windnet or awea.wind.home as a listserve, send a message to Tom Gray at tomgray@econet.org and explain your interest in wind energy. The Usenet "news groups" alt.energy.renewable and sci.energy are open to anyone with access to the internet.

For those with high speed modems, graphics capability, and the appropriate software, there are several sites on the World Wide Web offering wind energy topics with both text and graphics. One of the best is the Wind Turbine Workgroup of the Aerospace Institute's Department of Construction and Design at the Technical University of Berlin. The uniform resource locator or URL is http://rotor.fb12.tu-berlin.de/engwindkraft.html. This site includes information in English, German, and French and contains one of the WWW's most extensive selections on wind energy.

EcoNet
18 De Boom St.
San Francisco, CA 94107
phone: 415 442 0220
fax: 415 546 1794
(accessible via numerous gateways)

Home Power BBS
P.O. Box 520
Ashland, OR 97520-0520
BBS: 707 822 8640

Small Wind Turbine Manufacturers

Small wind turbines range in size from micro turbines, those less than 1,000 watts, to machines 10 meters in diameter driving generators up to 25 kW. The smallest turbines are suitable only for battery-charging and similar applications. As of the mid 1990s, only machines in the 7-meter size class and above are suitable for direct interconnection with an electric utility.

An expanded list of small wind turbine manufacturers is available on disk from Real Goods (1-800-919-2400).

Amp Air
P.O. Box 416
Poole, Dorset BH12 3LZ
United Kingdom
phone: +44 1202 74 9994
fax: + 44 1202 73 6653

Atlantis Windkraftanlagen
Glogauer Strasse 19/21
D 10999 Berlin, Germany
phone: +49 30 618 20 21
fax: +49 30 618 90 79

Bergey Windpower Co.
2001 Priestley Ave.
Norman, OK 73069
phone: 405 364 4212
fax: 405 364 2078
email: sales@bergey.com
http: //www.bergey.com/

Giacobone
Cerro Fitz Roy 1080
5800 Rio Quarto, Argentina
phone: +54 58 64 62 58
fax: +54 58 63 43 79

Hamilton Ferris
P.O. Box 126
Ashland, MA 01721
phone: 508 881 4601
fax: 508 881 3846

J. Bornay Aerogeneradores
Avda. de 1bi 76–78
ES 03420 Castalla, Alicante, Spain
phone: +34 6 556 00 25
fax: +34 6 555 07 52
email: bornay@alc.servicom.es

LMW Renewables
P.O. Box 279
NL 3700 AG Barneveld, The Netherlands
phone: +31 342 42 19 86
fax: +31 342 42 17 59

LVM Products
Old Oak Close
Arlesey, Bedfordshire SG15, 6XD, UK
phone: +44 146 27 33 336
fax: +44 146 27 30 466

Marlec Engineering Company
Rutland House
Corby, Northants NN17 1XY, UK
phone: +44 536 201 588
fax: +44 536 400 211
email: marlec@dial.pipex.com
http: //dialspace.dial.pipex.com/marlec

Northern Power Systems
New World Power Technology
1 Northwind Road
P.O. Box 659
Moretown, VT 05660-0659
phone: 802 496 2955
fax: 802 496 2953

Proven Engineering Products
Moorfield Industrial Estates
Kilmarnock KA2 0BA, Scotland
phone: +44 1563 543 020
fax: +44 1563 539 119
email: gordon.proven@post.almac.co.uk

SOMA
Sunrise Solar
49 Vista Ave.
Cococabana, New South Wales 2251
Australia
phone: +61 43 811 531
fax: +61 43 821 880

Southwest Windpower
2131 N. First St.
Flagstaff, AZ 86003-2190
phone: 520 779 9463
fax: 520 779 1485

SoWiCo Solartech
Holderdorp 68
D 49536 Lienen, Germany
phone: +49 54 83 14 91
fax: +49 54 83 81 66

Survivor Energy Systems
Synergy Power Corporation
19–27 Wyndham St., 20F Wilson House
Central Hong Kong, China
phone: 852 2846 3168
fax: 852 2810 0478
email: 100314.2615@compuserv.com
http: //solstice.crest.org/renewables/
synergy/index.html

Vergnet S.A.
6, rue Henri Dunant
45140 Ingre, France
phone: +33 2 38 22 75 00
fax: +33 2 38 22 75 22
email:vergnet@wanadoo.fr

Westwind Turbines
Venco Products
29 Owen Rd.
Kelmscott, Western Australia 6111
Australia
phone: +61 93 99 52 65
fax: +61 94 97 13 35
email: venwest@iinet.net.au

Wind Turbine Industries
Prior Lake Machine
16801 Industrial Circle, SE
Prior Lake, MN 55372
phone: 612 447 6064
fax: 612 447 6050

World Power Technologies
19 Lake Ave. North
Duluth, MN 55802
phone: 218 722 1492
fax: 218 722 0791
email: wpt@cp.duluth.mn.us
http: //www.webpage.com/wpt/

Medium-Sized Wind Turbine Manufacturers

There are numerous manufacturers that build reliable wind turbines for large-scale commercial applications such as wind power plants. By the late-1990s the following companies had become important suppliers of mid-sized wind turbines, those 10–50 meters in diameter. As mentioned in the text, many of these machines have also been installed as single turbines or in small clusters for farms, businesses, and cooperatives. Several manufacturers are continuing to build models in the 10–20 meter size class that are suitable for individual or light commercial applications.

There are more than 24,000 wind turbines 10–30 meters in diameter operating throughout the world, mostly in Europe and California. Analysts consider wind turbines of this size a proven technology. Another 5,000 wind turbines 30–40 meters in diameter have been installed during the early 1990s. This technology represented the state-of-the-art for utility-scale wind turbines as *Wind Power for Home & Business* went into its fifth printing. Manufacturers began introducing wind turbines greater than 40 meters in diameter during 1994.

For a complete list of U.S. manufacturers, write the American Wind Energy Association. For manufacturers outside the United States, contact the British, Canadian, Danish, Dutch, German, or European trade associations.

An expanded version of the following table listing more than one hundred different wind turbines from 10 meters in diameter to 60 meters in diameter is available on diskette from the publisher.

Bonus Energy A/S
Fabriksvej 4
DK 7330 Brande, Denmark
phone: +45 97 18 11 22
fax: +45 97 18 30 86

Ecotecnia
Amistat 23, 1st
ES 08005 Barcelona, Catalonia, Spain
phone: +34 32 25 76 00
fax: +34 32 21 09 39

Enercon
Dreekamp 5
D 2960 Aurich, Germany
phone: +49 49 41 927 0
fax: +49 49 41 927 199

GET-Gesselschaft für Energietechnik
Kieler Strasse 53
D 24768 Rendsburg, Schleswig-Holstein
Germany
phone: +49 43 31 55 051
fax: +49 43 31 55 944

Jacobs Energie
Hamburger Strasse 57
D 25746 Heide, Schleswig-Holstein
Germany
phone: +49 48 12 025
fax: +49 48 11 266

Lagerwey Windturbine
P.O. Box 279
NL 3770 AG Barneveld
The Netherlands
phone: +31 34 24 22 724
fax: +31 34 24 22 861

Markham & Company Limited
Broad Oaks Works
Chesterfield S41 0DS, UK
phone: +44 1246 276 121
fax: +44 1246 211 424

Micon
Milskowej 8
DK 8900 Randers, Denmark
phone: +45 86 46 76 00
fax: +45 86 46 77 18
email: mail@micon.dk

Mitsubishi Heavy Industries
1-1 Akunoura-Machi
85091 Nagasaki, Japan
phone: +81 958 28 40 92
fax: +81 958 28 61 74

NedWind
Remmerden 9
NL 3910 AC Rhenen, The Netherlands
phone: +31 31 76 19 004
fax: +31 31 76 12 129

Nordex
Svindbaek
DK 7323 Give, Denmark
phone: 45 75 73 44 00
fax: 45 75 73 41 47

Nordtank A/S
Nyballevej 8
DK 8444 Balle, Denmark
phone: +45 86 33 72 00
fax: +45 86 33 73 74

Riva Calzoni
Via Emilia Ponente 72
I 40133 Bologna, Italy
phone: 39 51 52 75 11
fax: 39 51 6 57 46 55

Sudwind
Prinzenstrasse 32-33
D 10969 Berlin, Germany
phone: + 49 30 61 69 26 40
fax: +49 30 61 69 26 77
email: suedwind.berlin@t-online.de

Tacke Windtechnik
Holsterfield 5a
D 48499 Salzbergen, Germany
phone: +49 59 71 97 08 0
fax: +49 59 71 97 08 88

Vestas DWT
Smed Hansens Vej 27
DK 6940 Lem, Denmark
phone: +45 97 34 11 88
fax: +45 97 34 14 84

Wind Energy Group, Ltd
Taywood House
Southall, Middlesex UB1 2QX, UK
phone: +44 181 578 2366
fax: +44 181 575 8318

Windmaster Nederland
Zuiveringweg 26
NL 8243 PZ Leystad, Flevoland
The Netherlands
phone: +31 32 02 54 940
fax: +31 32 02 54 576

Wind Technik Nord
Guner Weg 11
D 25920 Stedesand, Germany
phone: +49 46 62 14 14
fax: +49 46 62 14 25

Wind World A/S
Buttervej 60
DK 9990 Skagen, Denmark
phone: +45 98 44 40 11
fax: +45 98 44 57 56

Zond Systems
P.O. Box 1910
Tehachapi, CA 93561
phone: 805 823 6700
fax: 805 822 7880
email; zond@compuserv.com
http://www.zond.com

Table H

Characteristics of Selected Medium-Sized Wind Turbines

Manufacturer	Model	Rotor Dia. (m)	Swept Area (m²)	Power (kW)	No. of Blades	(1) Speed of Rotor	(2) Rotor Control	Overspeed Control
Enercon	E12	12.0	113	30	3	v	p	variable pitch
Jacobs Energie	Aeroman	14.8	172	33	2	c	p	variable pitch
Lagerwey	LW 18/80	18.0	254	80	2	v	p	variable pitch
Ecotecnia	28/225	28.0	616	225	3	c	s	pitchable tips
Windtechnik Nord	200/29	29.0	661	200	3	c	s	pitchable tips
GET	GDW 29	29.0	661	225	3	c	s	pitchable tips
Vestas	V29	29.0	661	225	3	c	p	variable pitch
Wind World	W3000	29.2	670	250	3	c	s	pitchable tips
Enercon	E-30	30.0	707	280	3	v	p	variable pitch
Micon	M750	31.0	755	400	3	c	s	pitchable tips
Riva Calzoni	M30S2	33.0	855	350	1	v	p	pitchable tips
Bonus	300 Mk II	33.2	866	300	3	c	s	pitchable tips
Sudwind	N3330	33.4	876	270	3	c	s	pitchable tips
Mitsubishi	MWT 450	39.0	1195	450	3	c	p	variable pitch
Wind Energy Group	MS3 400L	39.4	1216	400	2	c	p	variable pitch
Zond	Z40	40.0	1257	500	3	c	p	variable pitch
Enercon	E40	40.3	1276	500	3	v	p	variable pitch
GET	GET 41	41.0	1320	600	3	c	s	pitchable tips
Vestas	V42	42.0	1385	600	3	c	p	variable pitch
Jacobs	43/600	43.0	1452	600	3	c	s	pitchable tips
Nordex	N43	43.0	1452	600	3	c	s	pitchable tips
Nordtank	600/43	43.0	1452	600	3	c	s	pitchable tips
Tacke	TW 600	43.0	1452	600	3	c	s	mechanical brake
Windtechnik Nord	600/43	43.0	1452	600	3	c	s	pitchable tips
Micon	M1500	43.4	1479	600	3	c	s	pitchable tips
NedWind	NW44	43.8	1505	500	2	c	s+p	variable pitch
Bonus	600 Mk II	44.0	1521	600	3	c	s	pitchable tips
Ecotecnia	44/600	44.0	1521	600	3	c	s	pitchable tips
Vestas	V44	44.0	1521	600	3	c	p	variable pitch
Markham	VS45	45.0	1590	600	3	v	p	variable pitch
Wind World	W4500	45.4	1619	550	3	c	s	pitchable tips
Tacke	TW 600e	46.0	1662	600	3	c	s	mechanical brake
Zond	Z46	46.0	1662	700	3	c	p	variable pitch
Micon	M1800	48.0	1810	600	3	c	s	pitchable tips
WindMaster		48.0	1810	750	2	c	p	variable pitch
Bonus		50.0	1963	1000	3	c	s	pitchable tips
Nordex	N52	52.1	2132	800	3	c	s	pitchable tips

c=constant speed, v=variable speed, s=stall regulated, p=variable pitch.
Note: Some models from the same manufacturer have larger rotor diameters for the same power output. These turbines are intended for lower wind speed regimes.

Mechanical Wind Pumps

The companies below manufacture or import mechanical water-pumping windmills. Dempster and Aermotor still build their own wind wheels (rotors). All are back-geared and self-regulating. O'Brock and Topper offer catalogs full of windmills (both new and rebuilt), pumps, and paraphernalia. Mountain Crane and Windmill rebuild and install used water-pumping windmills.

Aermotor Windmill Corp.
P.O. Box 5110
San Angelo, TX 76902
phone: 915 658 2795
fax: 915 655 7147

Climax Windmills
P.O. Box 265244
Three Rivers, Transvaal 1935
Republic of South Africa
phone: +27 16 23 12 65
fax: +27 16 23 31 05

Dempster Industries
P.O. Box 848
Beatrice, NE 68310
phone: 402 223 4026
fax: 402 228 4389

FIASA (Fabrica de Implementos
    Agricolas SA)
Hortiguera 1890
Buenos Aires, 1406
Argentina
phone: +54 1 923 1081 15

Koenders Mfg. Co. Ltd.
P.O. Box 171
Englefeld, Saskatchewan S0K 1N0
Canada
phone: 306 287 3139

Mountain Crane and Windmill
P.O. Box 1187
Diamond Springs, CA 95619
phone: 800 494 6364 or 916 622 0947
fax: 916 644 3008

O'Brock Windmill Distributors
9435 12th St.
North Benton, OH 44449
phone: 330 584 4681
fax: 330 584 4682
email: windmill@cannet.com
(Send $3.00 for catalog)

Southern Cross Corp.
632 Ruthven St.
Toowoomba
Queensland QLD 4350, Australia
phone: +61 76 38 4988
fax: +61 76 38 5898

Topper Co.
P.O. Box 30369
San Angelo, TX 76903-3069
phone: 800 775 3277 or 915 658 3277
fax: 915 655 3086

Windtech International
P.O. Box 1509
Coffeyville, KS 67337
phone: 316 251 7897
fax: 316 251 3622

Dealers

These firms sell wind turbines. Some also install and service them.

Alternative Energy Engineering
P.O. Box 339
Redway, CA 95560-0339
phone: 800 777 6609 or 707 923 2277
fax: 707 923 3009
email: energy@alt-energy.com

Energy Transfer Systems
P.O. Box 121
Acton, CA 93510
phone: 805 269 5410

Jade Mountain
P.O. Box 4616
Boulder, CO 80306
phone: 800 442 1972 or 303 449 6601
fax: 303 449 8266
email info@jademountain.com
http://www.jademountain.com

Kansas Wind Power
13569 214th Road
Holton, KS 66436
phone: 913 364 4407
fax: 913 364 4407

Lake Michigan Wind & Sun
E. 3971 Bluebird Rd.
Forestville, WI 54213
phone: 414 837 2267
fax: 414 837 7523
email: lmwands@aol.com

Mountain Pass Windpower
711 N. C St.
Livingston, MT 59047
phone: 406 222 1707

Real Goods Trading Corp.
555 Leslie St.
Ukiah, CA 95482-5507
phone: 800 919 2400 or 707 468 9214
fax: 707 468 0301
email: realgood@realgoods.com
http: //www.realgoods.com

Towers

As a rule, tower manufacturers don't sell to individuals. They prefer to deal directly with the manufacturer of the wind turbine. By handling sales this way they can assure themselves that the tower is matched properly to the wind turbine. Unarco Rohn is one of the few suppliers of guyed lattice and free-standing truss towers for small wind turbines in the United States. Most wind machines greater than 10 meters in diameter are supplied with their own tower.

Rohn
P.O. Box 2000
Peoria, IL 61656
phone: 309 697 4400
fax: 309 697 5612

NRG Systems
P.O. Box 509
Hinesburg, VT 05461-0509
phone: 800 448 9463 or 802 482 2272
fax: 802 482 2272

handwritten annotations:
NET METERING IN PA. CONSIDER 10 KW
399 KW/150 KW OVER SOMEBODY
35 A DIFFERENT STORY
AMERICAN WIND ENERGY ASSOC.
202-383-2500
" NET METERING
LOCAL PUBLIC UTILITIES COMMISSION. < KW ?"
GET A LAWYER

Recording Anemometers

Both NRG and Zond Systems build a full line of recording anemometers. Simple wind-run odometers cost about $150. More sophisticated instruments record minute by minute wind speed, direction, and turbulence. NRG, Zond, and Second Wind offer serial data loggers that can collect data from multiple wind speed and direction sensors. These loggers require a computer with the requisite software to analyze the stored data. NRG also manufacturers a line of hinged anemometer masts from 10-meters to 40-meters tall (30- to 140-feet in English units).

Neat Systems
12 Strathmore Ave.
Dunblane, Perthshire FK15 9HX
United Kingdom
phone: +44 786 82 26 70
fax: +44 786 82 55 45

8 JUNE 99 JULIA VERY HELPFUL

NRG Inc.
110 Commerce St.
Hinesburg, VT 05461-0509
phone: 802 482 2255 or 800 448 WIND
fax: 802 482 2272
email: sales@nrgsystems.com

8 JUNE 99

R. M. Young Co.
2801 Aero Park Dr.
Traverse City, MI 49684
phone: 616 946 3980
fax: 616 946 4772
email: young@traverse.com

8 JUNE 99

Second Wind
366 Somers St.
Somerville, MA 02144
phone:617 776 8520
fax: 617 776 0391
email: sales@secondwind.com

Zond Systems
P.O. Box 1910
Tehachapi, CA 93561
phone: 805 823 6700
fax: 805 822 7880
email: zond@compuserv.com

POWER AT HUB HEIGHT

Periodicals

Windpower Monthly, the wind industry's principal trade publication, is an unaffiliated observer of wind energy worldwide. *Home Power* magazine is the principal source for information on small wind turbines and the issues facing them, and, as the name implies, serves homeowners using stand-alone or off-the-grid power systems. The editors at *Windpower Monthly* also publish *Wind Stats,* a quarterly technical journal, and *Naturlig Energie,* a Danish magazine for the Association of Windmill Owners. Both are good sources for statistical data on wind turbine performance. *Wind Energy Weekly* features news about the commercial side of the wind industry in the United States. The *Windletter* is the association's monthly membership newsletter. The

British and European wind energy associations jointly publish *WinDirections,* a quarterly source for tracking developments in Europe. Natta's *RENEW* is another good source for monitoring activity in the United Kingdom. *Wind Energie Aktuell* and *Neue Energie* (in German), monthly magazines of the German wind energy associations, are good sources for developments in Germany. *DEWI Magazin* is a quarterly technical publication of the German Wind Energy Institute, which includes abstracts in English. *Wind Engineering* is the industry's sole peer-reviewed technical journal. *Independent Energy Magazine* periodically carries articles on trends in the commercial wind industry. *Systèmes Solaires* (in French) devotes at least one issue per year to wind energy developments in Francophone countries. *Windmillers' Gazette* focuses on water-pumping windmills with articles on both historical and contemporary technical issues.

DEWI Magazin
Eberstrasse 96
D 2940 Wilhelmshaven Niedersachsen
Germany
phone: +49 44 21 48 080
fax: +49 44 21 48 0843

Home Power Magazine
P.O. Box 520
Ashland, OR 97520-0520
phone: 916 475 3179
fax: 916 475 0836
email: hp@homepower.org
http://www.homepower.org

Independent Energy Magazine
1421 Sheridan Rd.
Tulsa, OK 74112
phone: 800 922 3736 or 918 835 3161
fax: 918 831 9776

Neue Energie
Bundesverband WindEnergie (BWE)
Natruper Strasse 70
D 49090 Osnabruck, Germany
phone: +49 541 96 19 185
fax: +49 541 96 19 186
email: iwb_bwe_@t-online.de

RENEW
Network for Alternative Technology and
   Technology Assessment
Faculty of Technology
Milton Keynes, Bucks MK7 6AA, UK
phone: +44 190 865 4638
fax: +44 190 865 4052

Systèmes Solaires
146, rue de l'Université
F 75007 Paris, France
phone: +33 1 44 18 00 80
fax: 33 1 44 18 00 36
email: sunnic.watt@utopia.EUnet.fr

Wind Energie Aktuell
Bundesverbandes WindEnergie
Lutherstrasse 14
D 30171 Hannover Niedersachsen
Germany
phone: +49 511 28 23 63
fax: +49 511 28 23 77

Wind Energy Weekly/Windletter
American Wind Energy Association
122 C St., NW, 4th Floor
Washington, DC 20001
phone: 202 383 2520
fax: 202 383 2505
email: windmail@mcimail.com

Wind Engineering
Multi-Science Publishing Ltd.
107 High Street
Brentwood, Essex CM14 4RX, UK
phone: +44 127 722 4632
fax: +44 127 722 3453

Windmillers' Gazette
P.O. Box 507
Rio Vista, TX 76093
phone: 817 755 1110

Anchors and Guying Hardware

A.B. Chance
210 N. Allen St.
Centralia, MO 65240
phone: 314 682 5521

Joslyn Manufacturing Co.
3700 S. Morgan St.
Chicago, IL 60609
phone: 312 927 1420

Preformed Line Products
P.O. Box 91129
Cleveland, OH 44101
phone: 216 461 5200

Tools and Work Belts

Klein Tools
7200 McCormick Rd.
Chicago, IL 60645

SINCO Products
P.O. Box 361
East Hampton, CT 06245

Photovoltaics and DC Appliances

Alternative Energy Engineering, Kansas Wind Power, Jade Mountain, and Real Goods offer catalogs with hard to find accessories for life off-the-grid. These catalogs include photovoltaic panels, batteries, inverters, DC appliances, energy-efficient refrigerators, compact fluorescent lights, and much more. Real Goods also publishes the *Solar Living Sourcebook,* a comprehensive, regularly updated guide to the latest developments in renewable energy equipment and energy-efficient appliances.

Alternative Energy Engineering
P.O. Box 339
Redway, CA 95560-0339
phone: 800 777 6609 or 707 923 2277
fax: 707 923 3009
email: energy@alt-energy.com
http://www.asis.com/aee

Jade Mountain
P.O. Box 4616
Boulder, CO 80306
phone: 800 442 1972 or 303 449 6601
fax: 303 449 8266
email: info@jademountain.com
http: //www.jademountain.com

Kansas Wind Power
13569 214th Road
Holton, KS 66436
phone: 913 364 4407
fax: 913 364 4407

Real Goods Trading Corp.
555 Leslie St.
Ukiah, CA 95482-5507
phone: 800 919 2400 or 707 468 9214
fax: 707 468 0301
email: realgoods@realgoods.com
http: //www.realgoods.com

Inverters

Vanner Weldon Dynamote
4282 Reynolds Dr.
Hilliard, OH 43026
phone: 614 771 2718

PowerStar
1050 E. Duane Ave., #D
Sunnyvale, CA 94086-2626
phone: 408 774 6810
fax: 408 774 6818

Statpower Technologies
7725 Lougheed Hwy.
Burnaby
British Columbia V5A 4V8
Canada
phone: 604 420 1585
fax: 604 420 1591

Trace Engineering
5916-195th N.E.
Arlington, WA 98223
phone: 206 435 8826
fax: 206 435 2229

Batteries

C&D Power Systems
3043 Walton Rd.
Plymouth Meeting, PA 19462
phone: 215 828 9000
fax: 215 834 3899

Powersafe Standby Batteries
North Haven, CT 06473
phone: 203 777 0037
fax: 203 773 1010

Yuasa-Exide
P.O. 14145
Reading, PA 19612-4145
phone: 800 538 3627

Johnson Controls
Specialty Battery Div.
900 E. Keefe Ave.
Milwaukee, WI 53212
phone: 414 961 6500
fax: 414 961 6506

Saft Nife
711 Industrial Blvd.
Valdosta, GA 31601
phone: 912 247 2331, or 800 556 6764
fax: 912 247 8486

Trojan Battery
12380 Clark St.
Santa Fe Springs, CA 90670
phone: 310 946 8381, or 800 423 6569
fax: 310 941 6038

Plans

Centre for Alternative Technology
Machynlleth
Powys, Wales SY20 9AZ, UK
phone: +44 165 470 2400
fax: +44 165 470 2782
Internet: cat@gn.apc.org

Kragten Design
Populierenlaan 51
5492 SG Sint-Odenrode, The Netherlands
phone and fax: +31 41 38 75 770

Lindsay Publications
P.O. Box 12
Bradley, IL 60915-0012
phone: 815 935 5353

Scoraig Wind Electric
Dundonnell, Ross Shire, Scotland IV23 2RE
United Kingdom
phone: +44 18 54 633 286
email: hugh.piggott@enterprise.net

I. Bibliography

General Interest

Cole, Nancy, and Skerrett, P.J. 1995. *Renewables Are Ready: People Creating Renewable Energy Solutions.* Chelsea Green Publishing, White River Junction, VT. Case studies and resource information on scores of successful renewable energy projects in communities throughout North America.

Gipe, Paul. 1995. *Wind Energy Comes of Age.* John Wiley & Sons. New York. A chronicle of wind energy's progress from its rebirth during the oil crises of the 1970s, through a troubled adolescence in the 1980s in California's mountain passes, to its maturation on the plains of northern Europe in the 1990s. The book argues that wind energy is no longer an "alternative" source of energy, but rather, because of improvements in performance, reliability, and cost-effectiveness, a fully commercial technology for generating electricity.

Leckie, Jim, et al. 1975. *More Other Homes and Garbage.* Sierra Club Books. San Francisco. One of the best overall books on alternative energy available. Runs the gamut from wind to bioconversion.

Marier, Donald. 1981. *Wind Power for the Homeowner.* Rodale Press, Emmaus, PA. A readable introduction to wind energy; intended for homeowners.

Park, Jack, and Schwind, Dick. 1977. *Wind Power for Farms, Homes, and Small Industry.* Packed with useful information. Available from National Technical Information Service, U.S. Dept. of Commerce, 5285 Port Royal Rd., Springfield, VA 22161.

Perez, Richard. 1985. *The Complete Battery Book.* Tab Books. Blue Ridge Summit, PA. All you ever wanted to know about batteries, and then some.

Schaeffer, John, et al. 1994. *Solar Living Source Book*, Eighth Edition. Chelsea Green Publishing. White River Junction, VT. Catalogue and discussion of energy-efficient products, including compact fluorescent lights, photovoltaic panels, inverters, batteries, cables, and much more.

Wilson, Alex. 1990. *Consumer Guide to Home Energy Savings.* American Council for an Energy-Efficient Economy, 1001 Connecticut Ave. NW, Washington, DC 20036. As the name suggests, a detailed guide on how to reduce household energy consumption; includes the energy efficiency of major appliances.

Wind Resources and Siting

Battelle Pacific Northwest Laboratory. March 1980. *A Siting Handbook for Small Wind Energy Conversion Systems.* Useful when siting your own wind turbine. PNL-2521 Rev. 1. Available from the National Technical Information Service, U.S. Dept. of Commerce, 5285 Port Royal Rd., Springfield, VA 22161.

Battelle Pacific Northwest Laboratory. March 1987. *Wind Energy Resource Atlas of the United States.* This is the updated version of Battelle's classic work mapping the U.S. wind resource. The update incorporates new data collected since the original atlas was published in 1980. DOE/CH 10094-4. Available from the American Wind Energy Assoc., 122 C St., NW, 4th Floor; Washington, DC 20001. Or the National Technical Information Service, U.S. Dept. of Commerce, 5285 Port Royal Rd., Springfield, VA 22161.

Risø National Laboratory. 1989. *European Wind Atlas.* Published for the European Community. 656 pp. Available from Risø, P.O. Box 49, 4000 Roskilde, Denmark.

Risø National Laboratory. 1992. *Wind Atlas Analysis and Application Program.* Available from Risø, P.O. Box 49, 4000 Roskilde, Denmark.

Designing, Building, or Installing Your Own

For tinkerers and those wanting to learn more about installing small wind turbines, Michael Hackleman's books are hard to beat. Jack Park's book is useful for those wanting a broader background in the mechanics of wind turbines. Lindsay's reprints offer several designs suitable for backyard experimenters.

Bergey Windpower Co. *BWC 1500 or BWC Excel Installation Manual.* 2001 Priestley Ave., Norman OK 73069. No doubt the best installation manual in the business today. Many details on installing guyed towers as well as the use of screw, expanding, and rock anchors.

Hackleman, Michael. 1974. *Wind and Windspinners.* Earthmind, P.O. Box 63, Ben Lomond, CA 95005. Describes how to build your own Savonius rotor; written in down-to-earth language.

_____. 1975. *The Home-Built, Wind-Generated Electricity Handbook.* Earthmind, P.O. Box 63, Ben Lomond, CA 95005. Explains how to find, lower, and rebuild pre-Rea wind generators, plus much more.

Park, Jack. 1981. *The Wind Power Book.* Cheshire Books. Palo Alto, CA. The best overall book on the technology and design of wind machines. An excellent reference for experimenters and professionals alike.

Powell, F.E. 1910. *Windmills and Wind Motors.* Lindsay Publications, P.O. Box 12, Bradley, IL 60915-0012. One of the many Lindsay reprints for do-it-yourselfers.

Kurtz, Edwin, and Shoemaker, Thomas. 1986. *The Lineman's and Cableman's Handbook.* 7th edition. McGraw-Hill. New York. An extremely useful handbook on setting poles, stringing conductors, and installing earth anchors.

Solar Photovoltaics

Stand-alone Photovoltaics Systems: A Handbook of Recommended Design Practices. March 1990. Photovoltaic Design Assistance Center. Sandia National Laboratories, Albuquerque, NM.

The Design of Residential Photovoltaic Systems. December, 1988. Photovoltaic Design Assistance Center. Sandia National Laboratories. Albuquerque, NM.

Davidson, Joel. 1987. *The New Solar Home Book: The Photovoltaic How-To Handbook.* Aatec Publications. Ann Arbor.

Potts, Michael. 1993. *The Independent Home: Living Well with Power from the Sun, Wind, and Water.* Chelsea Green Publishing. White River Junction, VT.

Strong, Steven and Scheller, William. 1993. *The Solar Electric House: Energy for the Environmentally-Responsive, Energy-Independent Home.* Distributed by Chelsea Green Publishing, White River Junction, VT.

Water Pumping

Planning for an Individual Water System. 1982. 4th edition. American Association for Vocational Instructional Materials. Athens, GA. A good, easy-to-understand reference on rural water systems.

Hays, Dick, and Allen, Bill. 1983. *Windmills and Pumps of the Southwest.* Eakin Press. Austin, TX. Advice from an oldtime windmiller on installing and operating a water-pumping windmill.

Index

CHELSEA GREEN

Sustainable living has many facets. Chelsea Green's celebration of the sustainable arts has led us to publish trend-setting books about organic gardening, solar electricity and renewable energy, innovative building techniques, regenerative forestry, local and bioregional democracy, and whole foods. The company's published works, while intensely practical, are also entertaining and inspirational, demonstrating that an ecological approach to life is consistent with producing beautiful, eloquent, and useful books, videos, and audio cassettes.

For more information about Chelsea Green, or to request a free catalog, call toll-free (800) 639–4099, or write to us at P.O. Box 428, White River Junction, Vermont 05001.

Chelsea Green's titles include:

  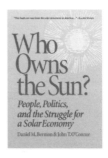

The Straw Bale House

The Independent Home:
  Living Well with Power
  from the Sun, Wind,
  and Water

Independent Builder:
  Designing & Building a
  Home Your Own Way

The Rammed Earth House

The Passive Solar House

The Sauna

The Real Goods Solar
  Living Sourcebook

The Flower Farmer

Passport to Gardening:
  A Sourcebook for 21st
  Century Gardeners

The New Organic Grower

Four-Season Harvest

Solar Gardening

The Contrary Farmer's
  Invitation to Gardening

Forest Gardening

Whole Foods Companion

Who Owns the Sun?

Hemp Horizons

Beyond the Limits

Loving and Leaving the
  Good Life

The Man Who Planted
  Trees